Springer Tracts in Modern Physics
Volume 124

Editor: G. Höhler
Associate Editor: E. A. Niekisch

Editorial Board:
S. Flügge H. Haken J. Hamilton
W. Paul F. Steiner J. Treusch P. Wölfle

Springer Tracts in Modern Physics

* denotes a volume which contains a Classified Index starting from Volume 36

B. Sitar   G. I. Merson
V. A. Chechin   Yu. A. Budagov

Ionization Measurements
in High Energy Physics

Revised and Enlarged English Edition
With 184 Figures

Springer-Verlag
Berlin Heidelberg GmbH

Dr. Branislav Sitar
Comenius University, Lediny 35, 84103 Bratislava, Slovakia

Dr. Gabriel I. Merson
Dr. Valery A. Chechin
P. N. Lebedev Physical Institute of the Russian Academy of Science,
Leninsky Prospect 53, 117924 Moscow, Russia

Dr. Yury A. Budagov
Joint Institute for Nuclear Research, Head Post Office, P. O. Box 79,
101100 Moscow, Russia

Translator:
Dr. G. B. Pontecorvo
JINR, Dubna, Laboratory of Nuclear Problems, Head Post Office,
P. O. Box 79, 10100 Moscow, Russia

Manuscripts for publication should be addressed to:
Gerhard Höhler
Institut für Theoretische Teilchenphysik der Universität Karlsruhe, Kaiserstrasse 12,
D-76131 Karlsruhe, Germany

*Proofs and all correspondence concerning papers in the process of publication should
be adressed to:*
Ernst A. Niekisch
Haubourdinstrasse 6, D-52428 Jülich, Germany

Title of the original Russian edition: Ionisation Measurements in
High Energy Physics © Energoatomizdat, Moscow 1988

ISBN 978-3-662-14923-2

Library of Congress Cataloging-in-Publication Data. Ionization measurements in high energy
physics / B. Sitar ... [et . al.]. p. cm. – (Springer Tracts in modern physics; v. 124). Includes
bibliographical references and index.
ISBN 978-3-662-14923-2   ISBN 978-3-540-46870-7 (eBook)
DOI 10.1007/978-3-540-46870-7
1. Ionizing radiation–
Measurement. 2. Nuclear counters. I. Sitar, B. (Branislav), 1946– . II. Series: Springer Tracts in
modern physics; 124. QC1.S797 vol. 124 [QC795.42] 530 s–dc20 [539.7'22'0287] 91-31754

This work is subject to copyright. All rights are reserved, whether the whole or part of the
material is concerned, specifically the rights of translation, reprinting, reuse of illustrations,
recitation, broadcasting, reproduction on microfilm or in any other way, and storage in data
banks. Duplication of this publication or parts thereof is permitted only under the provisions of
the German Copyright Law of September 9, 1965, in its current version, and permission for use
must always be obtained from Springer-Verlag Berlin Heidelberg GmbH.
Violations are liable for prosecution under the German Copyright Law.

© Springer-Verlag Berlin Heidelberg 1993
Originally published by Springer-Verlag Berlin Heidelberg New York in 1993
Softcover reprint of the hardcover 1st edition 1993

The use of general descriptive names, trademarks, etc. in this publication does not imply, even
in the absence of a specific statement, that such names are exempt from the relevant protective
laws and regulations and therefore free for general use.

Typesetting: Springer T_EX-in-house system
Production Editor: P. Treiber
56/3140-5 4 3 2 1 0 – Printed on acid-free paper

Preface

This book is a revised English version of the Russian edition by Budagov, Merson, Sitar and Chechin, published in 1988. The original text has been significantly extended and updated and a new chapter (Chap. 4), written by Sitar, has been added.

The book deals with ionization damping of fast charged particles in matter and with experimental methods of ionization measurements based on this phenomenon, which are applied in high energy physics for the identification and separation of charged particles according to their ionizing powers.

The primary subjects of the book are proportional, drift and streamer chambers, which are now widely used in experiments at high-energy particle accelerators. Traditional methods of ionization measurements with cloud, spark, and ionization chambers are also discussed.

The book contains illustrations and handbook information which should be useful in planning and performing ionization measurements and analysis of the data obtained.

The authors hope that this book will be useful for scientific workers, engineers, post-graduate and undergraduate students specializing in high energy and cosmic ray physics, experimental nuclear physicists and those working on ionizing radiation. We take this opportunity to thank our colleagues for helping us with useful advice. We will also be grateful to any reader who finds occasion to comment on this book.

Bratislava and Moscow
June 1993

B. Sitar
G.I. Merson
V.A. Chechin
Yu.A. Budagov

Contents

0. Introduction

The ionization of matter by charged particles is a phenomenon of particular significance of experimental nuclear physics and high energy physics, since it serves as the basis of the operation of most detectors of elementary particles. The registration of ionization led to the discovery of radioactivity and of cosmic rays, and thus initiated the science dealing with the structure of matter. Furthermore, advances in experimental technique, accompanied by the development of essentially novel ionization detectors and measurement methods, led to the discovery of the extremely diverse world of elementary particles. Investigations in this field have already been going on for nearly a century.

Theoretical and experimental studies of ionization damping of charged particles in a medium, which began at the same time as the creation of the foundation of modern physics, i.e. of relativity and quantum theories, have played a remarkable role in the history of physics of the twentieth century. As quantum concepts were being established, experiments on the passage of fast particles through matter together with atomic spectroscopy, provided the observational basis of new physical ideas. These experiments can be considered as significant as investigations of the scattering of high energy particles for understanding of their internal structures. It is, therefore, no surprise that the passage of ionizing particles through matter drew the attention of such outstanding physicists as Rutherford, Thomson, Bohr, Fermi, Bethe, Landau and others, whose names are immortalized in concepts and formulas in this field of physics.

The operation of the majority of detectors of charged particles is based on the ionization of matter. It may be worth reminding the reader that an ionizing particle travelling at a speed v close to the velocity of light c, loses less than 10^{-12} J per g/cm^2 of matter in a detector. To transform such a minute energy into a macroscopic signal an appropriate amplification mechanism is required. Amplification is achieved by some avalanche-like process in the metastable operating medium of the detector (Wilson cloud chamber, diffusion cloud chamber, bubble chamber, nuclear emulsion), or under the action of an accelerating electric field inside the detector (gas-discharge counters, proportional, drift, spark and streamer chambers), and/or by means of an external electronic device, e. g. an amplifier or a photomultiplier (ionization chamber, proportional, scintillation, semiconductor and Čerenkov detectors).

The most important property of a majority of detectors is the dependence of their response, i.e., of the *observed ionization effect J* upon the ionization produced by a charged particle. They are the J, v detectors which will be discussed here. The observed ionization effect in track detectors corresponds to the number of structure elements of a track (ions, droplets, grains, bubbles, streamers etc.), and in electronic detectors to the amplitude of the output electric signal. In many cases the observed ionization effect can be described in terms of ionization damping, namely the average specific energy losses of a particle ($-dE/dx$, where E is the particle energy and x is the path length) due to excitation and ionization (*the specific ionization energy losses*), or the average number of inelastic ionizing collisions per unit path length, dN_1/dx (*the specific primary ionization*).

According to theoretical predictions confirmed by experimental data, as the energy of a particle, $E = \gamma mc^2$, increases, where c is the speed of light, $\gamma = (1 - v^2/c^2)^{-1/2} = (1 - \beta^2)^{-1/2}$ is the particle's Lorentz factor and m is its mass, the inelastic collision cross section of such a fast ($v \gtrsim 0.2c$) particle with an atom falls proportionally to β^{-2} and passes through an *ionization minimum* close to $\gamma \simeq 4$. At relativistic velocities ($\gamma \gtrsim 10$), owing to the Lorentz transformation of the electromagnetic field of a particle, the radius of the region near the particle trajectory where the field is sufficiently high for excitation and ionization of atoms of the medium increases; so the inelastic collision cross section rises approximately as $\ln \gamma$. At sufficiently high γ, polarization of the atoms by the electromagnetic field of the particle hinders further increase of the interaction region, and leads to saturation of the relativistic rise of the cross section (*the Fermi plateau*). Such screening is stronger in dense matter and has been termed the *density effect*.

The density effect reduces the relativistic rise of energy losses ($-dE/dx$) due to ionization, and in the region of very large γ, this rise is only related to the enhancement of the kinematically possible maximum energy, $\omega_{\max}$, transferred in collisions to the atomic electrons. In some cases the energy transfers are limited by the conditions of observation, but not kinematically. So, for example, δ-electrons with kinetic energies $\varepsilon_e > T_0$ (where T_0 is an energy which depends on the size of the registration region), carry a major part of their energy away. In such cases one is said to deal with specific average *restricted energy losses* $(-dE/dx)_{T_0}$, which, like the collision cross section in the relativistic energy region, tend toward the Fermi plateau because of saturation. This also explains the saturation of the most *probable energy losses* $\Delta_0(x)$ corresponding to the maximum in the distribution of ionization losses in a layer of matter of thickness x.

Our understanding of the ionization damping of fast charged particles in matter has undergone a long evolution. We now have a thorough grasp of its details and peculiarities; this has allowed us to advance from investigations of this phenomenon to using it practically for identifying high energy particles by their ionizing powers. The identification (separation) of relativistic particles is more reliable the greater the *relativistic rise*, R_J, of ionization (the ratio of J on the Fermi plateau and in the ionization minimum), the more precise the ionization measurements (*the ionization resolution*), and the larger the relative difference in mass of the particles being identified. The magnitudes of R_J and of γ_{pl} corresponding to saturation of the energy dependence of dJ/dx, are (usually, but

Table 0.1. Relativistic rise of ionization R_J and saturation threshold, γ_{pl}, in various ionization detectors

State of the medium	Detector	Measured quantity	Material	R_J	γ_{pl}
Gaseous matter (1 bar, 0 °C)	Gas ionization detectors filled with noble and/or multiatomic gases and their mixtures (Wilson cloud chambers, ionization, proportional, drift chambers etc.)	The most probable ionization in a gas layer of $x = 2$ cm	He	1.55	260
			Ne	1.59	360
			Ar	1.59±0.05	380
			Kr	1.71±0.09	510
			Xe	1.75±0.05	570
			C_3H_8	1.30±0.03	80
Condensed matter	Bubble chambers	Bubble density	H_2	1.01±0.01	10
			C_3H_8	1.13±0.03	100
			$CBrF_3$	1.32±0.05	50
	Nuclear emulsions	Grain or blob density	AgBr	1.06–1.15	70
	Scintillation detectors	Signal amplitude	$(CH)_n$	1.01–1.02	5
			NaI(Tl)	1.11–1.14	20
	Semiconductor detectors	Signal amplitude	Si	1.1	10

The results for He and Ne were obtained by Monte-Carlo simulation

not always) enhanced as the atomic number, Z, increases and as the density of the medium decreases (Table 0.1). The properties of the detector also play an important role in determining the weights, $\mathcal{P}(\omega)$, of the contributions to the observed ionization effect, provided by individual ionizing collisions and depending on the energy transfer ω. Concerning measurement precision, for identification of relativistic particles the relative standard ionization measurement uncertainty,

σ_J, must be significantly smaller than R_J: $\sigma_J \lesssim (R_J - 1)/10$. Methods for improving the accuracy of ionization measurements are analyzed in Chaps. 3 and 5.

Generally speaking, ionization measurements only permit determination of the Lorentz factor of a particle. To estimate the mass of a particle of known charge ze (where e is the charge of the electron) one must also know one more independent characteristic, for example, a momentum, p, or an energy, E, which are determined respectively by the deflection of the particle in a magnetic field, and by the energy liberated in total-absorption calorimeters. Modern ionization detectors such as multiwire proportional, drift and streamer chambers installed in a magnetic field are capable of simultaneous measurement of ionizing powers, coordinates and the momenta of particles. This permits a reconstruction in space of the interaction picture, and a complete kinematical analysis of events. Thus, such detectors are established in high-energy physics and the long term prospects for their wide spread application are good. For this reason many operating large-scale detectors and those under construction for use at high energy accelerators, include large ionization identifiers with high speed readout capabilities for data acquisition and processing.

During recent years new methods have been developed to identify relativistic particles. They are based on the recording of Vavilov-Čerenkov radiation rings (so-called ring imaging Čerenkov counters) and on detection of X-ray transition radiation (TR detectors or TRD). Such detectors exhibit certain advantages in the solution of a number of problems, but they are probably less universal than ionization identifiers. For instance, they are insensitive to slow interaction products and require additional devices for measurement of particle momenta. The scopes of application of ionization and other identification methods are presented in Fig. 0.1.

This book is devoted to the methods and results of ionization measurements in gas detectors. The same problems have been recently discussed by Boos et

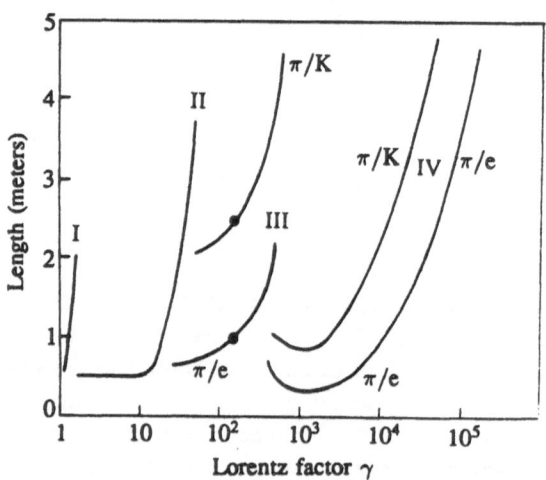

Fig. 0.1. Detector length L necessary for particle identification versus the particle Lorentz factor γ. The points on the curves for ionization detectors correspond to normal pressure (Willis 1978). Curves correspond to measurements of the time of flight (*I*), Čerenkov radiation (*II*), ionization (*III*), transition radiation (*IV*)

al. (1979), and Miakonkaya and Chasnikov (1983) for bubble chambers, nuclear photoemulsions and solid state detectors. We have tried to detail the methods of ionization measurements which are carried out nowadays with the help of various ionization detectors. It is useful to review the history of studies of the ionization damping of fast charged particles in matter, and to pay a tribute to the traditional methods of ionization measurements with the aid of cloud, spark and ionization chambers, and to the methods of determining the ionizing powers of particles according to the efficiencies of gas-discharge detectors. However, most attention is given to new ionization measurement methods, namely, energy loss $(-dE/dx)$ measurements in drift and streamer chambers, the method of cluster counting in a drift chamber with longitudinal electron drift, and the light proportional drift chamber.

The book consists of six chapters. Chapter 1 develops the principal concepts of the description of charged particle passage through matter and presents the theory of ionization damping in a polarizable medium. Relationships are derived for the differential and total inelastic cross sections of fast charged particles colliding with atoms; for the specific ionization (total, restricted and doubly restricted) energy loss; for the specific primary ionization; and for the density effect in matter. The connection between these quantities and observed ionization effects in actual particle detectors is analyzed. Also, methods are described for computation of the fluctuations of energy loss in absorbers of various thicknesses.

Chapter 2 is devoted to the description of physical phenomena in the gas filling of an ionization detector, of electron and ion transport processes capable of distorting the measured ionization, and of physical processes responsible for amplifying and recording the ionization signal.

Chapter 3 describes the ionization measurement technique in proportional and drift detectors. Here a classification is given of multilayer ionization identifiers; methods for choosing their parameters and the gas mixture are considered. Factors influencing the precision of ionization and coordinate measurements are analyzed. The methods of data analysis in multilayer proportional detectors are listed. The promising method of cluster counting in a proportional chamber with longitudinal electron drift is discussed in detail.

In Chap. 4 the spatial resolution of multilayer drift chambers and the electronics for acquisition of ionization and coordinate information from the chambers are considered. Space and ionization measurements in gas detectors are closely related and are based on the same set of data, the coordinate information usually being given priority. Therefore, we have considered it necessary in this chapter to analyze the coordinate resolution, the count rate capability, and the methods of providing for stable operation of multilayer drift chambers. We have also analyzed methods of data-handling provided by high precision gas coordinate detectors, in which the coordinate is really measured by recording ionization tracks.

Chapter 5 is devoted to ionization measurements in gas track detectors, i.e. in cloud, spark and streamer chambers. Methods of ionization determination based on track structure parameters, and possibilities of automation of ionization measurements in track detectors are described.

Chapter 6 deals with ionization detectors used relatively seldom in high energy physics, but which could turn out to be the most suitable means to solve certain problems of particle identification or separation. These are ionization chambers, gas-discharge detectors of reduced efficiency (i. e., low-pressure spark chambers, spark chambers containing electro-negative admixtures, or hodoscopic neon flash tubes), and gas scintillation detectors involving light amplification, so-called light (or electro-luminescence) proportional drift chambers.

Certain issues, dealt with in this book, which are related to the operation of spark, proportional, drift and streamer chambers have already been discussed in the monographs by Allkofer (1969), Rice-Evans (1974), Ferbel (1987), Kleinknecht (1988) and in excellent reviews by Sauli (1977) and Charpak, Sauli (1984) cited at the end of this introduction. Since then new data have been obtained and new approaches have evolved. Their description is also included in the book.

References

Allkofer O. C. 1969: *Spark Chambers* (Thiemig, München)

Boos E. G, Vinitsky A. H., Ovsov Yu. V., San'ko L. A. 1979: *The automation of ionization loss measurements in bubble chambers* (in Russian). (Nauka, Alma-Ata)

Charpak G., Sauli F. 1984: Ann. Rev. Nucl. Part. Sci. **34**, 285

Ferbel T. 1987: *Experimental Techniques in High Energy Physics* (Addison-Wesley, New York)

Kleinknecht K. 1988: *Detektoren und Teilchenstrahlung* (Teubner, Stuttgart)

Miakonkaya G. S., Chasnikov I. Ya. (1983): *Charged Particle Identification According to Ionization in Dense Targets* (in Russian) (Nauka, Alma-Ata)

Rice-Evans P. 1974: *Spark, Streamer, Proportional and Drift Chambers* (Richelieu, London)

Sauli F. 1977: Preprint CERN No. 77–09

Willis W. J. 1978: Phys. Today **31**, 32

1. Ionization Effects in a Polarizable Medium

1.1 Ionization Energy Loss by Charged Particles in a Medium

Classical Theory of Energy Loss. The theoretical approach to the description of charged particle deceleration in a medium depends basically on the range of velocity variations of the particle or on its Lorentz factor. Here we shall consider "fast" particles with velocities $v \gtrsim 0.2c$. c is speed of light. Their Lorentz factors will be assumed to be not too great, so that losses due to bremsstrahlung and spin effects can be neglected ($\gamma \lesssim 10^2$–10^8 depending on the particle mass). This section is devoted to development of the main theoretical ideas concerning ionization energy loss in the indicated velocity range. Certain aspects of this problem have been dealt with in reviews (Bethe 1933; Uehling 1954; Fano 1963; Crispin, Fowler 1970; Ahlen 1980; Allison, Cobb 1980; Bichsel 1988) and monographs (Merzon et al. 1983; Bohr 1948; Bethe, Ashkin 1953; Rosse 1952; Sternheimer 1961; Andersen, Ziegler 1977; Bichsel 1972). The purpose of this section, and of the main part of Chap. 1, is to present a certain "theoretical minimum" necessary for understanding and calculating ionization deceleration processes of fast particles.

In the earliest publications (Thomson 1903; Darwin 1912; Bohr 1948) deceleration in a medium was calculated by considering the results of collisions of non-relativistic particles with free electrons. To describe Coulomb scattering of the particles on electrons, the Rutherford cross section

$$\sigma_{\mathrm{R}} = 2\pi \int\limits_{0}^{\varrho_{max}} \varrho\, d\varrho = \frac{2\pi z^2 e^4}{m_{\mathrm{e}} v^2} \int\limits_{\omega_{min}}^{\omega_{max}} \frac{d\omega}{\omega^2} = \frac{8\pi z^2 e^4}{v^2} \int\limits_{q_{min}}^{q_{max}} \frac{dq}{q^3} , \qquad (1.1)$$

was applied, where m_{e} is the electron mass; $q = \sqrt{2m_{\mathrm{e}}\omega}$ is the momentum transferred by the particle to the electron; e is the charge of an electron and ze is the particle's charge. The impact parameter ϱ and the energy ω transferred to the electron are related by

$$\omega(\varrho) = \frac{2z^2 e^4}{m_{\mathrm{e}} v^2}[\varrho^2 + (ze^2/m_0 v^2)^2]^{-1} , \qquad (1.2)$$

where $m_0 = mm_{\mathrm{e}}/(m + m_{\mathrm{e}})$ is the reduced mass of the particle of mass m

and of the electron. The integration limits, ω_{max} and q_{max}, are determined by the collision kinematics, $\omega_{max} = \omega(0) = 2m_0^2v^2/m_e$. It is assumed that $v \gg v_e$, where v_e is the characteristic velocity of the atomic electrons. The total energy lost by a particle per unit length in a medium of atom density n_a is

$$-\frac{dE}{dx} = n_a Z 2\pi \int\limits_0^{\varrho_{max}} \omega(\varrho)\varrho\, d\varrho = \frac{4\pi n_a z^2 e^4}{m_e v^2}\mathcal{L}_R\,, \tag{1.3}$$

where Z is the number of electrons in the atom. $\mathcal{L}_R$ is the stopping number; it can be written in three equivalent forms corresponding to (1.1)

$$\mathcal{L}_R = Z \int\limits_{\omega_{min}}^{\omega_{max}} \frac{d\omega}{2\omega} = Z \int\limits_0^{\varrho_{max}} \frac{\varrho\, d\varrho}{\varrho^2 + (ze^2/m_0v^2)^2} = Z \int\limits_{q_{min}}^{q_{max}} \frac{dq}{q}\,. \tag{1.4}$$

It can be seen from (1.4) that the energy loss diverges as $\omega_{min} \to 0$, $q_{min} \to 0$, or as $\varrho_{max} \to \infty$.

Subsequent theoretical publications on the loss of energy by particles can be classified according to the choice of effective integration limits in (1.4), corresponding to the adopted physical interaction pattern of particles with a medium. Thus, for instance, Thomson, who utilized the Rutherford planetary model, considered ϱ_{max} to be equal to the atomic radius.

In Bohr's (1913) paper (written several months before the publication of his famous article "On the constitution of atoms and molecules") the binding of electrons in an atom was taken into account in the case of sufficiently large impact parameters, ϱ. To this end all collisions were separated into "long-range" and "short-range" collisions in accordance with the parameter ϱ, and in each particular case respective simplifications were applied. Energy transfers for small ϱ were described as before by the Rutherford formula (1.4). In the case of large ϱ the interaction of the particle's electromagnetic field with the electrons was dealt with by analogy with dispersion and absorption of radiation. In doing so each electron was considered similar to a harmonic oscillator of natural frequency, ν_s, and the energy $\omega_s(\varrho)$ transferred to it by the interacting particle was calculated to be

$$\omega_s(\varrho) = \frac{2z^2e^4}{m_e v^2\varrho^2}Q(\varrho\nu_s/v)\,, \tag{1.5}$$

where $Q(x) = x^2[K_1^2(x) + K_0^2(x)]$, and $K_1(x)$ and $K_0(x)$ are modified Bessel functions. $Q \ll 1$ when $\varrho \gg v/\nu_s$; therefore, unlike (1.2), $\omega_s(\varrho)$ tends toward zero more rapidly than ϱ^{-2} as $\varrho \to \infty$, which results in a convergent integral for $\mathcal{L}_R$. The region $\varrho \gg v/\nu_s$ corresponds to adiabatic perturbation of a bound electron, since the collision time ϱ/v is much greater than the characteristic oscillator time, ν_s^{-1}. It is important that relations (1.2) and (1.5) transform into each other, i.e. are "matched", at a certain $\varrho = \varrho_1$ (where $ze^2/m_0v^2 \ll \varrho_1 \ll v/\nu_s$), since $Q(x \ll 1) \approx 1$. Using (1.2) for $0 \le \varrho \le \varrho_1$, and (1.5) for $\varrho_1 \le \varrho < \infty$ and incorporating these into (1.4), Bohr obtained

$$\mathcal{L}_{\text{Bohr}}(0 \leq \varrho \leq \varrho_1) = Z \ln \left(\frac{\varrho_1 m_0 v^2}{z e^2} \right) ;$$

$$\mathcal{L}_{\text{Bohr}}(\varrho_1 \leq \varrho < \infty) = \sum_{s=1}^{Z} \ln \left(\frac{1,123 v}{\varrho_1 \nu_s} \right) .$$

(1.6)

Summing these expressions yields

$$\mathcal{L}_{\text{Bohr}} = \sum_{s=1}^{Z} \ln \left(\frac{1,123 m_0 v^3}{z e^2 \nu_s} \right) .$$

(1.7)

Bohr's formula (1.7) actually corresponds to the choice in (1.4) of $\varrho_{\max} \approx v/\nu_s$, i.e., to the limit impact parameter of adiabaticity.

Bohr's method of separating the collisions into two (and sometimes into three) groups, and of further summing individual contributions, was subsequently applied in practically all theoretical works on ionization energy loss. Bohr's approach, based on classical mechanics, is applicable for sufficiently low particle velocities, i.e. for $v \ll z e^2/\hbar$, where $\hbar$ is the Planck constant, while still assuming that $v \gg v_e$ (v_e is the mean velocity of atomic electrons). Since for electrons of the K-shell, $v_e \sim Z e^2/\hbar$, Bohr's formula is only applicable for deceleration of slow multicharged particles in light substances.

Quantum Theory of Ionization Loss. A consistent quantum non-relativistic computation of the stopping number in the Born approximation ($z e^2/\hbar v \ll 1$) was performed by Bethe (1930). In the case of electromagnetic interaction of a fast particle with an atom, nearly all the lost energy is spent in ionization and excitation of the atom. Indeed, the ratio between a fast particle's energy losses due to elastic and inelastic collisions with atoms of mass M_a is of the order of magnitude of $m_e Z/M_a \ll 1$. Furthermore, in the case of inelastic collisions, only a small part ($\sim m_e/M_a$) of the energy lost by the particle is spent in enhancement of the recoil atom kinetic energy. Consequently, neglecting corrections of the order of $m_e Z/M_a$, one may consider the atomic electrons to be bound to an infinitely heavy atom. Instead of using the impact parameter ϱ, Bethe, unlike Bohr, classified collisions according to the momentum, q, transferred to the atom that represents an observable quantum-mechanical quantity:

$$-\frac{dE}{dx} = n_a \sum_{k=1}^{k_{\max}} \omega_k \sigma_k = n_a \sum_{k=1}^{k_{\max}} \int_{q_{\min}}^{q_{\max}} \omega_k \frac{d\sigma_k}{dq} dq ,$$

(1.8)

where $d\sigma_k/dq$ is the excitation differential cross section of the atom to the level of energy E_k; $\omega_k = E_k - E_0$ is the excitation energy; and $q_{\min} = \omega_k/v$. For brevity we shall also usually use the term "excitation" to imply ionization; in this case the sum in (1.8) is replaced by an integral over the levels of the continuous spectrum. Bethe calculated (1.8) with the additional assumption that $v \gg v_e$, this may be violated for the K- and L-electron shells of heavy atoms. Luckily, the main results obtained in the Born approximation hold for the deceleration of heavy particles ($m \gg m_e$) down to $v \sim v_e$. Further improvements can be made by the introduction of corrections to the binding energies of the K- and L-electrons (Walske 1956). Since $\omega_{max} = 2m_0^2 v^2/m_e$, the condition $v \gg v_e$ signifies that $\omega_{max} \gg \bar{\omega}_a$ (where $\bar{\omega}_a$ is the mean binding energy of electrons in the atom), and that in the case of energy transfers of $\omega \sim \omega_{max}$ the electrons can be considered free. In addition to the above, this condition permits capture of atomic electrons by the particle to be neglected.

Bethe divided the range of q values into three intervals applying different simplifications to each interval. Matching the solutions at intermediate q values yields (for particles other than the electron):

$$\mathcal{L}_{Bethe} = Z \ln(2m_0 v^2/I) , \tag{1.9}$$

where I is the mean logarithmic ionization potential of the medium (Sect. 1.3). If the ionizing particle is an electron, and if in the collision it transfers to the atomic electron a significant part of its initial energy, then it is necessary to take into account the exchange effect and to set $\omega_{max} = m_e v^2/4$. The result is that

$$\mathcal{L}_e = Z \ln \left(\frac{m_e v^2}{2I} \sqrt{\frac{e}{2}} \right) . \tag{1.10}$$

The conditions for transformation into each other of the classical, (1.7), and quantum, (1.9), expressions for the stopping number, $\mathcal{L}$, at intermediate values of $ze^2/\hbar v \sim 1$, were investigated by Bloch (1933). Bloch derived the general expression for $\mathcal{L}$, from which the results obtained by Bohr, (1.7), and by Bethe, (1.9), follow as limiting cases for high and low $ze^2/\hbar v$ values, respectively. Bloch's equation is written

$$\mathcal{L}_{Bloch} = Z \left[\ln \left(\frac{2m_0 v^2}{I} \right) + \psi(1) - \psi \left(1 + i\frac{ze^2}{\hbar v} \right) \right] .$$

Here $\psi(x)$ is the logarithmic derivative of the gamma-function. To obtain (1.7) from this equation, it is necessary to take into account that when $ze^2/\hbar v \gg 1$,

$$\psi(1) - \psi \left(1 + i\frac{ze^2}{\hbar v} \right) \simeq \ln \left(\frac{1.123 \hbar v}{2ze^2} \right) .$$

The stopping number for a heavy relativistic particle was calculated by Bethe (1932) and by Möller (1932). Results in the general case depend on the particle mass and spin. However, for $E \ll (m/2m_e) \cdot mc^2$ this dependence drops out and (1.9) is replaced by the Bethe-Bloch formula:

$$-\frac{dE}{dx} = \frac{4\pi n_a e^4}{m_e v^2} Z \left\{ \ln \left(\frac{2m_e v^2 \gamma^2}{I} \right) - \beta^2 - \frac{\delta_E}{2} \right\} , \tag{1.11}$$

where δ_E is the correction due to the density effect; $\beta = v/c$. For ultrarelativistic electrons and positrons, i.e., when $\gamma \gg 1$ (Uehling 1954) we have

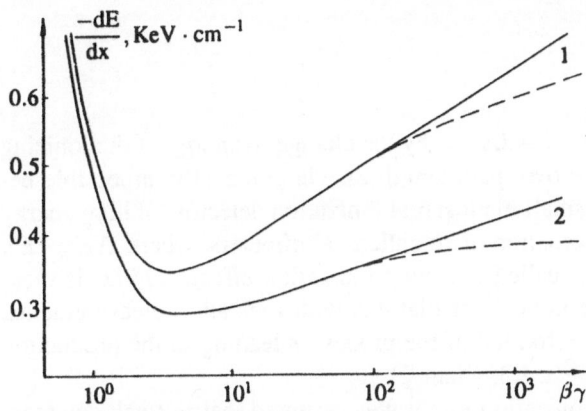

Fig. 1.1. Dependence of total (*1*) and limited (*2*) energy loss on $\beta\gamma$ in helium under normal conditions: the dashed lines indicate an influence of the medium density effect; $T_0 = 100\,\text{keV}$

$$\left(-\frac{dE}{dx}\right)_{e^-} = \frac{4\pi n_a e^4}{m_e v^2} Z \left\{ \ln\left(\frac{m_e c^2 \gamma^{3/2}}{\sqrt{2}\,I}\right) + \frac{1}{16} - \frac{\delta_E}{2} \right\} ; \qquad (1.12)$$

$$\left(-\frac{dE}{dx}\right)_{e^+} = \frac{4\pi n_a e^4}{m_e v^2} Z \left\{ \ln\left(\frac{\sqrt{2}\,m_e c^2 \gamma^{3/2}}{I}\right) + \frac{23}{24} - \frac{\delta_E}{2} \right\} ; \qquad (1.13)$$

The presence in (1.11–13) of terms proportional to $\ln\gamma$ results, for $\gamma \gtrsim 4$, in a logarithmic enhancement of $-dE/dx$ as the particle energy increases (the "relativistic rise"). However, for sufficiently large $\gamma \gtrsim I/\omega_p$ [where $\omega_p/\hbar = (4\pi n_a Z e^2/m_e)^{1/2}$ is the plasma frequency of the medium], screening of the particle's electromagnetic field becomes essential owing to polarization of the medium. This leads to significant suppression of the relativistic rise in $-dE/dx$ (Fig. 1.1). The corresponding correction δ_E occurring in (1.11–13) was computed by Fermi (1940), Landau and Lifshitz (1960) and others. Representative calculations of δ_E for a large number of substances were performed by Sternheimer (1981).

Ionization energy loss results from random collisions with individual atoms. Owing to the statistical nature of this process, the energy loss on any arbitrary part of the particle path, x, undergoes fluctuations, most noticeably when $x < l_c$ (where l_c is the characteristic distance between collisions). Energy loss fluctuations were considered by Landau (1944), Vavilov (1957), Blunk and Leisegang (1950) and others; a review of the respective theoretical results is presented in Sect. 1.6.

Observed Ionization Effect. So far only the total energy loss, taking into account all kinematically permissible energy transfers, $\omega \leq \omega_{max}$, has been considered. This quantity determines the range, R_p, of a particle in a medium:

$$R_p = \int\limits_0^E \frac{dE}{\left(-\frac{dE}{dx}\right)} \ .$$

Straightforward measurement of $-dE/dx$ by the change in energy of the ionizing particle, $\Delta E = (-dE/dx)\Delta x$ over path length Δx, is practically impossible because the ratio $\Delta E/E$ is too small in most real ionization detectors of high energy charged particles (with the exception of so-called calorimeters, where $\Delta E \sim E$). In these detectors only the so-called observed ionization effect, dJ/dx, is measured, and for comparing theoretical calculations with ionization measurements, a physical analysis must be performed of the processes leading to the production of the observed ionization (Sect. 1.3, Chap. 2).

In early publications on ionization loss it was assumed that in track detectors

$$\frac{dJ}{dx} = \frac{1}{w}\left(-\frac{dE}{dx}\right) \ . \tag{1.14}$$

This assumption was based on the fact that the average energy w required for the production of a single ion pair in a medium is practically independent of the particle velocity. In a real detector, however, collisions involving differing energy transfers are distinguishable, to various degrees. Thus, for example, in most detectors collisions with sufficiently large $\omega (\omega > T_0; \ T_0 \sim 10^3\text{--}10^5 \text{ eV})$ give no contribution to the observed ionization effect, since a δ-electron leaves the track or the fiducial volume of the detector taking with it from the region of observation the greater part of its energy. In addition, an energy transfer of $\omega < \omega_{\text{th}}$ (i.e., lower than the threshold energy $\omega_{\text{th}} \approx 10^2\text{--}10^3 \text{ eV}$) does not result in the formation of a track structure element. A better quantity for comparison with ionization measurement data in such detectors is not the total, but the "restricted" energy loss, $(-dE/dx)_{T_0}$, in which $\omega < T_0$, or even the "doubly restricted" energy loss:

$$\left(-\frac{dE}{dx}\right)_{\omega_{\text{th}},T_0} = n_a \sum_{\omega_k=\omega_{\text{th}}}^{\omega_k=T_0} \omega_k\sigma_k = n_a \int\limits_{\omega_{\text{th}}}^{T_0} \omega\frac{d\sigma}{d\omega}\,d\omega \ . \tag{1.15}$$

If ω_{th} and T_0 are comparable with the ionization potentials of atoms of the medium, then the conventional Bethe method for calculating $-dE/dx$, which utilizes the sum rule (1.27), cannot be applied. Since ω_{th} and T_0 depend on the operating conditions of the detector, and can vary over a wide range, the task actually reduces to calculation of the cross sections σ_k first, followed by the numerical summation (1.15). Essentially the same problem arises in the analysis of ionization effects in another class of detectors sensitive not to the total, but to the primary ionization (for example, in a streamer chamber and in the low-efficiency gas-discharge counter). In this case it is necessary to calculate the total number of collisions, N, or the amount of ionization collisions N_1, per unit length:

$$\frac{dN}{dx} = n_{\mathrm{a}} \int\limits_0^\infty \frac{d\sigma}{d\omega}\, d\omega \equiv n_{\mathrm{a}}\sigma \ ;$$

$$\frac{dN_1}{dx} = n_{\mathrm{a}} \int\limits_{I_1}^\infty \frac{d\sigma}{d\omega}\, d\omega \equiv n_{\mathrm{a}}\sigma_{\mathrm{i}} \ ;$$

(1.16)

where I_1 is the first ionization potential.

In the general case, dJ/dx is determined by a certain weight function, $\mathcal{P}(\omega)$, which depends on the operation mode of the detector. It also determines the contribution of collisions involving the energy transfer, ω, to the ionization effect:

$$\frac{dJ}{dx} = n_{\mathrm{a}} \int\limits_0^{\omega_{\max}} \mathcal{P}(\omega)\frac{d\sigma}{d\omega}\, d\omega \ .$$

(1.17)

In Table 1.1, the observed ionization effect in various detectors is compared with the theoretically computed quantity. From the Table, one can see that, as a rule, the ionization effect is determined either by dN_1/dx or by $(-dE/dx)_{\omega_{\mathrm{th}}, T_0}$. A more detailed analysis reveals that only for the bubble chamber and for nuclear photoemulsion, it is necessary to apply the general formula (1.17) with a non-trivial function $\mathcal{P}(\omega)$. A physical analysis of the observed ionization effects and of their relation to the calculated quantities, is extremely important for a correct understanding of the operation of detectors of fast charged particles. For example, the discrepancy, noted at different times, between the results of ionization measurements and theoretical calculations for high energy particles, has usually been caused by too straightforward a comparison being made between the observed ionization effect and the ionization energy loss.

The main task of the theory of ionization effects produced by fast particles in a medium, consists in calculating the cross sections, $d\sigma/d\omega$, and in performing the respective integrations in the formulas (1.15) to (1.17).

The cross section $d\sigma/d\omega$ will be considered in Sect. 1.2. This cross section is utilized in Sect. 1.3 for calculating the doubly restricted energy loss (1.5); in Sect. 1.4, for the calculation of the number of collisions (1.16); and in Sect. 1.5 for calculating energy loss fluctuations.

1.2 Differential Collision Cross Section of Fast Charged Particles with Atoms

Basic Relations. The electromagnetic interaction of a fast charged particle with atoms can be dealt with by applying perturbation theory. In the first approximation this process is described by a single Feynman diagram (Fig. 1.2).

The solid lines in Fig. 1.2 correspond to a particle with momentums p and p' and energies E and E' before and after the collision respectively. The wavy line

Table 1.1. Ionization effects in detectors of charged particles

Detectors	Observed ionization effect	Computed quantity to be compared with	Limits of recorded energy transfer [keV]	
			ω_{th}	ω_{max}, T_0
Gas detectors:				
Ionization Chamber	Ionization produced by charged particles	$(-dE/dx)_{\omega_{th}, T_0}$	0.01–0.02	$10-10^2$
Proportional and drift chambers	Electric signal proportional to the charge produced in the gas	Δ_0–probable energy loss	0.01–0.02	$10-10^2$
Gas scintillation detector	Number of photons produced in excitation of atoms of the gas by primary particle or secondary electrons	$(-dE/dx)_{\omega_{th}, T_0}$	$\simeq 0.003$	$10-10^2$
Electroluminescence chamber	Number of photons emitted by atoms in the course of multiple collisions of accelerated secondary electrons in the gas	Δ_0–probable energy loss	0.01–0.02	$10-10^2$
Wilson cloud chamber (with expansions delayed by about 0.2 s)	Number of droplets along a track, with the exception of dense blobs	$(-dE/dx)_{\omega_{th}, T_0}$	0.01–0.02	$\simeq 1$
Wilson chamber (with preliminary expansion); diffusion, discharge-condensation and discharge-diffusion chambers	Number of droplets on a track	dN_1/dx	0.01–0.02	ω_{max}
Streamer chamber (with high voltage pulse delayed by less than 1 μs)	Number of streamers (avalanches) along a track	dN_1/dx	0.01–0.02	ω_{max}
Low efficiency gas-discharge detector	Efficiency of discharge formation	dN_1/dx	0.01–0.02	ω_{max}

Solid-state and liquid:

Scintillation detector	Number of photons produced by the primary particle and the secondary electrons	$(-dE/dx)_{\omega_{\mathrm{th}},T_0}$	$\simeq 0.003$	$10\text{--}10^2$
Semiconductor detector	Number of electron-hole pairs produced by the charged particle in the sensitive layer	$(-dE/dx)_{\omega_{\mathrm{th}},T_0}$	$\simeq 0.003$	$\simeq 10$
Nuclear photo-emulsion	Number of developed AgBr microcrystals along the particle track	$\int\limits_{\omega_{\mathrm{th}}}^{T_0} \mathcal{P}(\omega)\dfrac{d^2N}{dx\,d\omega}\,d\omega$	$\simeq 0.005$	$1\text{--}10$
Bubble chamber	Number of bubbles on the particle track in an over-heated liquid	$\int\limits_{\omega_{\mathrm{th}}}^{T_0} \mathcal{P}(\omega)\dfrac{d^2N}{dx\,d\omega}\,d\omega$	$0.2\text{--}1$	$\simeq 10$

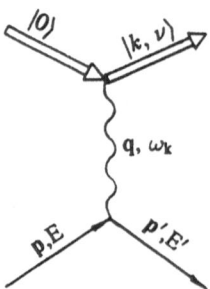

Fig. 1.2. Feynman diagram for a charged particle collision with an atom in the first Born approximation

represents a virtual photon that transfers to the atom a momentum $q = p' - p$ and an energy $\omega_k = E' - E$. The double lines correspond to transition of the atom from the initial $|0\rangle$ state to the final $|k, \nu\rangle$ state, where ν are quantum numbers indicating degenerate states of the atom with energy E_k. The cross section of this process can be written in the form (Fano 1963)

$$d\sigma_k = \frac{4\pi e^4}{v^2} \left\{ \frac{|F_k(q)|^2}{q^4} + \frac{|G_k(q)|^2}{(c^2 q^2 - \omega_k^2)^2} \right\} dq^2 . \tag{1.18}$$

Here

$$|F_k(q)|^2 = \sum_\nu \left| \left\langle k, \nu \left| \sum_{j=1}^{Z} \exp\left(\frac{-i q r_j}{\hbar} \right) \right| 0 \right\rangle \right|^2 \tag{1.19}$$

is the square inelastic form factor of the atom; summation is carried out over all the Z atomic electrons. Also,

$$2|G_k(q)|^2 = \sum_\nu \left| \left\langle k, \nu \left| \sum_{j=1}^{Z} \left[\exp\left(\frac{-i q r_j}{\hbar} \right) v_j \right. \right. \right. \right.$$
$$\left. \left. \left. + v_j \exp\left(\frac{-i q r_j}{\hbar} \right) \right] v_\perp \right| 0 \right\rangle \right|^2 ; \tag{1.20}$$

where $v_j = -i(\hbar/m_e) \cdot (\partial/\partial r_j)$ is the velocity operator of the j-th electron; and $v_\perp = v - (vq)q/q^2$ is the particle velocity component normal to q. The charge of the particle is $z = 1$.

We shall now make some comments to clarify relations (1.18–20):

a) Formulas (1.18–20) hold for sufficiently small $|q|$ such that

$$q^2/2m_e \ll m_e c^2 . \tag{1.21}$$

Under this condition, the atomic electrons conventionally considered to be non-relativistic in the initial state, are also non-relativistic in the final state. The atomic matrix elements, (1.19–20), are taken in a non-relativistic form. In addition, condition (1.21) allows exchange effects in the scattering of fast electrons to be neglected; the relatively small change in the particle momentum due to collision may also be neglected, therefore its motion is considered to be uniform and linear. If the momentum transfer, q, is greater than or equal to $m_e c$, then one may neglect instead the binding of electrons in the atom, since $m_e c \gg \sqrt{2 m_e I} \simeq p_{at}$ (where p_{at} are characteristic momenta of the atomic electrons, if the low-lying electron shells of heavy atoms are not taken into consideration). Such "hard" collisions depend on the particle's spin and mass, and their contributions must be taken into account in calculations of the total energy losses (1.11–13). However, for the ionization loss of restricted energy transfer, $\omega \le T_0 \lesssim 10^5$ eV, to be considered below, one can make use of relations (1.18–20), because the corresponding values of q satisfy the condition (1.21) (since $q^2/(2m_e) \le q^2_{max}/(2m_e) \approx T_0 \ll m_e c^2$). At a given energy, ω, the minimum transferred momentum is determined by kinematics:

$$q_{min}(\omega) = \omega/v \; .$$

b) The matrix elements in (1.19–20) depend on the direction of q; however, upon summation over ν this dependence drops out (in the case of a spherically symmetric atom). For transitions of an atom to the continuous spectrum it is necessary to substitute a continuous parameter, ω, for the index k in (1.18–20). In this case the cross section and the square matrix elements in (1.18) are replaced by differential quantities with respect to ω: for example, $\sigma_k \to d\sigma/d\omega$. Summation over ν, in this case, indicates integration over the angles of the outgoing atomic electron.

c) Formulas (1.18–20) correspond to the Coulomb gauge of potentials: $\text{div} A = 0$. In applying the Lorentz (Crispin, Fowler 1970) or the non-relativistic (Landau and Lifshitz 1982) gauges, expressions for $d\sigma_k/dq$ are obtained in another form, which can, of course, be reduced to the form of (1.18–20), taking into account conservation of current. The Coulomb gauge seems to be the most convenient one in this case, since it permits immediate identification in (1.18) of two terms exhibiting a definite physical meaning. The first term describes Coulomb ("longitudinal") interaction of the particle charge with the atom, and does not contain the velocity of light, c. The second term involves all the relativistic effects manifesting themselves in the collision, and describes the interaction of the particle and the atomic electrons through virtual ("transverse") photons. The contribution to (1.18) of the longitudinal interaction differs from the Rutherford cross section (1.1) only by the presence of the factor $|F_k(q)|^2$; this takes into account the electron binding, and gives the conditional probability of excitation of the atom to the level, k, when a momentum transfer to it of q occurs. The absence of a one-to-one correspondence between the momentum, q, and energy, ω, is due to a part of the momentum q being transferred to the atom as a whole; in other words this ambiguity could be explained by uncertainty in the momentum of the atomic electron. Note that, owing to the large mass of the atom, practically all the energy transferred to it is spent in its excitation or ionization.

The relativistic corrections in (1.18) are essential only when $q r_j/\hbar \lesssim 1$. Setting in (1.20) $\exp(-i q r_j/\hbar) \approx 1$, one can write (1.18) in the form

$$d\sigma_k = \frac{4\pi e^4}{v^2} \left\{ \frac{|F_k(q)|^2}{q^4} + \frac{v_\perp^2 \omega_k^2 |x_k|^2}{\hbar^2 (c^2 q^2 - \omega_k^2)^2} \right\} dq^2 , \tag{1.22}$$

where $|x_k|^2 = \sum_\nu |\langle k, \nu | \sum_{j=1}^Z x_j |0\rangle|^2$; x_j is one of the Cartesian coordinates of the j-th electron; and here $v_\perp^2 \approx v^2 - \omega_k^2/q^2$, since $qv = \omega_k$ owing to the quantity q/p being small. Instead of the matrix elements in (1.22), one usually introduces generalized and optical oscillator strengths, $f_k(q)$ and f_k, respectively (Bethe 1930):

$$f_k(q) = \frac{2m_e \omega_k}{q^2} |F_k(q)|^2 ; \tag{1.23}$$

$$f_k = \lim_{q \to 0} f_k(q) = \frac{2m_e \omega_k}{\hbar^2} |x_k|^2 . \tag{1.24}$$

Consequently, for transitions to the continuous spectrum,

$$\frac{d^2\sigma}{d\omega \, dq} = \frac{4\pi e^4}{m_e v^2 \omega} \left\{ f(\omega, q) + \frac{f(\omega)(v^2 q^2 - \omega^2)\omega^2}{(c^2 q^2 - \omega^2)^2} \right\} \frac{1}{q} . \tag{1.25}$$

The differential cross section of inelastic collision is

$$d\sigma/d\omega = \int_{\omega/v}^{\infty} \frac{d^2\sigma}{d\omega \, dq} \, dq = \frac{2\pi e^4}{m_e v^2 \omega} \left\{ 2 \int_{\omega/v}^{\infty} \frac{f(\omega, q) \, dq}{q} + f(\omega)[\ln(\gamma^2) - \beta^2] \right\} . \tag{1.26}$$

(Because the integrals converge rapidly, the upper integration limit $q_{max} = \sqrt{2m_e \omega_{max}}$ can be set equal to infinity.) The second term in (1.26) corresponds to the contribution of "transverse collisions"and gives the relativistic rise of the inelastic collision cross section.

Equations (1.18, 25) reflect an important peculiarity of the collision cross section of a fast particle with an atom: they contain factors related either only to the particle, or to the atom ($f_k(q)$, f_k). This is a direct consequence of the applicability of the Born approximation in such collisions. We note that the scattering of slow particles with $v \sim v_e$ depends on the properties of the composite atom-particle system. Consequently, the investigation of inelastic collisions of fast particles can serve as some kind of "spectroscopy" which permits us to relate the scattering properties to an important characteristic of the atom, $f_k(q)$.

Generalized Oscillator Strengths. To calculate $f_k(q)$ or $f(\omega, q)$, it is necessary to know the wave functions of the atom. An accurate calculation of $f(\omega, q)$ has only been performed for atomic hydrogen (Bethe 1930; Mott, Massey 1965) (Fig. 1.3).

The surface formed by the function $f(\omega, q)$ over the (ω, q)-plane is known as the *Bethe surface*; determination of its form represents one of the most important problems in the theory of inelastic collisions of fast charged particles with atoms.

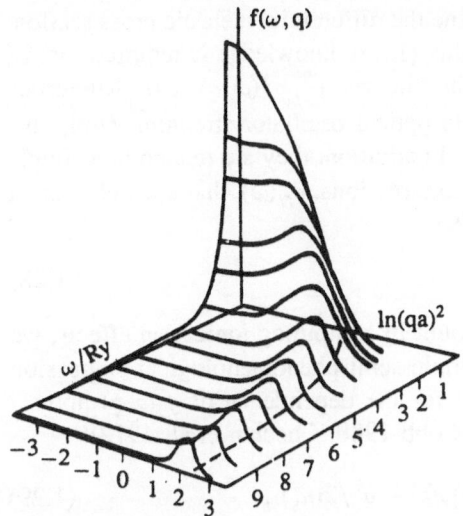

$f(\omega, q)$

$\ln(qa)^2$

ω/Ry

Fig. 1.3. The Bethe surface for atomic hydrogen: the dashed line in the plane indicates the position of the Bethe ridge $\omega = q^2/(2m_e)$; $Ry = m_e e^4/(2\hbar^2)$; $a_0 = \hbar^2/m_e e^2$

In this connection, a vast literature has been devoted to the Bethe surface (see reviews in Inokuti 1971; Inokuti et al. 1978; Bichsel 1988).

Usually Hartree-Fock wave functions are utilized for light atoms in theoretical work. The largest number of such calculations has been performed for helium. When $Z > 2$, they rapidly become more complicated, and their precision deteriorates (Powell 1976; Moiseiwitsch 1985). Experimental determination of $f(\omega, q)$ is based on the investigation of the energy spectrum of electrons scattered at a certain angle, which permits unambiguous determination of q and ω. Usually non-relativistic electrons are used, so that the second term in (1.25) can be dropped; $f(\omega, q)$ is found from comparison of (1.25) with the experimental data (Inokuti 1971).

Both theoretical and experimental studies of $f_k(\omega)$ have only been carried out for individual atoms and levels, k. At the same time, for computation of ionization effects in various substances, knowledge of $f(\omega, q)$ is required, generally, for a wide range of ω and q values. However, the requirements of the $f(\omega, q)$ determination accuracy are not too severe in this case, for two reasons. First, it is possible to restrict its determination to the precision of subsequent calculations of ionization effects, including the precision in determining the weight function $P(\omega)$. Second, ionization effects depend on the total contribution of collisions within a wide range of ω and q values, so details of the behavior of $f(\omega, q)$ are not essential. Application of the sum rule for $f(\omega, q)$ and $f(\omega)$ (Bethe 1930),

$$\int_0^\infty f(\omega, q)\, d\omega = \int_0^\infty f(\omega)\, d\omega = Z , \qquad (1.27)$$

permits the calculation of total energy loss $(-dE/dx)$ (or $(-dE/dx)_{T_0}$ for $T_0 \gg I$) in terms of a single parameter, the mean ionization potential, I. In the general case, however, when calculating ionization effects taking account of the weight function, $\mathcal{P}(\omega)$, it is necessary to determine the differential inelastic cross section, $d\sigma/d\omega$. However, in order to calculate this (1.26), knowledge is required, not of the function $f(\omega, q)$ itself, but only of the integral $\int_{\omega/v}^{\infty} f(\omega, q)\, dq/q$. Numerous calculations have been performed for the optical oscillator strengths $f(\omega)$, also occurring in (1.26) (Fano, Cooper 1969). In addition, they are related in a simple manner to the atomic photo-absorption cross sections, $\sigma_\gamma(\omega)$, that are well known from spectroscopic data (Berkowitz 1979):

$$f(\omega) = 2\hbar c Z n_a \sigma_\gamma(\omega)/\pi \omega_p^2 \;. \tag{1.28}$$

Taking these considerations into account in computing ionization effects, we shall take advantage of the following simple semi-phenomenological expression for $f(\omega, q)$; this accounts approximately for the dependence of $f(\omega, q)$ upon q and expresses it through $f(\omega)$ (Allison, Cobb 1980; Chechin et al. 1972):

$$f(\omega, q) = f(\omega)\theta(\omega - q^2/2m_e) + \mathcal{F}(\omega)\delta(\omega - q^2/2m_e) \;, \tag{1.29}$$

where $\mathcal{F}(\omega) = \int_0^\omega f(\omega')\, d\omega'$; $\delta(x)$ is the Dirac delta-function; $\theta(x) = 1$ for $x > 0$ and $\theta(x) = 0$ for $x < 0$. This form of $f(\omega, q)$ is based on a physical analysis of the Bethe surface (Fig. 1.3).

Function $f(\omega, q)$ differs noticeably from zero when $\omega \geq q^2/2m_e$. In this region two parts of $f(\omega, q)$ can be singled out:

1) A part corresponding to small q: $qr_a/\hbar \lesssim 1$ (where r_a is a characteristic dimension of the atom). Here $f(\omega, q)$ varies slowly as q grows, when $q^2/2m_e \lesssim \omega$, while when $q^2/2m_e \gg \omega$, it falls rapidly toward zero. This is related to long-distance collisions corresponding to small q, for which $f(\omega, q)$ is determined by the dipole properties of the atoms, i.e. by the function $f(\omega)$, and depends strongly on its electronic structure.

2) A part corresponding to large q: $qr_a/\hbar \gtrsim 1$. Here, above the curve $q^2/2m_e = \omega$, there exists on the Bethe surface a noticeable enhancement (the "Bethe ridge"), which becomes more and more pronounced with the growth of $q^2/2m_e$ above I. Such behavior of the "Bethe ridge" is common for all atoms and does not depend on the electronic structure. It is related to the fact that for large $q \gg \hbar/r_a$, when $q^2/2m_e \gg \hbar^2/m_e r_a^2 \approx I$, the collisions of a particle with an atom reduce to elastic scattering on the atomic electrons, which in this case can be considered free. Formula (1.29) takes these peculiarities into account. The first term is responsible for part (1), and the second for part (2) of the Bethe surface. It is important to point out that relation (1.29) satisfies the sum rule (1.27).

The form (1.29) chosen for $f(\omega, q)$ differs from its real form, which involves no delta-like or step-like singularities. Nevertheless, it reflects the main physical properties of the generalized oscillator strengths $f(\omega, q)$; their description in such a form leads to reasonable consequences in the calculations of inelastic collision cross sections with atoms and of ionization effects. We stress that (1.29) is used only for calculating the contribution of the Coulomb, "longitudinal", interaction to the differential cross section, (1.26).

20

Differential Inelastic Collision Cross Section. Substituting (1.29) into (1.26) we obtain

$$\frac{d\sigma}{d\omega} = \frac{2\pi e^4}{m_e v^2} \left\{ \frac{f(\omega)}{\omega} \left[\ln\left(\frac{2m_e v^2 \gamma^2}{\omega} \right) - \beta^2 \right] + \frac{\mathcal{F}(\omega)}{\omega^2} \right\} . \qquad (1.30)$$

When $\omega \lesssim I$, the first term in (1.30) is more important ; however, when $\omega \gg I$, the first term falls approximately like $\omega^{-4.5}$, while $\mathcal{F}(\omega)/\omega^2$ tends to $1/\omega^2$. Consequently, when $\omega \gg I$, (1.30) transforms as expected, into the Rutherford cross section (1.1).

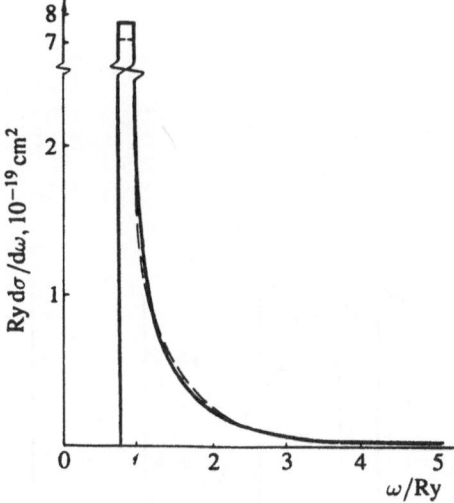

Fig. 1.4. Differential cross section of inelastic collisions of fast charged particles with a hydrogen atom for $\gamma = 4$. In the $\omega \leq Ry$ region the ratio of the total excitation cross section to $Ry/4 = \omega_\infty - \omega_1$ is shown

In Fig. 1.4 the cross section, $d\sigma/d\omega$, is presented, computed by (1.30) for atomic hydrogen, using the theoretical values of $f(\omega)$ (solid line), and the precise Möller formula (Möller 1932) (dashed line). For more complex atoms it is difficult to perform a comparison, since for these atoms there exist no reliable methods of calculating $d\sigma/d\omega$.

In a series of papers, the energy distributions, $d\sigma/d\varepsilon_e$ (where ε_e is the electron energy), were studied for electrons knocked out by a particle from the atom (Ogurtsov 1972). A simple relationship between $d\sigma/d\omega$ and $d\sigma/d\varepsilon_e$ exists, however, only for hydrogen and helium: $d\sigma/d\varepsilon_e = d\sigma/d\omega \, (\omega = \varepsilon_e + I_1)$.

In Fig. 1.5 an example is presented of the cross section $d\sigma/d\varepsilon_e$, measured in helium for various energies of the incident electrons. The solid lines are computed by (1.30) making use of (1.28) and the experimental data on the photo-absorption cross section, $\sigma_\gamma(\omega)$ (Berkowitz 1979), presented in Fig. 1.6 for the inert gases. The average energy ε_e of electrons knocked out from an atom depends weakly on the incident particle energy; for all atoms, $\overline{\varepsilon}_e \sim I$. From Figs. 1.5 and 1.6 it can be seen that (1.28) and (1.30) provide a good description of the cross section $d\sigma/d\omega$ within a wide range of ω values.

21

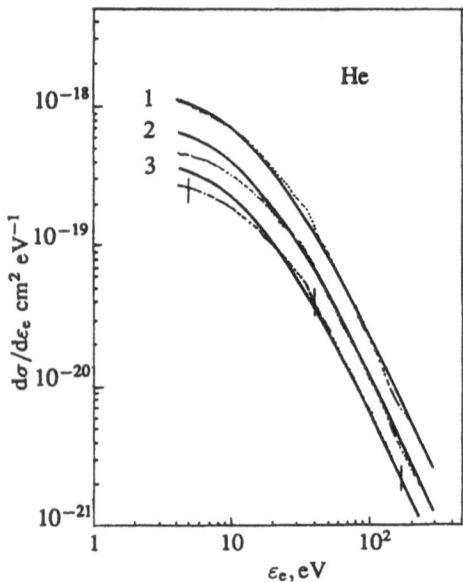

Fig. 1.5. Differential cross section of production of secondary electrons of energy ε_e in collisions of fast electrons with helium atoms at energies of $5 \cdot 10^2$ eV (*1*), 10^3 eV (*2*), $2 \cdot 10^3$ eV (*3*); the points represent experimental results (Opal et al. 1972)

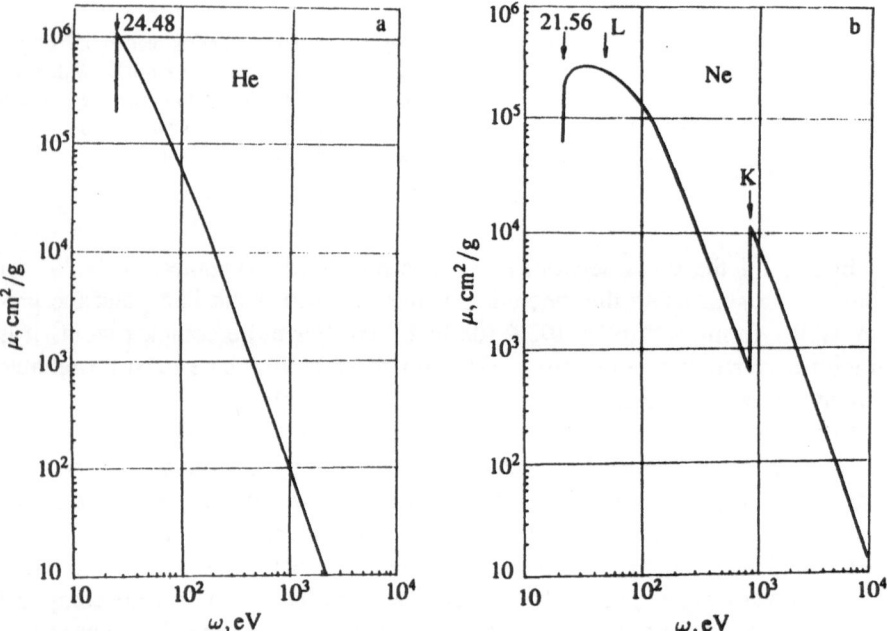

Fig. 1.6a–e. Dependence of the mass photoabsorption coefficient $\mu = n_a \sigma_\gamma(\omega)/\varrho$ on ω in the inert gases (ϱ is the density)

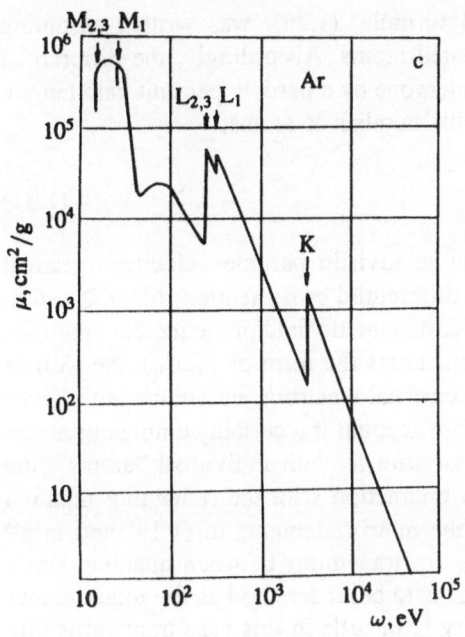

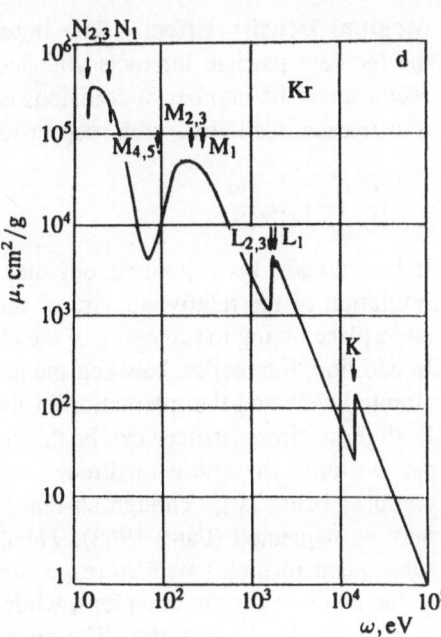

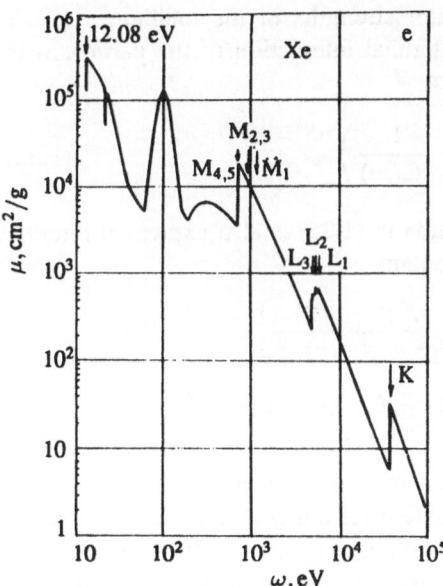

Medium Density Effect. The initial formula, (1.26), was written assuming the incident particle interacts with isolated atoms. Accordingly, the differential spectrum of the number of collisions undergone by a particle per unit path length, $d^2N/dx\,d\omega$ is additive with respect to the number of atoms:

$$\frac{d^2N}{dx\,d\omega} = n_\mathrm{a}\frac{d\sigma}{d\omega} \,. \tag{1.31}$$

It has already been pointed out that at relativistic particle velocities, gradual saturation of the relativistic rise of the differential cross section, for $\gamma \gtrsim \omega/\omega_\mathrm{p}$, takes place owing to screening of the electromagnetic field of the incident particle. In addition, interaction between the atoms alters the form of $f(\omega, q)$: the shift of atomic levels and the interaction of the external electrons are significant effects. Both these circumstances can be taken into account if a certain complex of atoms (in the limit, the entire medium) is dealt with as with individual "atoms", the complex being large enough so that its interaction with the remaining medium may be neglected (Fano 1963). Then, the matrix elements in (1.18) and in all subsequent formulas will be responsible for transitions between quantum states of the medium (or the complex), while Z is to be understood as the total number of electrons in the complex. The quantity $|F(\omega, q)|^2$ in this case transforms into a dynamical factor of the medium, $S(\omega, q)$, related to the permittivity of the medium, $\varepsilon(\omega, q)$. The generalized oscillator strengths of the medium, $\tilde{f}(\omega, q)$, determining the contribution of the longitudinal interaction of the particle with the medium (1.25), are written in the form

$$\tilde{f}(\omega, q) = \frac{2m_\mathrm{e}\omega}{q^2}S(\omega, q) = \frac{2\omega Z}{\pi\omega_\mathrm{p}^2}\mathrm{Im}\left(-\frac{1}{\varepsilon(\omega, q)}\right) \,. \tag{1.32}$$

The contribution of the transverse interaction in (1.25) is also expressed through $\varepsilon(\omega, q)$. Ultimately, instead of (1.25) we obtain

$$\begin{aligned}
\frac{d^2\sigma}{d\omega\,dq} &= \frac{2e^2}{\pi\hbar^2v^2n_\mathrm{a}q}\mathrm{Im}\left\{-\frac{1}{\varepsilon(\omega, q)} + \frac{v^2q^2 - \omega^2}{c^2q^2 - \varepsilon(\omega, q)\omega^2}\right\} \\
&= \frac{2e^2q}{\pi\hbar^2n_\mathrm{a}}\mathrm{Im}\left\{\left(1 - \frac{1}{\varepsilon(\omega, q)\beta^2}\right)(c^2q^2 - \varepsilon(\omega, q)\omega^2)^{-1}\right\} \,.
\end{aligned} \tag{1.33}$$

The task of calculating $\varepsilon(\omega, q)$ is even more complicated than calculating $f(\omega, q)$, since the quantum-mechanical problem for many atoms must be considered. Some information on $\mathrm{Im}\,\varepsilon^{-1}(\omega, q) \propto S(\omega, q)$ can be derived from experiments on the inelastic scattering of electrons in a medium, which are similar to the corresponding experiments with gases for determination of $f(\omega, q)$. As in the case of $f(\omega, q)$, the available theoretical and experimental data on $\tilde{f}(\omega, q)$ are insufficient for their straightforward application in calculations of ionization effects. Therefore, we shall use a form of $\tilde{f}(\omega, q)$ similar to (1.29) (Asoskov et al. 1982):

$$\tilde{f}(\omega, q) = \tilde{f}(\omega)\theta(\omega - q^2/2m_e) + \tilde{\mathcal{F}}(\omega)\delta(\omega - q^2/2m_e) ,$$

$$\tilde{f}(\omega) = \lim_{q \to 0} \tilde{f}(\omega, q) = \frac{2\omega Z}{\pi \omega_p^2} \mathrm{Im} \left(-\frac{1}{\varepsilon(\omega, 0)} \right) . \tag{1.34}$$

The arguments in favor of using (1.34) almost coincide with the previous ones in favor of (1.29); thus, for instance, $\tilde{f}(\omega, q)$ satisfies the sum rule (1.27). Integrating (1.33) over q, with substitution of (1.34) into the first term, and dropping the dependence on q in $\varepsilon(\omega, q)$ in the second term (which is justified, as in the derivation of (1.30), by the rapid convergence of the integral over q) we obtain

$$\begin{aligned}
\frac{d\sigma}{d\omega} &= \frac{2\pi e^4}{m_e v^2} \left\{ \frac{\tilde{f}(\omega)}{\omega} \left[\ln \left(\frac{2m_e v^2}{\omega|1 - \beta^2 \varepsilon|} \right) \right. \right. \\
&\quad \left. \left. + \frac{\mathrm{Re}\,\varepsilon - \beta^2 |\varepsilon|^2}{\mathrm{Im}\,\varepsilon} \arctan \left(\frac{\beta^2 \,\mathrm{Im}\,\varepsilon}{\beta^2 \mathrm{Re}\,\varepsilon - 1} \right) \right] + \frac{\tilde{\mathcal{F}}(\omega)}{\omega^2} \right\} \\
&= \frac{e^2}{\pi \hbar v^2 n_a} \mathrm{Im} \left[\left(\beta^2 - \frac{1}{\varepsilon} \right) \ln \left(\frac{2m_e v^2}{\omega(1 - \beta^2 \varepsilon)} \right) \right] + \frac{2\pi e^4}{m_e v^2} \frac{\tilde{\mathcal{F}}(\omega)}{\omega^2} , \tag{1.35}
\end{aligned}$$

where $\varepsilon = \varepsilon(\omega, 0) = \varepsilon(\omega)$.

This formula, which generalizes (1.30) to the case of a condensed medium, is the basic formula for calculating ionization effects. The second term in (1.35) corresponds, in the case of a transparent medium, to the Vavilov-Čerenkov radiation, while the other terms correspond to the so-called polarization losses. In calculating $d\sigma/d\omega$, the form of the function $\varepsilon(\omega)$ is determined in terms of the photo-absorption cross section, $\sigma_\gamma(\omega)$, with the aid of the known relationship

$$\mathrm{Im}\,\varepsilon(\omega) = \frac{n_a \hbar c}{\omega} \sigma_\gamma(\omega) ; \tag{1.36}$$

and the Kramers-Kronig relation for $\mathrm{Re}\,\varepsilon(\omega)$,

$$\mathrm{Re}\,\varepsilon(\omega) = 1 + \frac{2}{\pi} \fint_0^\infty \frac{\omega' \mathrm{Im}\,\varepsilon(\omega')\, d\omega'}{\omega'^2 - \omega^2} \tag{1.37}$$

where $\fint$ stands as usual for the principal value of the integral. We note that $\tilde{f}(\omega) = f(\omega)/|\varepsilon(\omega)|^2$, where $f(\omega)$ may be expressed through $\sigma_\gamma(\omega)$ using (1.28). Combining this with (1.36) results in

$$f(\omega) = \frac{2\omega Z}{\pi \omega_p^2} \mathrm{Im}\,\varepsilon(\omega) . \tag{1.38}$$

Thus, the procedure considered above for calculating the differential cross section, $d\sigma/d\omega$, of inelastic collisions of charged particles with atoms, is based on the semi-phenomenological form of the generalized oscillator strengths (1.29) and (1.34), which postulate a certain dependence of $f(\omega, q)$ and of $\tilde{f}(\omega, q)$ upon the momentum transfer, q. The dependences upon ω of $f(\omega)$ and $\varepsilon(\omega)$ are determined by the photo-absorption cross section, $\sigma_\gamma(\omega)$, with the aid of (1.28), (1.36) and (1.37). This calculation scheme has been increasingly frequent applied recently, for determining ionization effects and their fluctuations; it is sometimes termed the "photo-absorption ionization model" (Allison and Cobb 1980; Bichsel 1988).

1.3 Ionization Energy Loss

Average Restricted Energy Loss. The methods for calculating total and restricted ionization energy loss, $(-dE/dx)$ and $(-dE/dx)_{T_0}$ (for $T_0 \gg I$), are very similar to each other, but differ from the calculations of doubly restricted loss, $(-dE/dx)_{\omega_{th}, T_0}$, for $\omega_{th} \sim I$ and $T_0 \gg I$, and of restricted loss, for $T_0 \lesssim I_k$. In the first case it is possible to use the sum rule for $f(\omega, q)$, (1.27); the result will only contain a single parameter, the average ionization potential, I, which is sometimes considered as a parameter to be fitted. In the second case it is necessary to perform the numerical integration (1.15) with arbitrary ω_{th} and T_0; the calculations are not more complicated than the calculation of the overall ionization energy loss. In most of the cases the calculations have been performed of $(-dE/dx)$ and of $(-dE/dx)_{T_0}$, but not of $(-dE/dx)_{\omega_{th}, T_0}$. As pointed out in Sect. 1.1, this is due to the simple assumption, which is not always correct, that the observed ionization effect and the ionization energy loss for $T_0 \gg I$, are proportional.

We shall proceed by considering the most simple case of the average restricted loss for $m_e c^2 \gtrsim T_0 \gg I$, the general expression for which is obtained by the integration (1.26):

$$\left(-\frac{dE}{dx}\right)_{T_0} = n_a \int_0^{T_0} \omega \frac{d\sigma}{d\omega} \, d\omega = \frac{4\pi n_a e^4}{m_e v^2} \mathcal{L}_{T_0} \,, \tag{1.39}$$

where

$$\mathcal{L}_{T_0} = 2 \cdot \int_0^{T_0} d\omega \int_{\omega/v}^{\infty} \frac{dq}{q} f(\omega, q) + \int_0^{T_0} f(\omega) \, d\omega (\ln \gamma^2 - \beta^2) \,.$$

The contribution of various ω and q ranges to the integral in (1.39) is illustrated with the aid of Fig. 1.7, in which the density of the points is proportional to the contribution to (1.39) (Fano 1963). The lower integration limit, $Q_{min}[1 + Q_{min}/2m_e c^2] = \omega^2/2m_e v^2$, is shown by the solid line, and the Bethe ridge, $Q = \omega$, by the dashed line; $q^2/2m_e = Q[1 + Q/2m_e c^2]$.

If the atomic electrons were free and at rest, then all collisions would be concentrated on the Bethe ridge. Taking into account the momentum of the atomic electrons, $q_a \sim \sqrt{2m_e I}$, it is readily shown that the effective range of collisions in the (Q, ω)-plane is $\Delta Q^2 \approx \Delta \omega^2 = (Q - \omega)^2 \approx I\omega$. The intersection

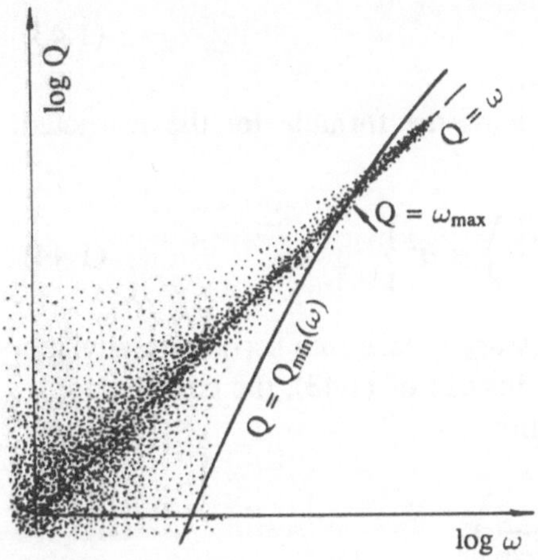

Fig. 1.7. Probability density distribution of collisions with various Q and ω

of the boundary $Q_{min}(\omega)$ with the Bethe ridge yields the kinematically possible limit $\omega \leq \omega_{max} = Q_{max} = 2m_e v^2 \gamma^2$. The existence of a range of "intermediate" Q, where the lower boundary, $Q_{min}(\omega)$, lies outside the effective (Q, ω) region (and can be ignored), simplifies the calculations considerably. A situation of that sort occurs for $I \ll \omega \ll 2m_e v^2$, and was dealt with by Bethe (1930). Consequently, the Bethe approach holds for $I \ll 2m_e v^2$ i.e. for $v \gg v_e$.

The calculation (1.39) is performed, following Bethe, by division into three parts of the region of integration over q. At low $q \leq q_1 \lesssim \hbar r_a$, instead of the generalized $f(\omega, q)$, one can make use of the optical oscillator forces, $f(\omega)$. Taking (1.27) into account, and assuming that $T_0 \to \infty$ (because of the rapid convergence of the integral over ω), we obtain

$$\mathcal{L}(q \leq q_1) = \frac{Z}{2} \left(\ln \frac{q_1^2 v^2 \gamma^2}{I^2} - \beta^2 \right) , \tag{1.40}$$

where

$$Z \ln I = \int_0^\infty f(\omega) \ln \omega \, d\omega . \tag{1.41}$$

In the intermediate q range ($q_1 \leq q \leq q_2$, $m_e c \gg q_2 \gg \sqrt{2m_e I}$), the order of integration over q and ω can be reversed, and (1.27) can be applied again; for $f(\omega, q)$ in this region:

$$\mathcal{L}(q_1 \leq q \leq q_2) = \frac{Z}{2} \ln \frac{q_2^2}{q_1^2} . \tag{1.42}$$

At large $q(q \geq q_2)$ the coupling of the electrons may be neglected by setting $f(\omega, q) = Z\delta(\omega - q^2/2m_e)$; this leads to the Rutherford expression for energy loss:

$$\mathcal{L}(q \geq q_2) = \frac{Z}{2} \ln \frac{2m_e T_0}{q_2^2} . \tag{1.43}$$

Summation of (1.40), (1.42), and (1.43) yields the formula for the restricted energy loss when $T_0 \gg I$:

$$\left(-\frac{dE}{dx}\right)_{T_0} = \frac{2\pi n_a e^4}{m_e v^2} Z \left\{ \ln \left(\frac{2m_e v^2 \gamma^2 T_0}{I^2} \right) - \beta^2 \right\} . \tag{1.44}$$

To obtain the total energy loss, it is necessary to take into account relativistic corrections to the Rutherford cross section; instead of (1.43), the corresponding contribution to the energy loss is of the form

$$\mathcal{L}(q \geq q_2) = \frac{Z}{2} \int\limits_{q_2^2/2m_e}^{Q_{\max}} dQ \left(\frac{1}{Q} - \frac{1}{2m_e c^2 \gamma^2} \right)$$

$$\approx \frac{Z}{2} \left[\ln \left(\frac{2m_e v^2 \gamma^2 2m_e}{q_2^2} \right) - \beta^2 \right] . \tag{1.45}$$

The sum of (1.40), (1.42) and (1.45) gives the Bethe-Bloch formula for the total energy loss of heavy particles (1.11) (for $\gamma \ll m/m_e$). Utilization of the form (1.29) for $f(\omega, q)$ also yields (1.44); this is readily checked by substitution of (1.30) into (1.39).

Influence of the Medium on the Ionization Energy Loss. Density Effect. From the preceding derivation of formula (1.11) it can be seen that equal contributions to the relativistic factor $Z(\ln \gamma^2 - \beta^2)$ are given by the regions of extremely small q (1.40) and by regions of extremely large q (1.45). In transition to restricted energy loss the second contribution drops out, and the density effect imposes a restriction on the relativistic rise, related to small q. This results, for sufficiently high γ, in the restricted loss, $(-dE/dx)_{T_0}$, reaching the Fermi plateau. The presence of this plateau reduces significantly the possibility of utilizing the relativistic rise of restricted loss for identification of charged particles, and for this reason the density effect has been under thorough investigation since 1940 (Fermi, 1940).

The most general approach to the description of the density effect consists in passing from interaction of a particle with individual atoms to its interaction, with complexes of atoms (the entire medium) (Sect. 1.2).

This modification is essential only for low momenta, $q < q_1$; therefore, setting $\varepsilon(\omega, q) \approx \varepsilon(\omega)$ in (1.33) and integrating over q we obtain

$$\left(-\frac{dE}{dx}\right)_{q<q_1} = \frac{e^2}{\pi v^2 \hbar^2} \operatorname{Im} \int\limits_0^\infty \left(\beta^2 - \frac{1}{\varepsilon(\omega)} \right) \ln \frac{q_1^2 v^2}{\omega^2 (1 - \beta^2 \varepsilon(\omega))} \omega \, d\omega . \tag{1.46}$$

This equation also expresses the force exerted upon a particle by its electromagnetic field in a medium (Landau and Lifshitz 1960). Equation (1.46) may also be derived by another approach applied by Fermi (1940), which calculates the total flux, $dS(\varrho_1)/dt$, of electromagnetic energy through a cylinder of radius $\varrho_1 \sim r_a$, with its axis directed along the particle trajectory. The field energy inside the cylinder being constant, so $dS(\varrho_1)/dt \approx (-dE/dx)_{q \leq q_1}$ for $\varrho_1 \approx \hbar/q_1$. In both the approaches the total energy losses are dealt with accurately, including losses due to Vavilov-Čerenkov radiation. An unambiguous identification of the contribution of this radiation in (1.46) is feasible only for a transparent medium, where $\text{Im}\,\varepsilon \to 0$. For all Čerenkov frequencies, $\beta^2 \text{Re}\,\varepsilon > 1$, and the phase of the logarithm in (1.46) is π, so that

$$\left(-\frac{dE}{dx}\right)_{\check{C}} = \frac{e^2}{v^2 \hbar^2} \int_{\omega_{\check{C}}} \left(\beta^2 - \frac{1}{\text{Re}\,\varepsilon}\right) \omega \, d\omega \ . \tag{1.47}$$

Taking into account that $\text{Im}\,\varepsilon$ is finite, substantially reduces the radiation, i. e. the energy flux far from the trajectory, because of both absorption and contraction of the Čerenkov bands. In this case it becomes difficult to separate the losses due to the Vavilov-Čerenkov radiation and those due to ionization of atoms, because they are dependent on the model chosen for $\varepsilon(\omega)$. This problem was investigated, for instance, by Budini et al. (1960), according to whom, the energy losses due to ionization of atoms are

$$\left(-\frac{dE}{dx}\right)_{\text{ion}} = \frac{2\pi n_a e^4}{m_e v^2} \cdot \left[\int_0^\infty \tilde{f}(\omega) \left(\ln \frac{q_1^2 v^2}{\omega^2 |1 - \beta^2 \varepsilon|} - \beta^2 \text{Re}\,\varepsilon\right) d\omega \right],$$

which represents a direct formal generalization of (1.40).

The most elegant and general computation of (1.46) was given by Landau and Lifshitz (1960) who utilized the analyticity of $\varepsilon(\omega)$ in the upper semi-plane of complex ω. Turning the contour of integration in (1.46) to the positive imaginary $\omega = i\nu$ axis, and taking into account that $\text{Im}\,\varepsilon(i\nu) = 0$ one can obtain:

$$\mathcal{L}_E(q \leq q_1) = \mathcal{L}(q \leq q_1) - \frac{Z}{2} \cdot \delta_E , \quad \text{where}$$
$$\delta_E = \ln \frac{\overline{\omega^2 + l^2}}{l^2} - \frac{l^2}{\omega_p^2 \gamma^2} \ . \tag{1.48}$$

and the index E refers to the energy loss dE/dx. Here averaging is performed with the weighting $\tilde{f}(\omega)/Z$:

$$\ln \overline{(\omega^2 + l^2)} = \int_0^\infty \tilde{f}(\omega) \ln(\omega^2 + l^2) \, d\omega/Z \ ; \tag{1.49}$$

and l is either a positive root of the equation $\varepsilon(il) = \beta^{-2}$, or zero, if no such root exists. The correction δ_E occurring in the contribution of small q is the same both for the total [(1.11)] and for the restricted energy loss:

$$\left(-\frac{dE}{dx}\right)_{T_0} = \frac{A_0}{\beta^2}\left(\ln\frac{2m_e v^2\gamma^2 T_0}{I^2} - \beta^2 - \delta_E\right) , \qquad (1.50)$$

where $A_0 = (2\pi n_a e^4/m_e v^2)\cdot Z = 0.1536(Z/A)\varrho$ MeV/cm. ϱ is the density of the substance in g/cm^3; A is the atomic mass.

It is very important that $\varepsilon(i\nu)$ is a monotonous function and that $\varepsilon(i0) \geq \varepsilon(i\nu) > 1$; therefore l and consequently, (1.48), depend weakly on the model representation of $\varepsilon(\omega)$. In this connection, a simple model of the medium in the form of a set of K harmonic oscillators is usually applied in calculations of the correction δ_E:

$$\varepsilon(\omega) = 1 - \sum_{j=1}^{K}\frac{f_j\omega_p^2}{Z(\omega^2 - \omega_j^2 + 2i\Gamma\omega)} , \qquad \Gamma \to 0 . \qquad (1.51)$$

Fermi (1940) for example, assumed that $K = 1$. In the most detailed calculations by Sternheimer, K was considered equal to the number of electron shells; f_j was considered equal to the number of electrons in a shell; and ω_j was chosen to be close to the ionization potential of the respective shell (Sternheimer 1981). In model (1.51), in accordance with (1.38), $f(\omega) = \sum_{j=1}^{K} f_j\delta(\omega - \omega_j)$. In this case,

$$\tilde{f}(\omega) = \frac{f(\omega)}{|\varepsilon(\omega)|^2} \approx \sum_{j=1}^{K} f_j\delta(\omega - l_j) .$$

Consequently,

$$\overline{\ln(\omega^2 + l^2)} = \frac{1}{Z}\sum_{j=1}^{K} f_j\ln(l_j^2 + l^2) ,$$

where $l_j^2 = \omega_j^2 + f_j\omega_p^2/Z$, while the correction is

$$\delta_E = \frac{1}{Z}\sum_{j=1}^{K} f_j\ln\left(\frac{l_j^2 + l^2}{\omega_j^2}\right) - \frac{l^2}{\omega_p^2\gamma^2} . \qquad (1.52)$$

For $\beta < \beta_0$, where $\beta_0 = 1/\sqrt{\varepsilon(0)}$, $l = 0$ and $\delta_E(0) = (1/Z)\cdot\sum_{j=1}^{K}\ln(l_j^2/\omega_j^2)$. This constant correction to the Bethe-Bloch formula is related to the shift of the atomic levels, ω_j, due to the influence of the medium and, as a rule, is insignificant. If in the Bethe-Bloch formula one utilizes $\tilde{I}$, the average logarithmic potential with account of the influence of the medium is

$$\ln\tilde{I} = \frac{1}{Z}\int_0^\infty \tilde{f}(\omega)\ln\omega\, d\omega = \frac{1}{Z}\sum_{j=1}^{K} f_j\ln l_j ,$$

then δ_E will be substituted by $\tilde{\delta}_E$:

$$\tilde{\delta}_E = \frac{1}{Z} \sum_{j=1}^{K} f_j \ln \left(\frac{l_j^2 + l^2}{l_j^2} \right) - \frac{l^2}{\omega_p^2 \gamma^2} . \tag{1.53}$$

In the case of such "renormalization" of the average ionization potential, the correction $\tilde{\delta}_E$ approaches zero when $\beta \leq \beta_0$.

As γ increases, the root l also increases. When $\gamma \gg l_j/\omega_p$, $l \approx \omega_p \gamma \gg l_j$; therefore

$$\tilde{\delta}_E \approx 2 \ln \left(\frac{\omega_p \gamma}{\tilde{I}} \right) - 1 , \tag{1.54}$$

and the restricted energy loss arriving at the Fermi plateau is

$$\left(-\frac{dE}{dx} \right)_{\substack{T_0 \\ \gamma \to 0}} = \frac{2\pi n_a e^4 Z}{m_e c^2} \ln \left(\frac{2m_e c^2 T_0}{\omega_p^2} \right) . \tag{1.55}$$

Unlike the restricted loss, the total loss continues to rise logarithmically with γ, although with a smaller derivative than before the onset of influence of the density effect (Fig. 1.1):

$$\left(-\frac{dE}{dx} \right)_{\gamma \to \infty} \to \frac{2\pi n_a e^4 Z}{m_e c^2} \left[\ln \left(\frac{2m_e c^2 \gamma}{I} \right)^2 - 1 \right] . \tag{1.56}$$

To represent the results of calculations of $\tilde{\delta}_E$ in a convenient manner, the following approximation formula was used in the work of Sternheimer:

$$\tilde{\delta}_E = \begin{cases} 0 & \log(\beta\gamma) \leq \log(\beta_0\gamma_0) = X_0 \\ 4{,}6 \log(\beta\gamma) + C + a(X_1 - \log(\beta\gamma))^m & X_0 \leq \log(\beta\gamma) \leq X_1 \\ 4{,}6 \log(\beta\gamma) + C & X_1 \leq \log(\beta\gamma) . \end{cases} \tag{1.57}$$

The parameters X_0, X_1, m are tabulated by Sternheimer et al. (1971) for various substances (for a certain choice of f_j and ω_j in (1.53)):

$$C = -2 \ln(\tilde{I}/\omega_p) - 1 ; \quad a = -(C + 4.6X_0)/(X_1 - X_0)^m .$$

With an uncertainty of up to 2 % in $(-dE/dx)_{T_0}$, the correction $\tilde{\delta}_E$ can be calculated with the aid of certain mean parameters $\overline{X}_0$, $\overline{X}_1$ and $\overline{m} = 3$, presented in Table 1.2 (Ahlen 1980) (Fig. 1.8).

Mean Ionization Potential. The only parameter in the Bethe-Bloch formula that characterizes the medium is the mean ionization potential, I. We have already noted that in a condensed medium I must be substituted by $\tilde{I}$, which depends on the aggregate state of the substance. For determination of $\tilde{I}$ numerous measurements of $(-dE/dx)_{T_0}$ have been performed in the region of $\beta < \beta_0$, where the density effect can be neglected. The potential $\tilde{I}$ is regarded in these experiments

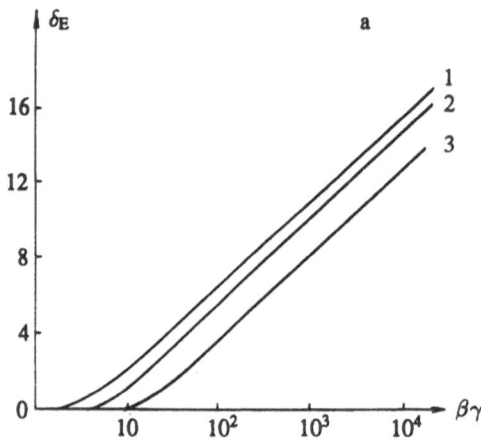

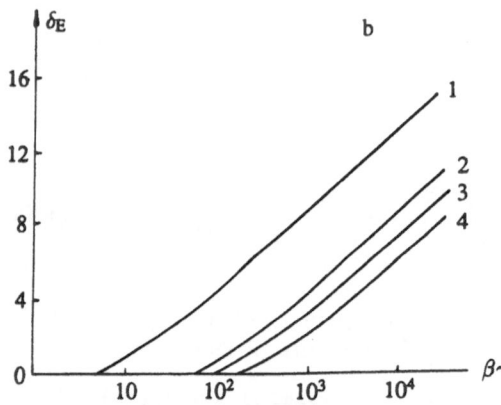

Fig. 1.8. Correction due to the density effect for condensed substances (a) and for gases (b): (a): 1 – Be; 2 – H_2O; 3 – nuclear photo-emulsion; (b): 1 – H_2 at $P = 10^7$ Pa; 2 – H_2; 3 – air; 4 – Ar (all gases are at $P = 10^5$ Pa except for 1)

as a fitted parameter. Results of these studies are presented by Janni (1982), Seltzer and Berger (1982) and, in part, in Table 1.3.

The first calculation of I for atomic hydrogen was performed by Bethe (1933): $I_H = 15.5\,\text{eV}$. For heavy atoms the Thomas-Fermi method is usually applied with $I/Z \approx 4.95\,\text{eV}$; more precise calculations with approximate wave functions yield $I/Z \approx 10\text{--}12\,\text{eV}$. For calculations one can make use of the following empirical dependence: $I = (11.2+11.7Z)\,\text{eV}$ for $Z \leq 13$; $I = (52.8+8.71Z)\,\text{eV}$ for $Z > 13$.

On the whole, there is good agreement between the theoretical and experimental values of I; thus, for instance, the enhancement of $\tilde{I}$ above I is noticeable in condensed substances: usually $\tilde{I}/I = 1.1\text{--}1.3$. The ionization potentials of chemical complexes and mixtures is found with the aid of the Bragg rule:

$$n_e \ln I = n_{1e} \ln I_1 + n_{2e} \ln I_2 + \dots , \qquad (1.58)$$

where n_{je} and I_j are the electron density and the ionization potential of the j-th element. Deviations from the Bragg rule due to influence of the chemical

Table 1.2. Parameters of the approximation formula (1.57) for the calculation of the correction $\tilde{\delta}_E$ due to the density effect

A. Solid substances and liquids

I	$|C|$	X_0	X_1
$100\,\text{eV} > I$	$|C| < 3.681$	0.2	2.0
$100\,\text{eV} > I$	$|C| \geq 3.681$	$0.321|C| - 1.0$	2.0
$100\,\text{eV} \leq I$	$|C| < 5.215$	0.2	3.0
$100\,\text{eV} \leq I$	$|C| \geq 5.215$	$0.326|C| - 1.0$	3.0

B. Gases at a pressure $P = 1.0133 \cdot 10^5$ Pa and a temperature $T_0 = 0\,^{\circ}\text{C}$

C	X_0	$X_{1_,}$
$10.0 > |C|$	1.6	4.0
$10.0 \leq |C| < 10.5$	1.7	4.0
$10.5 \leq |C| < 11.0$	1.8	4.0
$11.0 \leq |C| < 11.5$	1.9	4.0
$11.5 \leq |C| < 12.25$	2.0	4.0
$12.25 \leq |C| < 13.804$	2.0	5.0
$13.804 \leq |C|$	$0.325|C|-2.5$	5.0

C. Gases at a density $\eta \varrho_0$, where ϱ_0 is the density in normal conditions: $P = 1.0133 \cdot 10^5$ Pa, $T = 0\,^{\circ}\text{C}$

$X_0(\eta) = X_0 - 0.51 \log \eta$; $X_1(\eta) = X_1 - 0.5 \log \eta$; $a(\eta) = a$; $C(\eta) = C + 2.303 \log \eta$.

coupling of the outer electron shells may lead to an uncertainty of up to 1 % in the determination of $(-dE/dx)$. At present there exist sufficiently complete data on $\tilde{I}$ for many elements and substances, in various aggregate states, to permit calculation of $(-dE/dx)$ and $(-dE/dx)_{T_0}$ with an uncertainty of the order of 1 %. Calculations of the density effect yield about the same precision. As an example, the calculated restricted energy losses in inert gases under normal conditions are presented in Fig. 1.9a.

Doubly Restricted, $(-dE/dx)_{\omega_{\text{th}},T_0}$, and Restricted, $(-dE/dx)_{T_0}$, Energy Loss for $T_0 \lesssim I_K$. (I_K is the ionization potential of the K electron shell.) It was noted above that for describing the ionization effects in certain detectors one must determine the doubly restricted energy loss, $(-dE/dx)_{\omega_{\text{th}},T_0}$, where $\omega_{\text{th}} \sim I$. To compute it, we shall take advantage of (1.30) for the differential cross section, based on the semi-phenomenological form of the generalized oscillator forces (1.29). We shall take into account the fact that $(-dE/dx)_{\omega_{\text{th}},T_0} = (-dE/dx)_{T_0} - (-dE/dx)_{0,\omega_{\text{th}}}$, where

$$\left(-\frac{dE}{dx}\right)_{0,\omega_{\text{th}}} = \frac{2\pi n_a e^4}{m_e v^2} \int_0^{\omega_{\text{th}}} f(\omega) \left[\ln\left(\frac{2m_e v^2 \gamma^2 \omega_{\text{th}}}{\omega^2}\right) - \beta^2\right] d\omega \ . \tag{1.59}$$

Table 1.3. Properties of gases and vapors utilized in ionization detectors

Substance	Z	A [Eq. (1.50)]	ϱ [10^{-3} g/cm^3]	ω_p [eV]	I_1 [eV]	I [eV]	w [eV] [Eq. (1.14)]	$(-dE/dx)_{min}$ [keV/cm]	$(dN_t/dx)_{min}$ [cm^{-1}]	$(dN_1/dx)_{min}$ [cm^{-1}]
Helium He	2	4.003	0.165	0.263	24.59	38.5	42.3	0.322	7.6	3.3
Neon Ne	10	20.18	0.838	0.587	21.56	133.8	36.4	1.452	39.9	10.9
Argon Ar	18	39.95	1.662	0.787	15.76	181.6	26.3	2.541	96.6	24.8
Krypton Kr	36	83.8	3.478	1.113	14.00	340.8	24.05	4.750	197.5	33.0
Xenon Xe	54	131.3	5.485	1.363	12.13	508.8	21.9	6.862	313.3	44.8
Hydrogen H$_2$	2	2.016	0.0837	0.263	15.43	20.4	36.3	0.342	9.4	4.7
Nitrogen N$_2$	14	28.013	1.166	0.694	15.58	97.8	34.65	2.097	60.5	20.8
Oxygen O$_2$	16	31.999	1.331	0.741	12.08	115.7	30.83	2.360	76.5	23.7
Air	14.46	28.98	1.205	0.707		103.12	33.8	2.148	63.6	23.7
Methane CH$_4$	10	16.043	0.668	0.588	12.71	44.13	27.1	1.608	59.3	24.8
Ammonia NH$_3$	10	17.031	0.718	0.592	10.14	61.11	26.5	1.586	59.8	
Acetylene C$_2$H$_2$	14	26.038	1.093	0.699	11.41	61.42	25.75	2.339	90.8	31.5
Ethylene C$_2$H$_4$	16	28.954	1.174	0.746	10.51	53.51	25.8	2.696	104.5	40.4
Etane dioxide	18	30.070	1.250	0.788	11.50	48.07	24.38	2.870	117.7	40.5
Carbon dioxide CO$_2$	22	44.011	1.829	0.871	13.79	102.35	32.8	3.280	100.0	33.6
Nitrous dioxide N$_2$O	22	44.013	1.829	0.871	12.89	103.96	32.55	3.275	100.6	
Propan C$_3$H$_8$	26	44.097	1.833	0.958	11.07	49.69	23.45	4.138	176.5	67.6
Isobutane i-C$_4$H$_{10}$	34	58.125	2.416	1.099	10.55	50.56	23.2	5.402	232.8	83.6
Freon-13 CClF$_3$	50	104.46	4.345	1.314	12.91	133.4		7.732		
Freon-12 CCl$_2$F$_2$	58	120.91	5.030	1.416	12.31	147.7	29.55	8.887	300.7	
Freon-13B1 CBrF$_3$	68	148.91	6.194	1.533	11.83	195.0		10.153		
Sulfurhexafluoride SF$_6$	70	146.05	6.075	1.560	15.69	136.3		10.804		
Water H$_2$O	10	18.015	$0.986\ 10^{-2}$	$2.13\ 10^{-2}$	12.61	81.77	29.9	$1.98\ 10^{-3}$	0.067	0.025
Ethyl alcohol C$_2$H$_5$OH	26	46.070	$2.521\ 10^{-3}$	$3.44\ 10^{-3}$	10.47	62.99		$5.27\ 10^{-3}$		0.076
Acetone CH$_2$COCH$_3$	32	58.080	$3.179\ 10^3$	$3.81\ 10^{-3}$	9.69	64.89	28.5	$6.49\ 10^{-3}$	0.23	0.092
Benzene C$_6$H$_6$	42	78.114	$4.275\ 10^{-3}$	$4.37\ 10^{-3}$	9.25	61.42	23.3	$8.57\ 10^{-3}$	0.37	0.13
Methylal CH$_3$O$_2$CH$_2$	42	76.096	$4.165\ 10^{-3}$	$4.37\ 10^{-3}$	9.98	68.6		$9.06\ 10^{-3}$		
Carbon tetrachloride CCl$_4$	74	153.84	$8.420\ 10^{-3}$	$5.80\ 10^{-3}$	11.47	169.67	23.2	$1.38\ 10^{-3}$	0.59	

The data in the upper part of the table is for gases at $P = 1.0133 \cdot 10^5$ Pa (1 atm), 20°C; the lower part lists values for vapors at $P = 1.33 \cdot 10^2$ Pa (1 Torr), 20°C

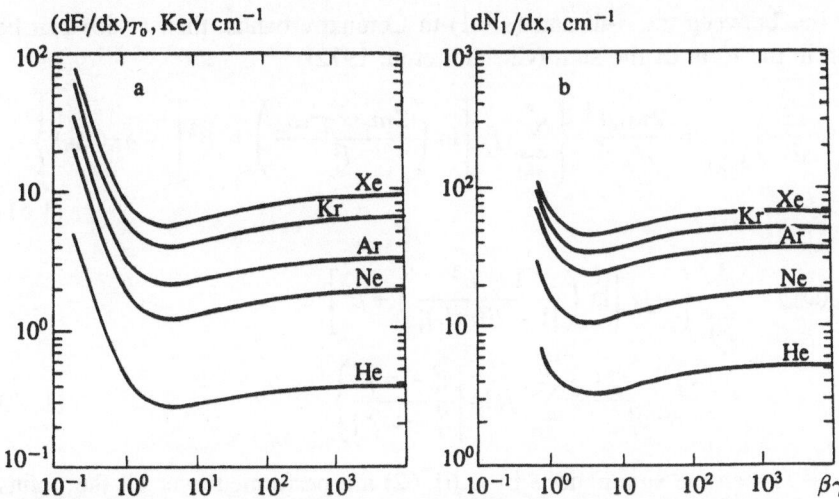

Fig. 1.9. Calculated dependence upon $\beta\gamma$ of restricted energy loss (**a**) and of primary ionization (**b**) in inert gases under normal conditions; $T_0 = 100\,\text{keV}$

Expression (1.59) is a natural generalization of the formula for restricted energy loss (1.44) and transforms into it for $\omega_{\text{th}} \gg I$; under this condition, the upper limit of integration in (1.59) can be set equal to infinity, and the mean logarithmic ionization potential, I, can be introduced. However, when $\omega_{\text{th}} \sim I$, numerical integration of (1.59) is necessary.

To take account of the density effect we shall make use of (1.35) instead of (1.30). Then

$$
\left(-\frac{dE}{dx}\right)_{0,\omega_{\text{th}}} = \frac{2\pi n_{\text{a}} e^4}{m_{\text{e}} v^2} \cdot \int\limits_0^{\omega_{\text{th}}} \tilde{f}(\omega) \left[\ln\left(\frac{2m_{\text{e}} v^2 \omega_{\text{th}}}{\omega^2 |1 - \beta^2 \varepsilon|}\right) \right.
$$
$$
\left. + \frac{(\text{Re}\,\varepsilon - \beta^2 |\varepsilon|^2)}{\text{Im}\,\varepsilon} \arctan\left(\frac{\beta^2 \text{Im}\,\varepsilon}{\beta^2 \text{Re}\,\varepsilon - 1}\right) \right] d\omega
$$
$$
= \frac{e^2}{\pi v^2 \hbar^2} \text{Im} \int\limits_0^{\omega_{\text{th}}} \left(\beta^2 - \frac{1}{\varepsilon}\right) \ln\left(\frac{2m_{\text{e}} v^2 \omega_{\text{th}}}{\omega^2 (1 - \beta^2 \varepsilon)}\right) \omega\, d\omega \ . \qquad (1.60)
$$

Application at $\omega_{\text{th}} \sim I$ of the analytical method of Landau and Lifshitz (1960) in (1.60) is not possible; in this case numerical calculations must be performed, which are always extremely cumbersome, since they require computation of the non-trivial complex function $\varepsilon(\omega)$ by the photo-absorption ionization model (Sect. 1.2). As in the case of restricted energy loss $(-dE/dx)_{T_0}$, the results of such calculations can be assumed to depend weakly on the model representation of $\varepsilon(\omega)$. If the model of discrete harmonic oscillators (1.51) is applied to $\varepsilon(\omega)$, then contributions to the integral are given only by points $\omega = l_j$ (first term) and Čerenkov frequency bands ($\omega_j/\hbar$, $\omega_j'/\hbar$) (second term), where ω_j' is the lower boundary of the j-th Čerenkov band, determined by the equation $\text{Re}\,\varepsilon(\omega_j') = \beta^{-2}$.

If ω_{th} lies between the s-th and $(s+1)$-th Čerenkov bands, then (1.60) can be written in the form of the sum (Chechin et al. 1972)

$$\left(-\frac{dE}{dx}\right)_{0,\omega_{th}} = \frac{2\pi n_a e^4}{m_e v^2}\left\{\sum_{j=1}^{s} f_j\left[\ln\left(\frac{2m_e v^2 \gamma^2 \omega_{th}}{l_j^2}\right) - \beta^2\right] - \tilde{\delta}_E(\omega_{th})\right\},$$

(1.61)

where

$$\tilde{\delta}_E(\omega_{th}) = \sum_{j=1}^{s}\left\{-f_j\left[\ln\left(\frac{1-\beta^2}{|1-\beta^2\varepsilon(l_j)|}\right) + \beta^2\right]\right.$$
$$\left. + \frac{(\omega_j^2 - \omega_j'^2)}{\gamma^2 \omega_p^2} + \sum_{i=1}^{K} f_i \ln\left|\frac{l_i^2 - \omega_j^2}{l_i^2 - \omega_j'^2}\right|\right\}.$$

(1.62)

If $\omega_{th} \gg I$, then the summations in (1.61, 62) are performed over all the points l_j and over all Čerenkov bands; then (1.61) reduces to (1.50), and (1.62) reduces to the Sternheimer correction, (1.53).

In performing calculations by (1.61, 62) the discrete oscillator forces corresponding to excitation of the atom can be taken from spectroscopic data. The oscillator forces, f_j, for the continuous spectrum are found from the continuous distribution $f(\omega)$, obtained from measurements of the photo-absorption cross section for atoms. To this end $f(\omega)$ is divided into K parts ($K \approx 10$) to each of which there correspond the forces of equivalent discrete oscillators, $f_j = \int_{\omega_{j,l}}^{\omega_{j,u}} f(\omega)\,d\omega$, where $\omega_{j,l}$ and $\omega_{j,u}$ are the lower and upper boundaries, respectively, of the j-th partition interval. The oscillator energies may be expressed as

$$\ln \omega_j = f_j^{-1}\int_{\omega_{j,l}}^{\omega_{j,u}} f(\omega)\ln\omega\,d\omega.$$

Relativistic Rise of Ionization Loss. An extremely important characteristic (from the point of view of applying ionization detectors for identification of relativistic charged particles) is the relative enhancement of the ionization effect from the minimum at $\gamma \approx 4$ up to the Fermi plateau, as $\gamma \to \infty$, the so-called *relativistic rise* $R_J = (dJ/dx)(\gamma \to \infty)/(dJ/dx)(\gamma \approx 4)$. If (dJ/dx) is related to $(-dE/dx)_{T_0}$, then the relativistic rise of restricted energy loss, $R_E(T_0)$, is implied.

In the case of restricted energy loss, the growth of T_0 is accompanied by an enhancement of the role played by short-range ("Rutherford") collisions; their contributions yield no relativistic rise, being proportional to β^{-2}. As a result, the relativistic rise, $R_E(T_0)$, diminishes. In the case of doubly restricted energy loss the growth of ω_{th} is accompanied by a rapid decrease in the contribution of collisions with the outer electron shells; this is strongly affected by the density effect, and reaches the Fermi plateau at relatively low values of the Lorentz factor $\gamma_{pl}^1 \sim I_1/\omega_p$. Here the relativistic rise increases, since the contribution of the low-lying electron shells is more pronounced; for these, the respective Lorentz factor $\gamma_{pl}^K \sim I_K/\omega_p$ is significantly higher and consequently the relativistic rise is greater. However, when $\omega_{th} > I_K$, and the contribution of long-range collisions is strongly suppressed, the relativistic rise begins to decrease owing to the dominant contribution of short-range collisions. Such behavior of the relativistic rise is essential in analyzing the operation of threshold detectors, such as the bubble chamber or a nuclear photo-emulsion (Asoskov et al. 1982). For gas detectors it may turn out, in certain cases, that $\omega_{th} \approx 0$ while $T_0 \lesssim I_K$. If so, the contribution of low-lying electron shells is suppressed and the relativistic rise decreases (Fig. 1.10).

From (1.61) and (1.62) it can be seen that the contribution of oscillators with $\omega_j > \omega_{th}$ is present only in the definition of the function $\varepsilon(\omega)$; if $\omega_{s+1} \gg \omega_{th}$, then its dependence on $\varepsilon(\omega)$ may be dropped assuming

$$\varepsilon(\omega) \approx \varepsilon_{ef}(\omega) = 1 - \sum_{j=1}^{s} f'_1 \omega_p'^2 / Z(\omega^2 - \omega_j^2 + 2i\Gamma\omega) , \qquad (1.63)$$

where $f'_j = Zf_j/\mathcal{F}_s$, $\omega_p'^2 = \mathcal{F}_s\omega_p^2/Z$. In this case $(-dE/dx)_0$, ω_{th} can be written in a form similar to the Bethe-Bloch formula with the Sternheimer correction corresponding to the "effective" dielectric constant ε_{ef}, see (1.63) (Alakoz et al. 1975):

$$\left(-\frac{dE}{dx}\right)_{0,\omega_{th}} = \frac{2\pi n_a e^4}{m_e v^2} \mathcal{F}_s \left[\ln\left(\frac{2m_e v^2 \gamma^2 \omega_{th}}{\bar{I}'^2}\right) - \beta^2 - \tilde{\delta}'_E \right] ,$$

$$\ln \tilde{I}' = \frac{1}{Z}\sum_{j=1}^{s} f'_j \ln l_j , \quad \tilde{\delta}'_E = \frac{1}{Z}\sum_{j=1}^{s} f'_j \ln\left(\frac{l_j^2 + l'^2}{l_j^2}\right) - \frac{l'^2}{\omega_p'^2\gamma^2} ; \qquad (1.64)$$

$$\varepsilon_{ef}(il') = \beta^{-2} .$$

Thus, for instance, at $\gamma \gg \tilde{I}'/\omega_p'$

$$\left(-\frac{dE}{dx}\right)_{0,\omega_{th}} \simeq \frac{2\pi n_a e^4}{m_e c^2} \mathcal{F}_s \ln\left(\frac{2m_e c^2 \omega_{th}}{\omega_p'^2}\right) . \qquad (1.65)$$

Relations (1.63–65) have a simple physical meaning: oscillators with frequencies $\omega_j \gg \omega_{th}$ are not excited and therefore take no part in decelerating the particle.

Observed Ionization Effect and Ionization Energy Loss. In Sect. 1.1 it was pointed out that the ionization effect observed in detectors of total ionization depends on the doubly restricted energy loss (1.15). On which physical phenomena is this based? That is, which physical phenomena allow the approximate relation $dJ/dx \approx (1/w) \cdot (-dE/dx)_{\omega_{th},T_0}$ to hold? Here w is independent of v and its physical meaning is the average energy spent in the medium for the production of a single free electron (Christophorou 1970).

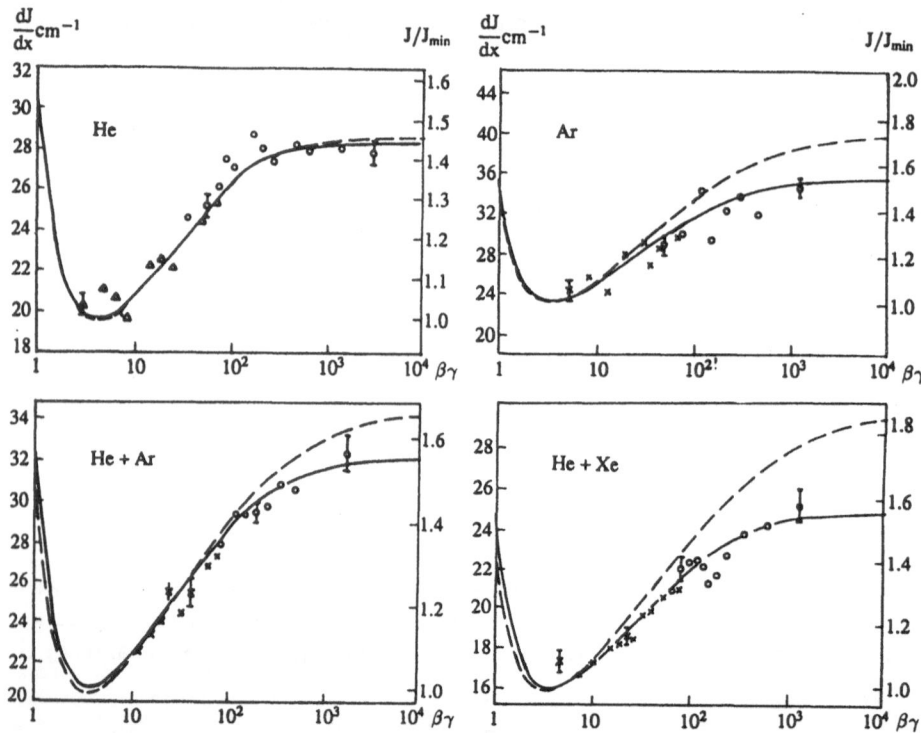

Fig. 1.10. Dependence of track density in a Wilson chamber operating in the droplet-counting mode on $\beta\gamma$: ———— computation by formula (1.64) for $\omega_{th} \simeq 1\,\text{keV}$; $\cdots\cdots$ computation by (1.50) for $T_0 = 1\,\text{keV}$ (Merson 1972); the points are experimental data ($\bullet$ – muons, $\circ$ – electrons) (Kepler et al. 1958); $dJ/dx = (-dE/dx)_{T_0}/w$

The energy lost by a particle in a substance is spent on: (a) overcoming the binding energy of electrons in the atom, (b) the kinetic energy of the knocked-out electrons, and (c) Vavilov-Čerenkov radiation.

In ionizing the outer electrons, an energy of the order of $2I_1$ is required for the production of an ion pair, since the average kinetic energy of each knocked-out electron is $\bar{\varepsilon}_e \sim I_1$ (Sect. 1.2). Accordingly, ionization of the K- and L-shells requires energies of $2I_K$ and $2I_L$, and so on. This energy is further redistributed among the atoms in portions of the order of $2I_1$ as a result of secondary processes. Indeed, electrons knocked out from the inner shells ionize atoms, spending an energy of about $2I_1$ on each secondary electron. Transition of the atoms to the final state is accompanied by a cascade of Auger electrons and by emission of fluorescence quanta, which transfer their energy to secondary electrons and to photo-electrons; the latter ionize the outer electron shells again. The process of energy division continues until the energy lost by the particle is spent up in portions of about $2I_1$ in ionization of the outer electrons. The electrons, thus produced, that have an energy lower than the lowest excitation potential $I^* \lesssim I_1$, lose their energy and are thermalized as a result of elastic collisions with the atoms. Estimates reveal that secondary ionization takes place at distances of the order of 10^{-4} cm from the particle trajectory, i.e. within the track width. In any case, a certain part of the initially released energy leaves the track trajectory and yields no contribution to dJ/dx. Firstly, δ-electrons (as well as the high-energy fluorescence radiation) with an energy $\varepsilon_e \gtrsim T_0$ carry nearly all their energy away from the track premises. This can be taken into account by introducing the efficiency of energy released by δ-electrons (and quanta) inside the track, i.e. by introducing a certain weight function, $\mathcal{P}(\omega)$ (1.17), which falls to zero in the region of $\omega \gtrsim T_0$. However, because of the weak, logarithmic dependence of the energy loss on T_0, it is sufficient to consider the restricted energy loss $(-dE/dx)_{T_0}$, which implies that T_0 indicates a certain effective boundary. Secondly, because gases are transparent to Vavilov-Čerenkov radiation, it propagates far from the particle trajectory. Its registration in gas detectors is possible only because of photo-absorption by admixtures, or because of the photo-effect occurring in the walls. However, the relative contribution of the Vavilov-Čerenkov radiation to the total loss is small, even in the region of the Fermi plateau: about 5 % in H_2 and He, and less than 1 % in heavier gases.

As a result, in energy transfers of $\omega \lesssim T_0$, nearly all the lost energy is released in the immediate vicinity of the particle trajectory, since the energy required for the production of a single secondary electron is, on average, about $2I_1$. Therefore, in total ionization detectors, in which the observed ionization effect is proportional to the number of secondary electrons, dN_t/dx, $dJ/dx \propto dN_t/dx \approx (1/w)[-dE/dx]_{T_0}$, $w \approx 2I_1$ (Table 1.3). The independence (or to be more precise, the weak dependence) of w on v follows from the fact that the kinetic energy of electrons knocked out from the outer shells is independent (weakly dependent) on v and is close to I_1.

Some detectors which have been discussed in this section – (gas proportional, drift and ionization chambers, gas scintillation detectors, and the Wilson cloud chamber operating in the droplet-counting mode) – are considered in greater detail in Chaps. 2–5.

1.4 Primary Ionization and the Number of Collisions

Ionization Cross Section. The Bethe Formula. Determination of the primary ionization is one of the main problems in the theory of atomic collisions, and it is important not only for high energy physics, but also for plasma physics, astrophysics, physics of the upper atmosphere etc.. The relativistic rise of the specific primary ionization in gases is utilized for identification of high energy particles in a manner similar to the utilization of the relativistic rise of ionization energy loss. This method is used at present, for instance, in spectrometers based on large streamer chambers (Sect. 4.4).

As in the case of the differential collision cross section (Sect. 1.2), accurate calculation of the ionization cross section, σ_i, has been performed only for atomic hydrogen. According to Bethe (1933)

$$\sigma_i = r \frac{2\pi e^4}{m_e v^2 I_1} \left(\ln \frac{2 m_e v^2 \gamma^2}{I_1} + s - \beta^2 \right), \tag{1.66}$$

where the dimensionless parameters r and s depend on Z, and for hydrogen have the following values: $r = 0.285$, $s = 3.04$. In calculations for heavier substances approximate wave functions of atoms were used, and the macroscopic and other methods were applied (Inokuti 1971; Moiseiwitsch 1985). Available theoretical and experimental data do not give a general outlook of the dependence of σ_i upon Z and γ, even less so if the density effect is taken into account. We shall therefore make use of the semi-phenomenological formula (1.30) for $d\sigma/d\omega$ and integrate it over ω within the limits $I_1 \leq \omega < \infty$. The result can be represented in the form (1.66), where

$$\frac{r}{I_1} = \int\limits_{I_1}^{\infty} \frac{f(\omega)\, d\omega}{I_1} \; ; \quad \frac{r}{I_1} s = \int\limits_{I_1}^{\infty} \frac{f(\omega)}{\omega} \left(\ln \frac{I_1}{\omega} + 1 \right) d\omega + \frac{\mathcal{F}(I_1)}{I_1}. \tag{1.67}$$

(Owing to the rapid convergence of integrals in the calculation of the ionization cross section, and the number of collisions, one can set the upper limit of integration, T_0, equal to infinity.)

The results of calculations for inert gases by formulas (1.66), (1.67) and (1.68) are presented in Table 1.4; these utilize the data on photo-absorption coefficients. From the last column of Table 1.4 it can be seen that the calculated and observed cross sections differ, on average, by less than 6 %. In accordance with (1.66) the quantity $\sigma_i \beta^2$ is a linear function of $\ln \beta^2 \gamma^2 - \beta^2$:

$$\sigma_i \beta^2 = \frac{A_1}{n_a} [\ln(\beta^2 \gamma^2) - \beta^2 + A_2] \tag{1.66a}$$

(the so-called *Fano plot*). The parameters A_1/n_a and A_2 (or r and s) can be obtained by an appropriate fit of the experimental dependence of $\sigma_i(\beta^2)\beta^2$ (Table 1.5) (Rieke, Prepejchal 1972). Formula (1.66a) and Table 1.5 permit estimation of σ_i and $dN_1/dx = n_a \sigma_i$ in the γ-range at which the density effect is absent.

Table 1.4. Comparison of calculated and experimental ionization cross sections for hydrogen and the inert gases

Atom	Lorentz factor	Ionization cross section $[10^{-19}\ cm^2]$		$\sigma_{exp}/\sigma_{calc}$
	γ	Calculation (1.66, 67)	Experiment (Chechin et al. 1972)	
H	4.0	0.89	0.92	1.03
He	4.0	1.41	1.43	1.01
Ne	4.0	4.38	4.57	1.04
Ar	3.0	9.9	9.07	0.92
Ar	5.4	10.2	9.46	0.93
Kr	3.0	13.3	12.6	0.96
Kr	5.4	13.7	13.2	0.97
Xe	3.0	17.9	17.3	0.96
Xe	5.4	18.5	18.1	0.98

Table 1.5. Parameters describing the dependence of the ionization cross section on $\ln(\beta^2 \gamma^2) - \beta^2$; $\sigma_i = (A_1/n_a \beta^2)[\ln(\beta^2 \gamma^2) - \beta^2 + A_2]$

Species	$10^{20} \cdot A_1/n_a$	A_2
O_2	7.9	9.2
NO	8.1	9.8
H_2O	6.1	9.9
PH_3	8.6	10.0
H_2S	9.4	8.4
CO_2	10.8	9.7
BF_3	13.2	9.1
CF_4	19.2	8.2
$(CN)_2$	13.6	9.9
$CH_4 + CD_4$	7.9	9.9
C_2H_2	9.8	10.3
C_2H_4	12.6	10.2
C_6H_6	32.9	9.3
Isobutane	26.6	10.0
n-Pentane	34.5	10.0
Neopentane	36.6	9.3
n-Hexane	43.1	9.7
Cyclopropane	19.9	10.0
Cyclohexane	41.2	9.7
CH_3OH	11.6	10.7
C_2H_5OH	18.6	9.8
Acetone	22.3	9.9
$(CH_3)_2O$	19.0	10.3
C_3H_8	20.4	9.6

Influence of the Medium Density Effect on the Primary Ionization. To take into account the density effect in the primary ionization, one can make use of the differential cross section (1.35):

$$\sigma_i = \frac{e^2}{\pi \hbar^2 v^2 n_a} \text{Im} \int\limits_{I_1}^{\infty} \left(\beta^2 - \frac{1}{\varepsilon} \right) \left(\ln \frac{2m_e v^2 \gamma^2}{\omega(1 - \beta^2 \varepsilon)} + 1 \right) d\omega + \frac{2\pi e^4}{m_e v^2} \frac{\tilde{\mathcal{F}}(I_1)}{I_1} .$$

$$(1.68)$$

Unlike the calculation of energy loss, here the analytical method of Landau and Lifshitz (1960) cannot be applied even when $I_1 \to 0$. One must, therefore, decide upon some concrete form of $\varepsilon(\omega)$, for instance, (1.51), and perform numerical summations similar to (1.61, 62). The specific primary ionization, dN_1/dx, the total and restricted numbers of collisions, dN/dx, and $(dN/dx)_{\omega_{th},\infty}$, can be calculated from the following formulas (Chechin et al. 1972):

$$\left(\frac{dN}{dx} \right)_{\omega_{th},\infty} = \frac{2\pi n_a e^4}{m_e v^2} \left[\sum_{j=s+1}^{k} \frac{f_j}{l_j} \left(\ln \frac{2m_e v^2 \gamma^2}{l_j} - \beta^2 + 1 \right) \right.$$

$$\left. + \frac{1}{\omega_{th}} \sum_{j=1}^{s} f_j - \delta_N(\omega_{th}) \right] ;$$

$$(1.69)$$

$$\delta_N(\omega_{th}) = \sum_{j=s+1}^{k} \left[-\frac{f_j}{l_j} \left(\ln \frac{1 - \beta^2}{|1 - \beta^2 \varepsilon(l_j)|} + \beta^2 \right) \right.$$

$$\left. + \frac{2(\omega_j - \omega_j')}{\gamma^2 \omega_p^2} + \sum_{i=1}^{k} \frac{f_i}{l_j} \ln \frac{(l_i - \omega_j)(l_i + \omega_j')}{(l_i + \omega_j)(l_i - \omega_j')} \right] ,$$

$$(1.70)$$

where $\omega_s < \omega_{th} < \omega_{s+1}$. Here $\delta_N(\omega_{th})$ is the correction for the density effect by which (1.69) differs from the Bethe formula (1.66); parameters r and s are determined by (1.67) with the substitution $f(\omega) \to \tilde{f}(\omega) \approx \sum_{j=1}^{k} f_j \delta(\omega - l_j)$. All the remaining notation is the same as in (1.61, 62).

To calculate (1.69, 70) one can use values of f_j and ω_j computed by the same method as for the energy loss (1.61), with one difference; the oscillator energies are now

$$\omega_j = f_j \left(\int\limits_{\omega_{j,l}}^{\omega_{j,u}} \frac{f(\omega) d\omega}{\omega} \right)^{-1} .$$

In spite of its being somewhat cumbersome, this method for calculating the primary ionization exhibits great universality, since it permits us to perform calculations for any medium where the photo-absorption cross sections, or at least the relative number of electrons on a shell, f_j/Z, and the respective ionization potentials of the shells, I_j, are known. The values of f_j and I_j for a wide range of substances have long been used in calculations of ionization energy loss taking account of the density effect (Sect. 1.3). Some of the results of calculations of dN_1/dx are presented in Fig. 1.9b.

Relativistic Rise of the Specific Primary Ionization and of the Restricted Number of Collisions. As in the case of $(-dE/dx)_{\omega_{th},T_0}$, an important characteristic of the dependence of $(dN/dx)_{\omega_{th},\infty}$ upon $\beta\gamma$ is the relativistic rise, $R_N(\omega_{th})$. The dependence of $R_N(I_1) \equiv R_{N1}$ upon the atomic number differs essentially from that of R_E. As Z increases, R_E also increases, since the mean ionization potential increases and, accordingly, the beginning of the Fermi plateau is shifted towards large γ: $\gamma \gtrsim \gamma_{pl} \approx I/\omega_p \propto \sqrt{Z}$. Collisions with various shells of the atom give a contribution proportional to I_j^{-1} to the primary ionization; therefore the contribution of the outer shells to dN_1/dx is more significant than to $(-dE/dx)_{\omega_{th},T_0}$. The arrival at the Fermi plateau is usually determined by

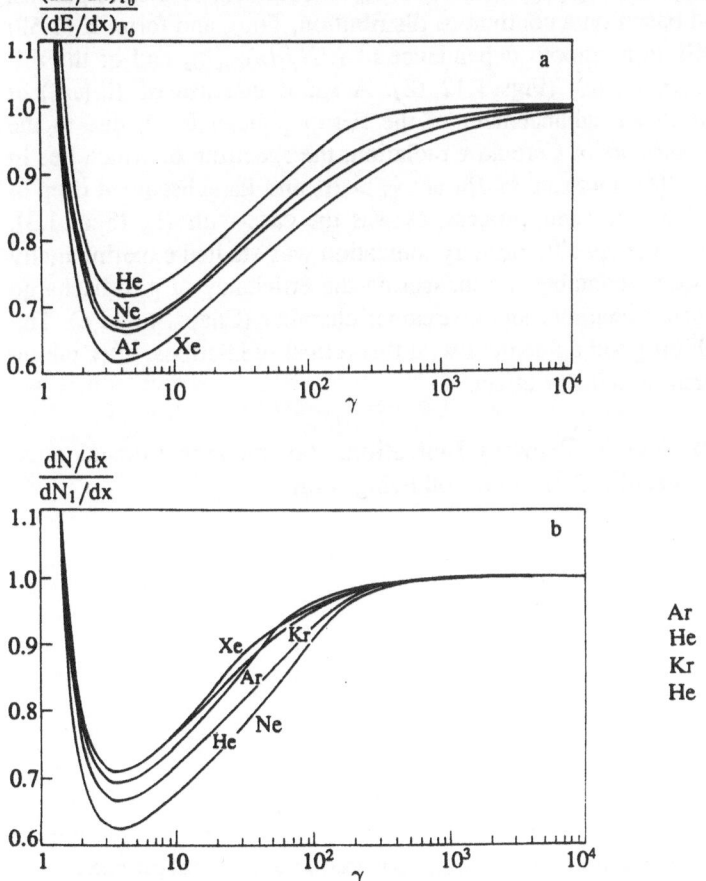

Fig. 1.11. Dependence on $\beta\gamma$ of restricted energy loss (**a**) and of primary ionization (**b**), normalized to unity on the Fermi plateau, in inert gases in normal conditions ($T_0 = 100\,\mathrm{keV}$)

such collisions for dN_1/dx, and occurs when $\gamma \gtrsim I_1/\omega_p \propto Z^{-1/2}$; consequently, R_{N1} falls as Z increases. For this reason the use of He, Ne and of He–Ne mixtures in detectors, is most expedient, from the point of view of identification of fast charged particles by the relativistic rise of the specific primary ionization. If, on the contrary, the observed ionization effect is related to energy loss, it is better to make use of the heavy gases Ar, Kr, Xe. These comments are illustrated in Fig. 1.11.

In real detectors the threshold energy, ω_{th}, may differ from I_1. Thus, fast secondary processes occurring in the gas – the formation of molecular ions, collisions of excited atoms with admixtures, photoionization, and the Vavilov-Čerenkov radiation – produce additional ionization of the gas and thereby lower, in effect, the threshold energy of the detector. On the other hand, introduction of electro-negative admixtures into a gas-discharge detector may suppress the development of an electric discharge in a region of low ionization density, which is equivalent to an enhancement of ω_{th}. In calculation of $(dN/dx)_{\omega_{th},\infty}$ by formulas (1.69, 70), a "step-like" dependence upon ω_{th} is obtained. A more rigorous, but tedious, method based on a continuous distribution, $f(\omega)$, and formula (1.68) with $I_1 \rightarrow \omega_{th}$, leads to a smooth dependence of $(dN/dx)_{\omega_{th},\infty}$ and of the relativistic rise, $R_N(\omega_{th})$, on ω_{th} (Figs. 1.12, 13). A small increase of $R_N(\omega_{th})$ at $\omega_{th} < I_1$ is related to an enhancement of the Fermi plateau level, due to the contribution by the photons of Čerenkov radiation, the spectrum of which lies in the region below I_1. The increase in R_N at $\omega_{th} > I_1$, and its subsequent drop at $\omega_{th} \gg I_1$, is caused by the same process, as was the case with R_E (Sect. 1.3). The relativistic rise of the specific primary ionization was studied experimentally by using a Wilson cloud chamber, by measuring the efficiency of gas-discharge counters, and of a spark chamber and a streamer chamber (Chaps. 4 and 6). The obtained results exhibit good agreement with theoretical predictions, after taking into account the medium density effect.

Approximation for Specific Primary Ionization. For numerical computation it is convenient to present (1.69) in the following form:

$$\left(\frac{dN_1}{dx}\right) = \frac{A_1}{\beta^2}[\ln \beta^2\gamma^2 - \beta^2 + A_2 - \delta_N(I_1)] , \qquad (1.71)$$

where

Fig. 1.12. Restricted number of collisions $(dN/dx)_{\omega_{th},\infty}$ in inert gases in normal conditions at the ionization minimum (a) and on the Fermi plateau (b); arrows indicate the absorption edges of the atomic shells. When ω_{th} is higher than the K-edges of absorption, $(dN/dx)_{\omega_{th},\infty} \propto \omega_{th}^{-2}$ (Ermilova et al. 1977)

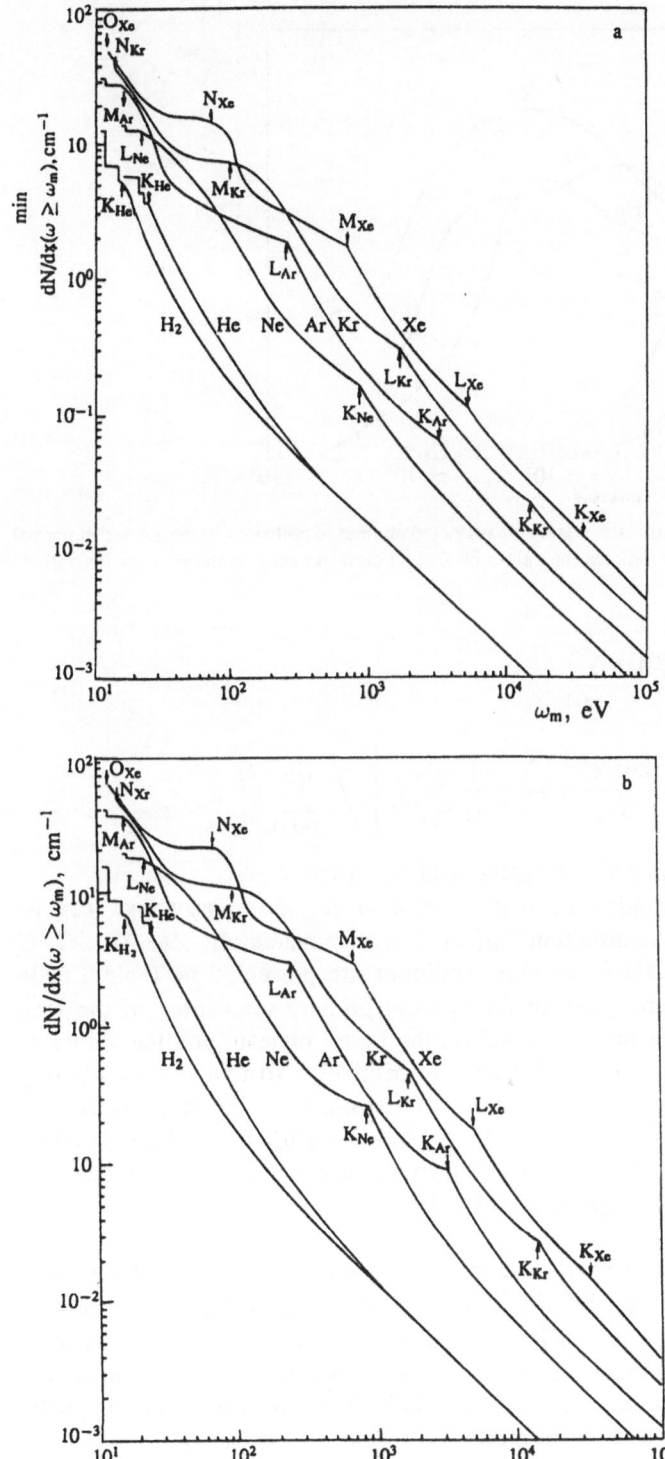

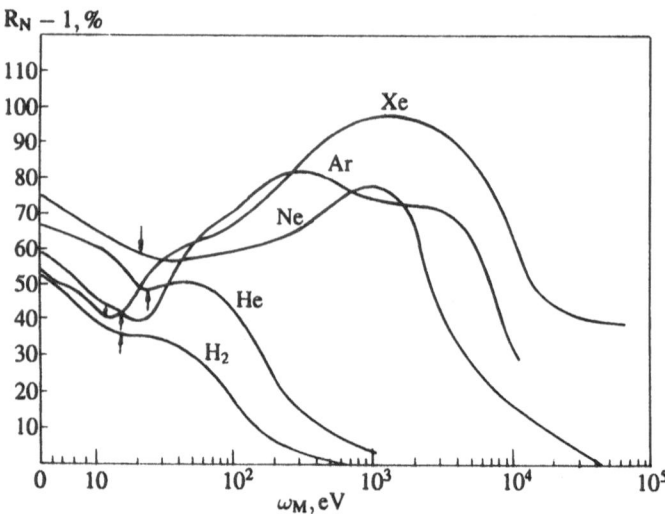

Fig. 1.13. Dependence of relativistic rise on the restricted number of collisions in inert gases in normal conditions upon ω_{th}; arrows indicate the values of $R_N(I_1)$ corresponding to the primary ionization

$$A_1 = (0.1536 \cdot 10^6 / A) \cdot \sum_{j=s+1}^{k} \frac{f_j}{l_j} \; ;$$

$$A_2 = \left[\sum_{j=s+1}^{k} \frac{f_j}{l_j} \left(\ln \frac{2 m_e c^2}{l_j} + 1 \right) + \frac{1}{I_1} \sum_{j=1}^{k} f_j \right] \bigg/ \sum_{j=s+1}^{k} \frac{f_j}{l_j} \; ,$$

where l_j is expressed in eV, ϱ in g/cm^3 and A_1 in cm^{-1}.

The same approximation formula [(1.57)] as for the energy loss, can be applied to calculate the correction, $\delta_N(I_1)$. The parameters A_1, A_2, X_0, X_1, C and m for the inert gases in normal conditions are presented in Table 1.6. In this table also values are given of the specific primary ionization, of the total collision numbers at the minimum and on the Fermi plateau, and the values of their relativistic rises, $R_N(I_1)$ and $R_N(0)$, respectively. To translate to arbitrary pressures and temperatures, it is necessary to perform the following substitution: $A_1 \to A_1 \eta_g$, $X_0 \to X_0 - 0.45 \cdot \log \eta_g$, $X_1 \to X_1 - b \cdot \log \eta_g$, $R_N \to R_N - d \cdot \log \eta_g$, where η_g is the ratio of the gas density to its density in normal conditions, and the parameters b and d are given in Table 1.6.

The Role of Transition Effects. The preceding discussion dealt with the motion of a charged particle in an infinite homogeneous medium. Near the interface of two homogeneous media (for instance, where the particle enters the working medium of the detector) there exist specific transition ionization effects related to the reshaping of the particle electromagnetic field. The transition electromagnetic field arising here leads to two different observable effects. In the transparency region of a given medium, the quanta of this field form the well-known transition

Table 1.6. The parameters in (1.71) for calculating the specific primary ionization in inert gases in normal conditions

Parameter	He	Ne	Ar	Kr	Xe
ϱ [10^{-3} g/cm^3]	0.18	0.9	1.78	3.71	5.85
A_1 [cm^{-1}]	0.24	0.84	1.83	2.55	3.45
A_2	11.6	10.9	11.5	11.3	11.3
X_0	1.88	1.31	1.50	1.47	1.40
X_1	3	3	3	3	3
$-C$	10.7	11.4	9.36	9.47	9.36
m	2.56	1.44	2.82	2.61	2.53
d	0.16	0.17	0.16	0.17	0.17
b	0.39	0.42	0.31	0.25	0.31
$(dN_1/dx)_{min}$ cm^{-1}	3.5	11.4	25.9	35.4	48.1
$(dN_1/dx)_{plateau}$ cm^{-1}	5.2	18.0	36.2	50.4	67.7
$R_N(\omega_{th} = I_1)$	1.49	1.58	1.40	1.42	1.41
$(dN/dx)_{min}$ cm^{-1}	5.3	12.7	29.3	35.4	48.1
$(dN/dx)_{plateau}$ cm^{-1}	8.0	19.3	40.7	50.4	67.7
$R_N(\omega_{th} = 0)$	1.51	1.52	1.39	1.42	1.41

radiation. On the other hand, in the frequency range where the medium exhibits absorption, the transition field yields a certain correction to the ionizing power of the particle in the vicinity of the interface. Indeed, it has long been known that in sufficiently thin samples of high-density matter surrounded by vacuum or gas there occurs an anomalous (suppressed) density effect. Its manifestation consists in an enhancement of the relativistic rise of: (a) the total energy loss of a particle traversing a thin layer of matter (Chechin 1985), (b) the K- (L- etc.) shell excitation cross section of atoms in experiments with thin foils ($\sim 10^{-3}$ cm) (Bak et al. 1986); and (c) local energy loss in the vicinity of the interface or in very thin samples ($\sim 10^{-4}$ cm) (Chechin and Ermilova, 1989). In the latter case, instead of the Fermi plateau, a significantly higher quantity will be obtained:

$$\left(-\frac{dE}{dx} \right)_{\substack{T_0 \\ \gamma \to \infty}} = \frac{2\pi n_a e^4 Z}{m_e c^2} \ln \frac{2m_e c^2 T_0}{\omega_p^2(\text{gas})}$$

where $\omega_p(\text{gas}) \ll \omega_p$ is the plasma frequency of the gas surrounding the sample.

The enhanced relativistic rise of energy loss in very thin samples of matter (surrounded by gas or vacuum) gives rise to the idea of utilizing this effect for particle identification. Unfortunately the total correction to the energy loss is so small ($\int \delta(dE/dx)\, dx \sim (e^2/\hbar c)\, I$), that only particle beams of sufficiently high intensity can be analyzed in this way.

1.5 Fluctuations and the Most Probable Energy Loss in Thin Samples of Matter

General Relations. Hitherto we have considered only the average energy loss $(-d\overline{E}/dx)$ and the average number of collisions $d\overline{N}/dx$ [1]. However, because of the statistical nature of the interaction process of a charged particle with matter, both the number of collisions and the energy transferred in the individual collisions exhibit fluctuations, and consequently, so do the ionization energy losses occurring in a sample of matter of thickness, x. We shall now take this into account, restricting our considerations to sufficiently thin samples of matter, in which the average energy loss is significantly lower than the particle energy: $\overline{\Delta E} \ll E$. In addition, we shall exclude from our study, rare, "catastrophic" collisions involving energy transfers comparable to the particle energy. Under such conditions the particle velocity and, consequently, the cross section of inelastic collisions, vary weakly with depth in the sample of matter, thus the subsequent collisions with atoms can be considered independent of each other. The operating substance in the detectors considered in the next chapter satisfy these conditions, and it thus represents a "thin" sample of matter. If, on the contrary, the energy loss, ΔE, is comparable with E, then collisions occurring in various parts of the detector cannot be considered independent; in addition, as the energy of the incident particle changes while it traverses the detector, its inelastic collision cross section also undergoes changes. Conditions corresponding to "thick" samples of matter are realized in total-absorption detectors, in the blocks of shielding from penetrating radiation, etc.. The methods for computing the energy loss, particle path ranges, and their fluctuations, differ considerably in this case from the methods presented below.

The probability $p(x, N)$ that a particle will undergo N collisions with atoms in a sample of matter of thickness, x, is determined, taking account of the above considerations, by the Poisson distribution, with an average $\overline{N} = x\,d\overline{N}/dx$:

$$p(x, N) = (\overline{N}^{N}/N!)\exp(-\overline{N}) . \tag{1.72}$$

To calculate the probability of energy loss, $f(x, \Delta)$, it is necessary to take into account that the probability, W_1 of an energy loss ω occurring in a single collision with an atom is

$$W_1(\omega) = \frac{d^2\overline{N}}{dx\,d\omega} \bigg/ \frac{d\overline{N}}{dx} = \frac{1}{\sigma}\frac{d\sigma}{d\omega} ,$$

where $\int_0^\infty W_1(\omega)\,d\omega = 1$; $W_1(\omega) = 0$ at $\omega < 0$. Consequently, in the case of N collisions, $W_N(\omega) = W_1(\omega)\,\mathcal{X}\,W_1(\omega)\,\mathcal{X}\,\ldots\,\mathcal{X}\,W_1(\omega)$ N times, where the symbol $\mathcal{X}$ indicates convolution over the variable ω:

---

[1] In this section the notation for the average energy loss and for the average number of collisions differs from the respective notation adopted in other sections of the book by a bar above the respective quantity.

$$W_1(\omega) \divideontimes W_1(\omega) = \int\limits_0^\omega W_1(\omega') W_1(\omega - \omega')\, d\omega' \, .$$

Taking into account that $W_0(\omega) = \delta(\omega)$, and utilizing (1.72), we obtain

$$
\begin{aligned}
f(x, \Delta) &= \sum_{N=0}^{\infty} \exp(-\overline{N}) \frac{\overline{N}^N}{N!} W_N(\Delta) \\
&= \exp(-\overline{N}) \left\{ \delta(\Delta) + x \frac{d^2 \overline{N}}{dx\, d\Delta} \right. \\
&\quad \left. + \frac{x^2}{2} \int\limits_0^\Delta \frac{d^2 \overline{N}(\Delta')}{dx\, d\Delta} \frac{d^2 \overline{N}(\Delta - \Delta')}{dx\, d\Delta}\, d\Delta' + \dots \right\} \, .
\end{aligned}
\tag{1.73}
$$

Formula (1.73) has a clear physical meaning: its first term gives the distribution of energy loss upon passage of the particle through a layer x without collisions; the second term gives the distribution in the case that a single collision takes place; and so on. Substituting into (1.73) an expression for $d^2 N/dx\, d\Delta$, it is possible to determine $f(x, \Delta)$ (the convolution method). Here the integrals entering into the convolutions, $W_N(\Delta)$, are found either numerically (Talman 1979; Bichsel 1988) or by Monte Carlo calculation (Ermilova et al. 1977; Allison and Cobb 1980). The convolution method is obviously applicable only for small x, when $\overline{N}$ is not too large. Otherwise, the calculations become too cumbersome, since convolutions with $N \gtrsim \overline{N}$ have to be determined. At the same time, as $\overline{N}$ increases, the details of the behavior of $W_1(\Delta)$ as a function of Δ become less and less essential, since they are smoothed out statistically in $W_N(\Delta)$. In these conditions the Landau method is more convenient, leading, in fact, to an approximate calculation of (1.73) for $\overline{N} \gg 1$ (Landau 1944).

The Landau Method. The Landau approach is based on a solution of the kinetic equation for $f(x, \Delta)$,

$$\frac{\partial f(x, \Delta)}{\partial x} = \int\limits_0^{T_0} w(\omega)[f(x, \Delta - \omega) - f(x, \Delta)]\, d\omega \tag{1.74}$$

with the initial condition $f(0, \Delta) = \delta(\Delta)$. Here

$$w(\omega) = \overline{N} W_1(\omega) = x d^2 \overline{N}/dx\, d\omega = x n_a\, d\sigma/d\omega \, .$$

The solution of equation (1.74) is found with the aid of the Laplace transformation,

$$\hat{L} f(x, \Delta) \equiv \int\limits_0^\infty f(x, \Delta) \exp(-p\Delta)\, d\Delta = \exp[-G(x, p)] \, , \tag{1.75}$$

where

$$G(x, p) = \int_0^{T_0} w(\omega)[1 - \exp(-p\omega)] \, d\omega .$$ (1.76)

The inverse Laplace transformation of (1.76) yields the following solution of (1.74)

$$f(x, \Delta) = \frac{1}{2\pi i} \int_{-i\infty}^{i\infty} \exp[p\Delta - G(x, p)] \, dp .$$ (1.77)

Relations (1.73, 77) are equivalent, which is readily verified by applying the Laplace transformation and performing the summation (1.73), taking account of the properties of the convolution of functions $\hat{L}W_N(\Delta) = [\hat{L}W_1(\Delta)]^N$. In turn, (1.73) can be obtained from (1.77) by expansion in powers of $G(x, p) - \overline{N}$. Equation (1.77) is exact. Regretfully, its integration with a reasonable spectrum of the number of collisions, $w(\omega)$, is very complicated and can be accomplished only in certain special cases. The main difficulty consists in that the integrals of rapidly oscillating functions in (1.76, 77) converge weakly; in this connection several methods have been proposed for their computation.

The Bohr, Landau, Vavilov and Blunk-Leisegang Solutions. In (1.77) the values of variable $p \sim 1/\Delta_{\text{eff}}$ are essential where Δ_{eff} represents the effective energy loss for which $f(x, \Delta)$ noticeably differs from zero. If $\Delta_{\text{eff}} \gg T_0$, i.e. if the average number of collisions in a sample is very large, such that $\overline{N} \gg \Delta_{\text{eff}}/T_0 \gg 1$, then the integrand in (1.76) can be expanded in powers of $p\omega \ll T_0/\Delta_{\text{eff}}$. Taking into account only the term which is linear in p yields $f(x, \Delta) \approx \delta(\Delta - \overline{\Delta})$, where $\overline{\Delta} = x(-d\overline{E}/dx) = \int_0^{T_0} w(\omega)\omega \, d\omega$; i.e. this signifies total neglect of fluctuations. The terms proportional to p and proportional to p^2 result in a Gaussian distribution for $f(x, \Delta)$, first obtained by Bohr (Bethe and Ashkin 1953):

$$f(x, \Delta) = \frac{1}{\sqrt{2\pi} \, \overline{\Delta E}} \exp\left[-\frac{(\Delta - \overline{\Delta})^2}{2\overline{\Delta E}^2}\right] ,$$ (1.78)

where

$$\overline{\Delta E^2} = \int_0^{T_0} w(\omega)\omega^2 \, d\omega .$$

As $x \propto \Delta_{\text{eff}}$ decreases and, accordingly, p increases, this expansion usually ceases to be valid. However, it can be applied for those collisions corresponding to sufficiently small energy transfers: $\omega \lesssim \omega_1$ for $\omega_1 p \sim \omega_1/\Delta_{\text{eff}} \ll 1$. To this end we shall single out in (1.76) the contributions of "far" ($\omega \leq \omega_1$) and "close"

$(\omega \geq \omega_1)$ collisions for which at $\omega_1 \gg I$, one can make use of the Rutherford collision spectrum (1.1):

$$w_R(\omega) = \xi/\omega^2 , \qquad (1.79)$$

where

$$\xi = x\frac{2\pi n_a Z e^4}{m_e v^2} = x\varrho \cdot 0.1536\frac{Z}{A}\frac{1}{\beta^2}$$

(ξ is expressed in MeV, and x in cm). Consequently,

$$G(x, p) = \int\limits_0^{\omega_1} w(\omega)\left(p\omega - \frac{p^2\omega^2}{2} + \frac{p^3\omega^3}{6} - \ldots\right) d\omega$$

$$+ \xi\int\limits_{\omega_1}^{T_0} [1 - \exp(-p\omega)]\, d\omega/\omega^2 . \qquad (1.80)$$

For samples of matter satisfying the conditions

$$I \ll \xi \ll T_0 , \qquad (1.81)$$

one may in (1.80) consider only the first terms of the sum and set $T_0 \to \infty$. We note that in accordance with Sect. 1.4 the average number of collisions $\overline{N} = x d\overline{N}/dx \sim 10\xi/I$. Consequently (1.81) signifies that $\overline{N} \gg 10$, i.e. the samples of matter considered are not too thin; application of the convolution method for these is quite difficult. Taking into account only the first term in the expansion (1.80) results in the Landau distribution

$$f(x, \Delta) = \frac{1}{\xi}\varphi(\lambda) , \qquad (1.82)$$

where $\varphi(\lambda)$ is the universal Landau function:

$$\varphi(\lambda) = \frac{1}{2\pi i}\int\limits_{-i\infty}^{i\infty} \exp(u\lambda + u\ln u)\, du ;$$

$$\lambda = \frac{\Delta - \Delta_0}{\xi} - 0.225 .$$

The expression for the most probable energy loss, Δ_0, taking account of the correction for the density effect, δ_E (Sternheimer, Peierls 1971), and corrections made by Maccabee and Papworth (1969) has the form

$$\Delta_0 = \xi\left[\ln\frac{2m_e v^2 \xi}{I^2(1 - \beta^2)} - \beta^2 + 0.198 - \delta_E\right] , \qquad (1.83)$$

which looks like the average restricted energy loss for $T_0 = 1.22\xi$. Taking account of the first two terms in the sum (1.80) gives the Blunk-Leisegang distribution (Blunk, Leisegang 1950).

Generalization of the Landau theory to the case of an arbitrary ratio, ξ/T_0, when a precise computation of the second integral in (1.80) is required, was first performed by Vavilov (1957):

$$f_V(x, \Delta) = \frac{\exp[k(1 + \beta^2 0.577)]}{\pi T_0} \int\limits_0^\infty \exp(kf_1) \cos(y\lambda_1 + kf_2) \, dy \,,$$

where subscript V indicates Vavilov, and

$$f_1 = \beta^2[\ln y - \text{Ci}(y)] - \cos y - y \, \text{Si}(y) \,;$$
$$f_2 = y[\ln y - \text{Ci}(y)] + \sin(y) + \beta^2 \text{Si}(y) \,;$$
$$\lambda_1 = k\lambda + k \ln k \,; \quad k = \xi/T_0 \,.$$

Si and Ci are the integral sine and cosine, respectively. When $\xi/T_0 \lesssim 0.01$, the Vavilov solution transforms into the Landau distribution; when $\xi/T_0 \gg 1$, the Vavilov solution transforms into a Gaussian distribution.

As pointed out above, all these solutions describe fluctuations of energy loss in short-range collisions ($\omega \gg I$) when the analytic expression (1.79) is known for the collision spectrum, $w_R(\omega)$. As to the long-range collisions, in the Landau distribution fluctuations of the corresponding energy loss are not taken into account at all [the term in (1.80) is linear in p)], while the Blunk-Leisegang correction only takes them into account approximately, in the form of a Gaussian distribution. An analysis of the relative contribution of the terms in the expansion (1.80) reveals that the Landau and Blunk-Leisegang theories are not applicable for $\xi \lesssim 10 \cdot I$ and for $\xi \lesssim 3 \cdot I$, respectively: because the terms of expansion (1.80) alternate in sign, the Landau theory lowers, while the Blunk-Leisegang theory enhances the width of the distribution $f(x, \Delta)$. Indeed, it has been noted for a long time that in the case of very low density samples of matter ($x \sim 1$–$10 \, \text{mg/cm}^2$), the experimental distributions of ionization energy loss are significantly broader than in the Landau theory, and narrower than predicted by the Blunk-Leisegang theory (Fig. 1.14).

Modification of the Landau Theory for Very Thin Samples of Matter. When $\xi \lesssim I$, expansion (1.80) cannot be applied; to obtain $f(x, \Delta)$ it is necessary to integrate (1.76,77) directly. The case of $\xi \sim I$ is of practical interest. Indeed, when $\xi \ll I$ (i.e. $\overline{N} \ll 10$), the convolution method is quite applicable, and, as can be seen from (1.73), the form of the distribution function, $f(x, \Delta)$, is determined directly by the collision spectrum. At the same time, when $\xi \sim I$, the average number of collisions is sufficiently high ($\overline{N} \sim 10$) so that one can hope to achieve a good description of $f(x, \Delta)$, even if a rough approximation is applied for the form of the collision spectrum.

Talman (1979) utilized the form (1.30) for the spectrum together with the additional assumption of a discrete distribution for the oscillator forces. This

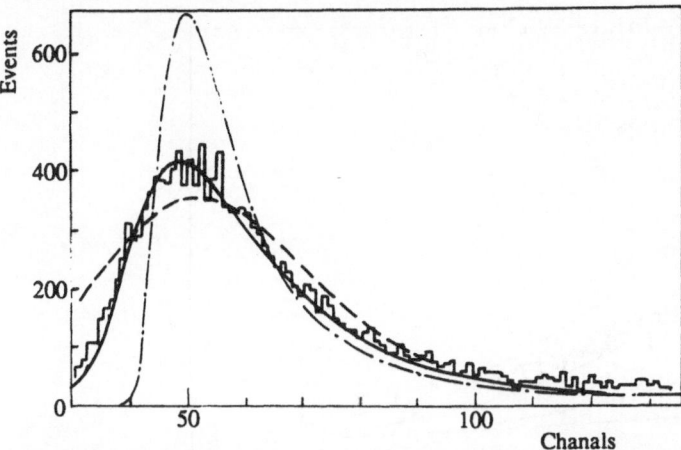

Fig. 1.14. Distribution of ionization energy loss for protons of momentum 2.1 GeV/c in 5 cm of [95 % Ar + 5 % CH$_4$]: histogram – experimental results; ――――― – calculation by harmonic-oscillator model; — · — · — · — – Landau theory, — — — — — — Blunk-Leisegang theory

approach was somewhat modified by Cobb et al. (1976) and Ermilova et al. (1977) with inclusion of the density effect. Substitution of (1.30) or (1.35) into (1.76) results in quite cumbersome expressions, so that the calculation of the inverse Laplace transformation (1.77) is performed numerically. The distributions thus obtained for ionization energy loss are in good agreement with the results of experiments performed with very thin samples of gas ($x \sim 1$–10 mg/cm^2).

When $\xi \gtrsim I$, the same good agreement is achieved with a much simpler model, in which all the electrons in the atom are regarded as independent harmonic oscillators with a natural frequency, $\omega_s/\hbar = I/\hbar$ (Chechin, Ermilova 1976). In this case the collision spectrum exhibits a δ-like behavior:

$$w(\omega) = \sum_{n=1}^{\infty} w_n \delta(\omega - nI) ,$$

where $w_1 = \xi(\ln a - 1)/I$; $w_{n \geq 2} = \xi/In(n-1)$; $\ln a = \ln(2m_e v^2 \gamma^2 / I) - \beta^2 - 0.577 - \delta_E$. Substitution of these expressions into (1.76) leads to a simple expression for $G(x, p)$:

$$G(x, p) = \frac{\xi}{I}[1 - \exp(-pI)] \ln \frac{a}{[1 - \exp(-pI)]} .$$

When $pI \sim I/\xi \ll 1$, $G(x, p) \simeq p\xi \ln(a/pI)$, which coincides with the Landau approximation. Consequently, $f(x, \Delta) = \sum_{n=1}^{\infty} C_n \delta(\Delta - nI)$, and the coefficients C_n are found from the recurrence relations

$$C_0 = \exp(-\xi \ln a/I) ; \quad C_1 = \xi C_0(\ln a - 1)/I ;$$

$$C_{n+1} = [\xi/(n+1)I] \left[C_n(\ln a - 1) + \frac{C_{n-1}}{1} + \ldots + \frac{C_0}{n} \right] . \tag{1.84}$$

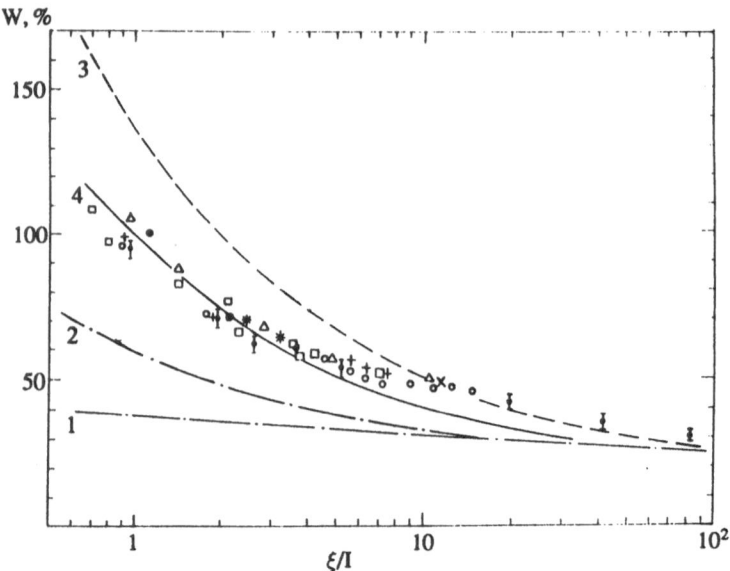

Fig. 1.15. Dependence of relative half-width of the ionization energy loss distribution in argon upon ξ/I at $\gamma = 4$: 1 and 2 – Landau theory; 3 – Blunk-Leisegang theory; 4 – harmonic-oscillator model. All curves, with the exception of 1, take into account the experimental resolution of the proportional detectors [(1.85)] (Chechin and Ermilova 1976)

The experimental and theoretical values for the relative half-width, W, of the energy loss distribution for various ξ/I are shown in Fig. 1.15 ($W = \delta/\Delta_0$ where δ is the width of the distribution at half-width). The good agreement with the experimental results for the distribution $f(x, \Delta)$ exhibited by the simplest model for $w(\omega)$, is explained by the large number of collisions in the gas sample ($\overline{N} \gtrsim 10$); here the details of the behavior of $w(\omega)$ in the $\omega \sim I$ region are averaged statistically in $f(x, \Delta)$. As gas samples become thinner ($\overline{N} \sim 1$), it clearly becomes necessary to make use of a better spectrum for the number of collisions, for example (1.30) or (1.35).

Until now all the characteristic ionization potentials of atoms and, accordingly, the energies transferred in long-range collisions were considered to be close to a certain mean energy I. At the same time, in the case of heavy atoms, the ionization potentials of low-lying shells are two or three orders of magnitude higher than I. It may turn out, then, that different approximations are applicable for different electron shells. Thus, for instance, when $I_1 \ll \xi \ll I_K \lesssim T_0$, the Landau theory can be utilized for the outer electrons, while the convolution method is applicable for the inner electrons. The overall distribution function results from the convolution

$$f(x, \Delta) = f_1(x, \Delta) \divideontimes f_K(x, \Delta) .$$

In Fig. 1.16 the ranges of applicability for various methods of computing $f(x, \Delta)$ are shown. Inside the range of applicability the accuracy of the respective ap-

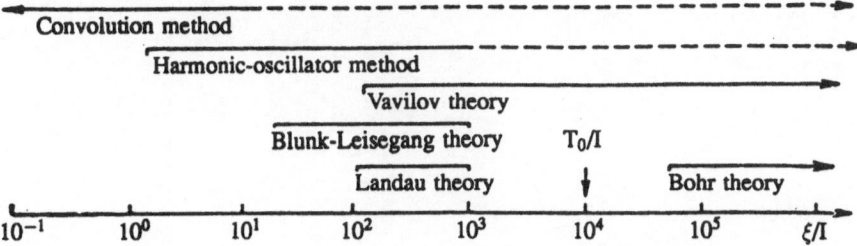

Fig. 1.16. Applicability ranges of various models for calculating $f(x, \Delta)$ at $T_0/I = 10^4$. The dashed lines indicate the regions of ξ/I for which the respective method can be applied, but its use becomes tedious

proximations are $\lesssim 10\%$. Sometimes discussed in the literature are corrections of the Blunk-Leisegang type to the Vavilov theory for $\xi \gtrsim I$ (Shulek et al. 1967). Since generally $T_0 \gg I$, it is usually not necessary to introduce such corrections: it is quite sufficient to make use either of the Blunk-Leisegang distribution (when $I \lesssim \xi \ll T_0$), or of the Vavilov distribution (when $I \ll T_0 \lesssim \xi$).

Simulation of the Energy Loss of Relativistic Particles in Thin Samples of Gas. As pointed out above, the most general method for calculating the fluctuations of energy loss, the convolution method, requires calculation of multiple integrals involving the spectrum of collision numbers, $w(\omega)$. In addition to numerical summation of such integrals, simulation of the energy loss is also utilized, which is actually equivalent to the calculation of the integrals by the Monte-Carlo method (Allison and Cobb 1980).

For Monte-Carlo calculations one must make use of the integral spectrum of the collision number, $x(d\overline{N}/dx)_{\omega_{\rm th},\infty}$ (Fig. 1.12). Then, events of a particle traversing the sample of gas are simulated in accordance with the following rules: (1) The number of inelastic collisions in a sample of thickness, x, is to follow a Poisson distribution with an average equal to $\overline{N} = x \, d\overline{N}/dx$. (2) Energy transfers in individual collisions must be distributed in accordance with the spectrum $(d\overline{N}/dx)_{\omega_{\rm th},\infty}$. To this end a pseudo-random number is generated by the computer in the interval $[(d\overline{N}/dx)_{1\,{\rm MeV},\infty}; \ d\overline{N}/dx]$, and then the corresponding energy transfer ω is determined. (3) The energy lost in a sample, Δ, is the sum of the energy losses which have occurred in N independent collisions. The statistics in each version of computation usually amounts to $2 \cdot 10^4 - 2 \cdot 10^6$ events.

The results of such simulations (Figs. 1.17, 18), reveal that the phenomenological model adopted for inelastic collision cross sections, (1.30, 35), which was applied in the Monte-Carlo method, provides good agreement with the experimental data from gas proportional detectors.

We note the important physical improvement introduced by the Monte-Carlo method in describing the dependence of probable energy loss, Δ_0, on ξ/I and γ, as compared with the Landau theory (Fig. 1.19, Table 1.6). The decrease in Δ_0 observed in experiments as compared with calculations by the Landau-Sternheimer theory (in particular on the Fermi plateau) is due to collisions with

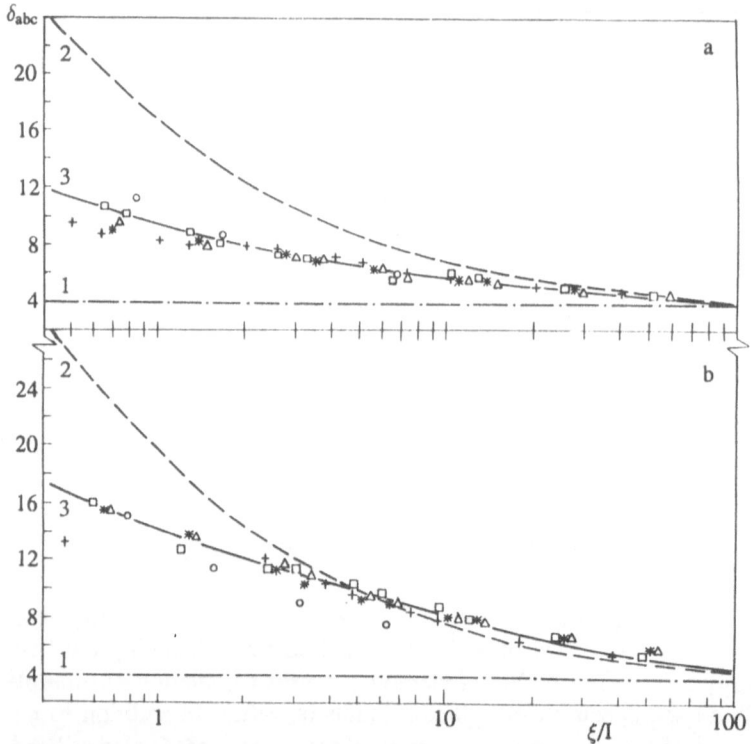

Fig. 1.17. Dependence of the absolute half-width δ of the energy loss distribution in inert gases upon ξ/I at the minimum ionization (a) and on the Fermi plateau (b): 1 – Landau theory; 2 – Blunk-Leisegang theory; 3 – approximation by the results of energy loss simulation for argon. The points are experimental results corresponding to various inert gases (o – He; + – Ne; □ – Ar; * – Kr; △ – Xe) (Ermilova et al. 1977)

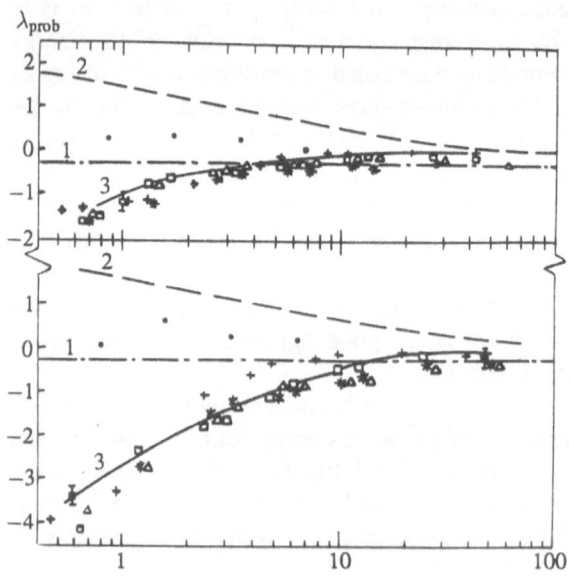

Fig. 1.18. Dependence upon ξ/I of the most probable energy loss in units of λ [(1.82, 83)] in inert gases; the notation is the same as in Fig. 1.17 (Ermilova et al. 1977)

electrons of the inner shells of the atom yielding no contribution to Δ_0 in very thin samples of gas. For example, in an argon sample 1 cm thick, $\Delta_0 \approx 1.22 \, \text{keV}$, while the ionization potential for the K-electrons of argon is $I_K = 3.2 \, \text{keV}$. Since polarization of the medium manifests itself for various shells of the atom at $\gamma \gtrsim I_j/\omega_p$, a partial or total switch-off of the contribution to Δ_0 of collisions with the K- and L-electrons results in the probable energy loss reaching the plateau at smaller values of γ, and results in a corresponding decrease in the relativistic rise of Δ_0.

When comparing theoretical computations with experimental data one must generally bear in mind that calculations concern the energy lost by a particle, while experimental data are related to the energy absorbed in the sensitive volume of a detector. Because of the emission of photons by excited atoms, and also because of δ- and photo-electrons with paths exceeding the dimensions of the detector, there occurs a leakage of a part of the energy from the detector. On the other hand, δ-electrons produced by the particle in the material placed in front of the detector may be absorbed by the sensitive volume. Estimations reveal that for samples of gas of thickness $x \sim 1$–$10 \, \text{cm}$, the change in Δ_0 related to these effects is not greater than 1 %.

The observed drop in the relativistic rise in thin gas samples as compared with the Landau theory, is extremely significant from the point of view of the possibility of applying gas proportional detectors for identification of charged particles. In this connection, other mechanisms to describe this phenomenon have been proposed. For example, such a fall has been related to the existence of a transition effect at the "detector wall – gas" boundary. This effect actually does exist but it is extremely small, since in gases under normal conditions it becomes noticeable only at distances of the order of $10^{-3} \, \text{cm}$ (Chechin 1975); this has been confirmed experimentally (Hasebe et al. 1978).

That the theoretical description of the fluctuations of energy loss in thin samples of matter is correct is very important for analyzing the operation of elementary particle detectors. Simulation of the energy loss utilizing realistic collision spectra [for instance, (1.30) and (1.35)] permits a detailed description of the data obtained in these detectors, taking account of such purely experimental effects as the influence of electron diffusion on charge collection in adjacent channels, the amplitude and time resolutions of the device, etc.. Thus, a close agreement is achieved between the results of calculations and of experiments (Fig. 1.19).

Fluctuations of Total and Primary Ionizations. In comparing the above-considered fluctuations of energy loss with fluctuations of the observed ionization effect, i.e. of the total ionization, it is necessary to take into account that the number, N_Δ, of secondary electrons for a given release of energy, Δ, also fluctuates with a dispersion $F N_\Delta$, where $F < 1$ is the Fano factor (Charpak 1970). (These fluctuations are smaller than Poisson fluctuations since they are not totally independent but are partially correlated in accordance with the condition $\Delta = \text{const.}$) Experimental uncertainties arising in the course of measuring the

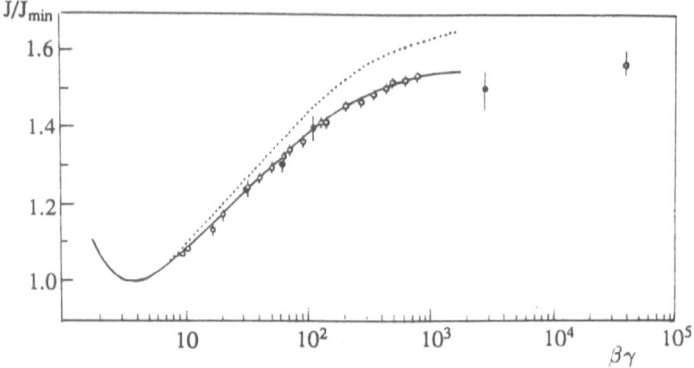

Fig. 1.19. Relativistic rise of a probable energy loss measured in the 128-layer detector, $128 \times 6\,\mathrm{cm}$ long, filled with $95\,\%$ Ar $+ 5\,\%$ CH$_4$; $P = 10^5$ Pa: - - - - - - calculation by (1.83); ———— results of energy loss simulation; points – data from experiments (Allison and Cobb 1980)

number of structure elements along a track, or in collecting and amplifying the charge in the detector, also contribute to the observed fluctuations of total ionization. Unlike the total ionization, the primary ionization fluctuates in accordance with the Poisson distribution (1.72).

Transformation of the primary ionization Poisson distribution into the asymmetric distribution followed by the total ionization, was observed experimentally: as the time of diffusion from the primary ionization blob is enhanced, secondary electrons start to contribute to the observed ionization effect (Fig. 1.20).

The total relative half-width (full width at the half maximum) is written,

$$W_t(x) = [W^2(x) + W_a^2(x)]^{1/2} , \tag{1.85}$$

where W and W_a are the relative half-widths of the ionization energy loss distribution and of experimental uncertainties, respectively. The latter weakly enhance the fluctuations of ionization, and they actually correspond to the energy loss distribution.

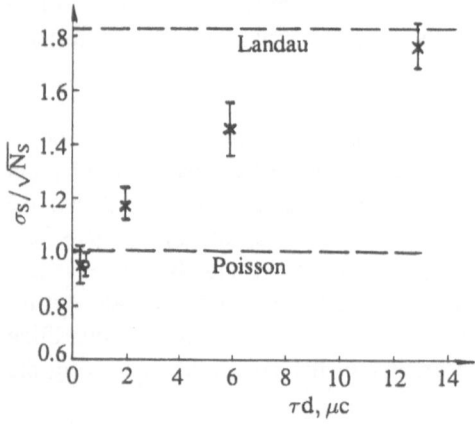

Fig. 1.20. Dependence of standard deviation σ_s of the number of streamers N_s along a 5 cm track length, upon the time delay τ_d of the high voltage pulse in a helium streamer chamber; $P = 4 \cdot 10^4$ Pa. The chambers were exposed to relativistic electrons (Davidenko et al. 1968)

In total ionization measurements, usually it is not the average quantity that is determined, but its most probable value, $N_0(x)$, which corresponds to the maximum of the ionization distribution in a sample of thickness, x. Unlike the mean ionization, the probable ionization is not distorted by the production of δ-electrons or fluorescence photons which leave the detector. The most probable ionization corresponds, within a good accuracy, to the probable energy loss:

$$N_0(x) \simeq \Delta_0(x)/w . \tag{1.86}$$

References

Ahlen S.P. 1980: Rev. Mod. Phys. **52**, 121. Sects. 1.1, 3

Alakoz A. V., Chechin, V. A., Kotenko L. P., Merson G. I., Ermilova V. C. 1975: Nucl. Instr. Meth. **124**, 41. Sect. 1.3

Allison W. W. M., Cobb J. H. 1980: Ann. Rev. Nucl. Sci. **30**, 253. Sects. 1.1, 2

Anderson H. H., Zeigler J. P. 1977: *Hydrogen Stopping Powers and Ranges in all Elements* (Pergamon, New York). Sect. 1.1

Asoskov V. S., Grishin V. M., Ermilova V. C., Kotenko L. P., Merson G. I., Chechin V. A. 1982: Trudy FIAN (Proc. P. N. Lebedev Phys. Inst., Moscow) **140**, 3 (in Russian). Sect. 1.2

Bak J. F., Peterson J. B. B., Uggerhoj E., Ostergaard K., Moller S. P., Sorensen A. H. 1986: Physica Scripta **33**, 147. Sect. 1.4

Berkowitz J. 1979: *Photoabsorption, photoionization and photoelectron spectroscopy* (Academic, New York). Sect. 1.2

Bethe H. 1930: Ann. Phys. **5**, 325. Sects. 1.1–3

Bethe H. 1932: Z. Phys. **76**, 293. Sect. 1.1

Bethe H. 1933: in *Handbuch der Physik*, 2nd ed., ed. by H. Geiger, K. Scheel, Vol. 24/1 (Springer, Berlin). Sects. 1.1, 3, 4

Bethe H., Ashkin, J. 1953: in *Experimental Nuclear Physics*, ed. by E. Segrè, Vol. 1 (Wiley, New York). Sects. 1.1, 5

Bichsel H. 1972: in *Amer. Instr. of Phys. Handbook*, ed. by D. E. Gray (McGraw-Hill, New York). Sect. 1.1

Bichsel H. 1988: Rev. Mod. Phys. **60**, 663. Sects. 1.1, 2, 5

Bloch F. 1933: Ann. Phys. **16**, 285. Sect. 1.1

Blunk O., Leisegang S. 1950: Z. Phys. **128**, 500. Sects. 1.1, 5

Bohr N. 1913: Phil. Mag. **25**, 10. Sect. 1.1

Bohr N. 1948: Kgl. Danske Vidersk. Selsk. Mat.-Fys. Medd. **18**, Nr. 8. Sect. 1.1

Budini P., Taffara L., Viola C. 1960: Nuovo Cimento **18**, 864. Sect. 1.3

Charpak I. 1970: Ann. Rev. Nucl. Sci. **20**, 195. Sect. 1.5

Chechin V. A. 1975: Dokl. Akad. Nauk SSSR **221**, 813 (English transl.: Sov. Phys.-Dok. **20**, 269). Sect. 1.5

Chechin V. A. 1985: Phys. Lett. A **108**, 453. Sect. 1.4

Chechin V. A., Ermilova V. C. 1976: Nucl. Instr. Meth. **136**, 551. Sect. 1.5

Chechin V. A., Ermilova V. C. 1989: Z. Phys. D **13**, 33. Sect. 1.4

Chechin V. A., Kotenko L. P., Merson G. I., Ermilova V. C. 1972: Nucl. Instr. Meth. **98**, 577. Sects. 1.2–4

Christophorou L. G. 1970: *Atomic and Molecular Radiation Physics* (Wiley, New York)

Cobb J. H., Allison W. W. M., Bunck J. N. 1976: Nucl. Instr. Meth. **133**, 315. Sect. 1.5

Crispin A., Fowler G. N. 1970: Rev. Mod. Phys. **42**, 290. Sects. 1.1, 2

Davidenko A. V., Dolgoshein B. A., Somov S. V., Semenov V. K. 1968: Zh. Eksp. Teor. Fis. **55**, 426 (Sov. Phys. – JETP **28**, 227). Sect. 1.5

Ermilova V. C., Kotenko L. P., Merson G. I. 1977: Nucl. Instr. Meth. **145**, 555. Sect. 1.5

Fano U. 1963: Ann. Rev. Nucl. Sci. **13**, 1. Sects. 1.1, 2

Fano U., Cooper J. W. 1968: Rev. Mod. Phys. **40**, 441. Sect. 1.2

Fermi E. 1940: Phys. Rev. **57**, 485. Sects. 1.1, 3

Hasebe N., Kikuchi J., Doke T., Nagata K., Nakamoto A. 1978: J. Phys. D **11**, 97. Sect. 1.5

Inokuti M. 1971: Rev. Mod. Phys. **43**, 297. Sects. 1.2, 4

Inokuti M., Itikawa Y., Turner J. E. 1978: Rev. Mod. Phys. **50**, 23. Sect. 1.2

Janni J. P. 1982: Atom. Data Nucl. Data Tables **27**, 147. Sect. 1.3

Kepler R. E., d'Andlau C. A., Freffer W. B., Hansen L. F. 1958: Nuovo Cimento **7**, 71. Sect. 1.3

Landau L. 1944: J. Phys. USSR **8**, 201. Sects. 1.1, 5

Landau L. D., Lifshitz E. M. 1960: *Electrodynamics of Continuous Media* (Addison-Wesley, Reading. Mass.). Sects. 1.1, 3, 4

Landau L. D., Lifshitz E. M. 1982: *Quantum Electrodynamics* (Pergamon, Oxford). Sect. 1.2

Maccabee H. D., Papworth D. C. 1969: Phys. Lett. A **30**, 241. Sect. 1.5

Merson, G. I. 1972: Yad. Fiz. **15**, 278 (English transl.: Sov. J. Nucl. Phys. **15**, 157). Sect. 1.3

Merson G. I., Sitar B., Budagov J. A. 1983: Fiz. Elem. Chastits At. Yadra **14**, 648 (English transl.: Sov. J. Part. Nucl. **14**, 270). Sect. 1.1

Moiseiwitsch B. L. 1985: Phys. Reports **118**, 133. Sects. 1.2, 4

Möller C. 1932: Ann. Phys. **14**, 531. Sects. 1.1, 2

Mott N. F., Massey H. S. W. 1965: *The Theory of Atomic Collisions* (Oxford University Press, London). Sect. 1.2

Ogurtsov G. N. 1972: Rev. Mod. Phys. **44**, 1. Sect. 1.2

Opal C. B., Beaty E. C., Peterson W. K. 1972: Atom. Data Nucl. Data Tables **4**, 209. Sect. 1.2

Powell C. J. 1976: Rev. Mod. Phys. **48**, 33. Sect. 1.2

Rieke F. F., Prepejchal W. 1972: Phys. Rev. A **6**, 1507. Sect. 1.4

Rossi B. 1952: *High Energy Particles* (Prentice-Hall, Englewood Cliffs). Sect. 1.1

Seltzer S. M., Berger M. J. 1982: Int. J. Sppl. Radiat. Isot. **33**, 1189. Sect. 1.3

Shulek P., Golovin B. M., Kulyukina L. A., Medved S. V., Pavlovich P. 1966: Yad. Fiz. **4**, 564 (English transl.: Sov. J. Nucl. Phys. **4**, 400). Sect. 1.5

Sternheimer R. M. 1961: in *Methods of Experimental Physics*, ed. by L. C. L. Yuan, C. S. Wu (Academic, New York). Sect. 1.1

Sternheimer R. M. 1981: Phys. Rev. B **24**, 6288. Sects. 1.1, 3

Sternheimer R. M., Berger M. J., Seltzer S. M. 1984: Atom Data Nucl. Data Tables **30**, 261. Sect. 1.3

Sternheimer R. M., Peirls F. 1971: Phys. Rev. B **3**, 3681. Sect. 1.5

Talman R. 1979: Nucl. Instr. Meth. **159**, 189. Sect. 1.5

Uehling E. A. 1954: Ann. Rev. Nucl. Sci. **4**, 315. Sect. 1.1

Vavilov P. V. 1957: Zh. Eksp. Teor. Fiz. **32**, 920 (English transl.: Sov. Phys. – JETP **5**, 749). Sects. 1.1, 5

Walske M. C. 1956: Phys. Rev. **101**, 940. Sect. 1.1

2. Physical Processes in Gas Ionization Detectors

2.1 Ionization of Gas in a Detector

The precision of ionization measurements and, consequently, the reliability of particle identification in gas ionization detectors depend on the following:

a) the inelastic collision cross section and its dependence upon $\beta\gamma$ (Chap. 1) which, basically, are determined by the statistical uncertainties of ionization measurements;

b) the relationship between the actual ionization occurring in the gas of the detector and the registered ionization effect. (This issue, already mentioned in Sect. 1.1, will be within the scope of this chapter and, partly, of Chap. 3.);

c) systematic errors in the determination of ionization (Chap. 3).

In the measurement of ionization in charged particle detectors three principal phases can be singled out in the formation of the ionization signal:

1) *Ionization* accompanied by the appearance of free electrons. This process, involving the formation of primary and secondary electrons (including δ-electrons), as well as fluorescence of the gas atoms, was considered in Sect. 1.1;

2) *Transport of electrons and ions*. These processes, which distort the information on the degree of ionization include electron thermalization, diffusion and drift of electrons and ions in the gas under the influence of an electric field. In transport processes collisions may take place between electrons, ions and atoms of the gas which are accompanied by additional ionization and excitation of atoms or by capture of electrons, thereby leading to inaccuracies in the measurements of the ionization effect. This error is proportional to the degree of ionization created by the particle in the gas and does not affect its relativistic rise;

3) *Registration* involving gas amplification and subsequent signal processing. At this stage, a distortion of the measured ionization may also occur owing to incomplete collection of the charge if the detector or the electronics are not fast enough or exhibit non-linearities (Chap. 3).

This chapter is devoted to transport and registration processes which not only depend on the composition and pressure of the gas, but also on the principle of operation and the time properties of the detector.

2.2 Transport of Electrons in a Gas

2.2.1 Thermalization of Electrons

In the course of thermalization, the electron energy ε_e is lowered as a result of elastic collisions with the gas from $\varepsilon_e \simeq I_1$ down to the thermal energy $\bar{\varepsilon}_e = (3/2)\varepsilon_c$ corresponding to the electrons being in a state of equilibrium thermal motion. Here $\varepsilon_c = kT/e$, where k is the Boltzmann constant, and T is the temperature of the gas. In the absence of an electric field ($\mathcal{E} = 0$) $T \approx 300\,\mathrm{K}$ and $\bar{\varepsilon}_e \approx 0.04\,\mathrm{eV}$. When $\mathcal{E} > 0$, the electron temperature and the average electron energy increase due to acceleration in the electric field; thus, for example, in argon in a field of $1\,\mathrm{keV/cm}$, $\bar{\varepsilon}_e \approx 4\,\mathrm{eV}$.

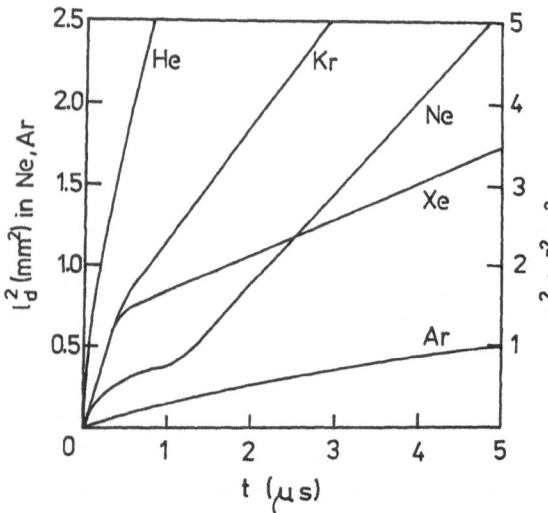

Fig. 2.1. Time dependence of the mean square of the electron displacement in a noble gas, resulting from thermalization and thermal diffusion under normal conditions (Davidenko et al. 1970). The regions where the slope is constant correspond to thermal diffusion

The thermalization time τ_t and the average displacement l_t of an electron from its initial position occurring in the time τ_t depend on the composition and pressure of the gas, as well as the field strength. For illustration we present in Fig. 2.1 the dependence on time of the total displacements, resulting from thermalization and diffusion, of electrons in noble gases ($\mathcal{E} = 0$, $P = 10^5\,\mathrm{Pa}$, where $T = 293\,\mathrm{K}$). The values of τ_t and l_t approximately correspond to the inflection points of the curves. Thermalization is most rapid in helium and slowest in argon, but the greatest displacement of electrons is observed in neon.

Introduction into the gas of small admixtures of molecules with low-lying excitation levels, for example, of N_2, N_2O or H_2O at about 100–200 Pa, leads to a significant enhancement of thermalization and a reduction of l_t (Davidenko et al. 1970), almost without affecting thermal diffusion.

The thermalization length in a proportional gas detector (down to electron energies of $\approx 4\,\mathrm{eV}$) has not been measured directly. Farr et al. (1978) and Sadoulet

(1982) have quoted the following values for argon, based on measurements by Lenard (1903): the mean absorption length is $\sim 0.5\,\mu$m for an M-shell electron (with a typical energy of 30 eV), $\sim 1\,\mu$m for an L-shell electron (400 eV), and $\sim 15\,\mu$m for a K-shell electron (4 keV). Most ($\sim 80\,\%$) of the collisions along the trajectory of a particle result in the creation of M-shell electrons with very small ranges ($\leq 1\,\mu$m). The probability that a δ-electron (with an energy of > 4 keV) will be produced is extremely small ($\sim 0.03\,\text{cm}^{-1}$ for $P = 10^5$ Pa), and it has practically no influence upon ionization measurements.

2.2.2 Diffusion of Electrons in the Absence of an Electric Field (Thermal Diffusion)

In gas track chambers (cloud, spark, streamer chambers) transport processes take place in the absence of an electric field or in very weak cleaning fields. In the absence of electric and magnetic fields thermalized electrons experience chaotic motion and are characterized by a *Maxwellian energy distribution*:

$$F(\varepsilon_e) = C_e \sqrt{\varepsilon_e} \exp(-\varepsilon_e/kT) . \tag{2.1}$$

where C_e is a constant. The mean electron energy $\bar{\varepsilon}_e \approx 0.04$ eV at $T = 300$ K and the mean electron velocity is $\bar{v}_e \simeq 10^7$ cm/s.

The collision frequency ν_e of electrons in the gas and their mean free path between collisions, l_e, are

$$\nu_e = n_a \bar{v}_e \sigma_e = \bar{\tau}_e^{-1} ; \quad l_e = 1/(\sigma_e n_a) , \tag{2.2}$$

where σ_e is the *elastic scattering cross section*; $\bar{\tau}_e$ is the average time between elastic collisions of an electron; $n_a = n_L P \cdot (273/T)$ is the number of atoms (molecules) per unit volume for a temperature T and pressure P, where $n_L = 2.69 \cdot 10^{25}\,\text{m}^{-3}$ is the Loschmidt constant ($P = 10^5$ Pa, $T = 273$ K).

In the kinetic theory of gases the *momentum transfer cross section* σ_m, known also as the transport cross section, or the diffusion cross section, is used more often. It represents a measure of the average forward directed momentum transferred by electrons (ions) in collisions with atoms (molecules) of the gas:

$$\sigma_m = \sigma_e(1 - \overline{\cos \theta_e}) ,$$

where θ_e is the angle between the direction of the electron motion and the direction of reference. The mean momentum transfer path (transport path) is given by

$$l_{mt} = 1/(n_a \sigma_m) . \tag{2.3}$$

In the case of isotropic scattering $\sigma_m = \sigma_e$. The cross sections σ_e and σ_m (Fig. 2.2) change noticeably with ε_e. The anomalously low cross section of electron scattering in certain gases at energies for 0.1 to 1.0 eV (Figs. 2.12, 16) is known as

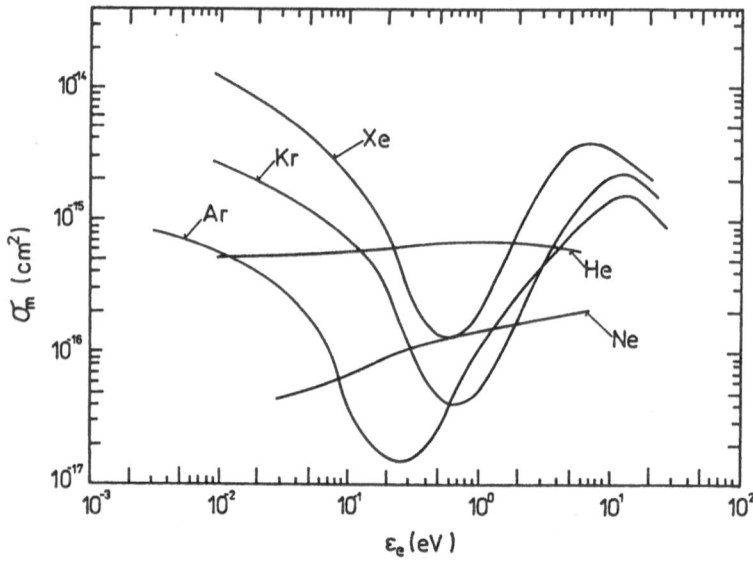

Fig. 2.2. The momentum transfer cross section σ_m for electrons in noble gases (Frost, Phelps 1964; Huxley, Crompton 1970)

Table 2.1. Diffusion coefficients for thermal electrons (Dolgoshein 1970) and electron transport cross sections (Allkofer 1969) in noble gases

Gas	He	Ne	Ar	Kr	Xe
D [cm^2/s]	310	3000	370		43
σ_m [10^{-16} cm^2]	5.3	0.45	3.8	22	72

the Ramsauer effect (Brown 1959; Hasted 1964). The diffusion coefficients and electron transport cross sections in noble gases are presented in Table 2.1.

Electrons and ions in a gas undergo diffusion towards areas of lower concentration. The diffusion current J_e flows in a direction opposite to the gradient of the electron (ion) concentration n_e; J_e is determined by the Fick diffusion law:

$$J_e = -(v_e/3n_a\sigma_m)\,\mathrm{grad}\,n_e = -D\,\mathrm{grad}\,n_e\ . \tag{2.4}$$

The constant D is called the *diffusion coefficient*. The equation for the rate of change of concentration with time follows from (2.4):

$$dn_e/dt = D\nabla^2 n_e\ . \tag{2.5}$$

Now, let $n(0)$ be the number of electrons (ions) at the origin of the reference frame at $t = 0$. After a period of time t the electrons will be found at the distances $r_1, r_2, \ldots r_n$. The mean square distance

$$\overline{r}^2 = \int\limits_0^\infty n_e r^2 \, dV \Big/ \int\limits_0^\infty n_e \, dV \, , \tag{2.6}$$

where n_e is the concentration of electrons in the volume $dV = 4\pi r^2 \, dr$. If the diffusion coefficient is constant in space, we obtain from (2.4) and (2.5) the equations

$$-\frac{\partial}{\partial t}(n_e \, dV) = -D\frac{\partial}{\partial r}\left(4\pi r^2 \frac{\partial n_e}{\partial r}\right) dr \tag{2.7}$$

and

$$\frac{\partial n_e}{\partial t} = D\left(\frac{2}{r}\frac{\partial n_e}{\partial r} + \frac{\partial^2 n_e}{\partial r^2}\right) \tag{2.8}$$

as well as the expressions for the electron concentration after the passage of time t:

$$n_e(t) = \frac{n(0)}{(4\pi Dt)^{3/2}} \exp(-r^2/4Dt) \tag{2.9}$$

and the mean square distance

$$\overline{r}^2 = 6Dt \, . \tag{2.10}$$

Expression (2.10) was derived by Einstein in 1905.

In the one-dimensional case (for example, for the projection of the electron density onto the x-axis)

$$n_e(t) = \frac{n(0)}{\sqrt{(4\pi Dt)^{1/2}}} \exp(-x^2/4Dt) \, , \tag{2.11}$$

while the root-mean-square deviation of the electrons from their initial sites σ_D or the electron diffusion displacement l_D is

$$\sigma_D(t) = l_D = \sqrt{2Dt} \, . \tag{2.12}$$

Distribution (2.11) is Gaussian (Fig. 2.3a). The diffusion of electrons in noble gases can be weakened by introduction of admixtures with large σ_m into the gas (Table 2.2). Thus, for example, helium-neon mixtures serve as more suitable fillings of spark and, in particular, of streamer chambers, than pure neon. Molecular admixtures with high σ_m, such as, for instance, saturated vapors of H_2O and C_2H_5OH, reduce the diffusion of thermal electrons to such an extent that well defined tracks are observed in a spark chamber up to about 50 μs after the passage of the particle.

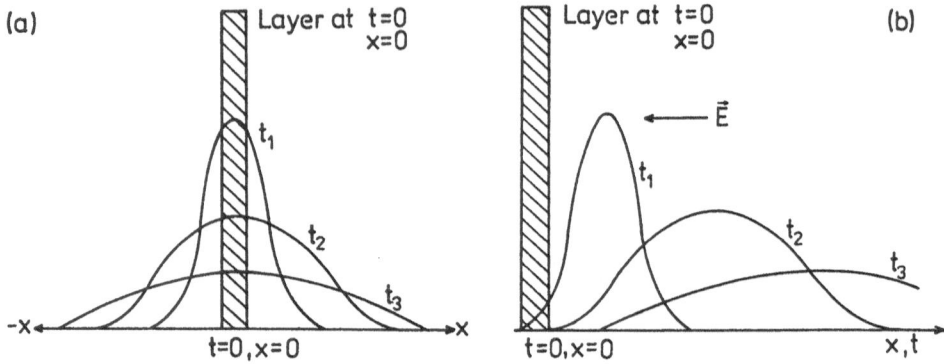

Fig. 2.3. Diffusion of an electron cloud in a gas in the absence of an electric field (a) and in an electric field (b) (Lewellyn-Jones 1957) (the *dashed area* represents the initial electron distribution)

Table 2.2. Momentum transfer cross sections σ_m for thermal electrons in molecular gases and in vapors ($\mathcal{E} = 0$) (Davidenko et al. 1970; Merzon et al. 1983)

Substance	N_2	O_2	H_2O	N_2O	CH_2	CO_2	C_3H_8	Iso-C_4H_{10}
σ_m [10^{-16} cm^2]	4	3	750	22	6	108	10	14

The thermal diffusion coefficient for a mixture of gases is

$$D = \left(\sum_j a_j / D_j \right)^{-1}$$

where a_j is the concentration (relative partial pressure), and D_j is the diffusion coefficient of the j-th component. The resulting diffusion displacement of electrons l_d, related to thermalization and thermal motion (Fig. 2.1), is

$$l_d = (l_t^2 + l_D^2)^{1/2} .$$

As a rule, thermalization somewhat enhances diffusion effects in comparison with purely thermal diffusion.

2.2.3 Motion of Electrons in an Electric Field

In gas-discharge detectors involving a constant power supply transport processes take place in the presence of an electric field. In an electric field of strength $\mathcal{E}$ the properties of the motion experienced by electrons in the gas differ from those at $\mathcal{E} = 0$.

a) The *electron temperature* increases and exceeds significantly (by a factor of 100 to 200) the temperature of the gas itself. Owing to its small mass an

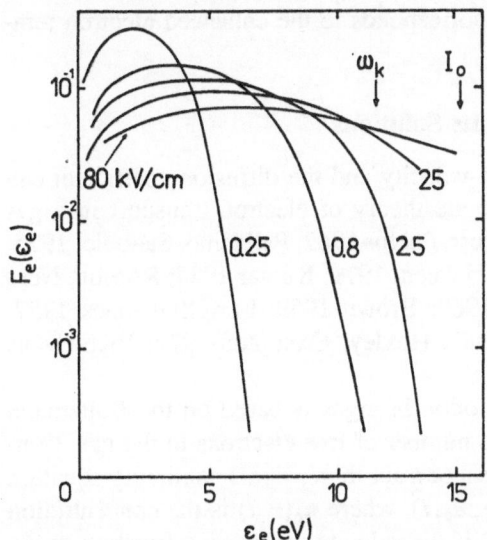

Fig. 2.4. Computed electron energy distribution in argon for various electric field strengths (Palladino, Sadoulet 1975)

electron is rapidly accelerated between collisions. In addition, in collisions with atoms (molecules) of the gas an electron losses only a small part of its energy, $\Delta\varepsilon_e/\varepsilon_e \sim m_e/M_a$ where m_e and M_a are the masses of the electron and atom (molecule), respectively. As a result, for example, the mean electron energy in a "typical" proportional detector increases up to 0.1–5.0 eV, and the distribution is no longer Maxwellian (Fig. 2.4).

b) In addition to the velocity of thermal motion v_e electrons also acquire an additional velocity w_d directed along the field lines. The *drift velocity* w_d is usually an order of magnitude smaller than v_e; it depends on the electric field strength $\mathcal{E}$, and on σ_m. From the simple kinetic theory developed by Townsend we obtain

$$w_d = \frac{e\mathcal{E}}{2m_e}\overline{\tau}_e = \frac{e\mathcal{E}}{2m_e}\frac{l_{mt}}{\overline{v}_e} = \frac{e\mathcal{E}}{2m_e\overline{v}_e n_a \sigma_m} \, , \qquad (2.13)$$

where $\overline{\tau}_e$ is the average time between collisions of the electrons and e is the charge of the electron. Expression (2.13) provides an approximate value for w_d; to be more precise (Rice-Evans 1974),

$$w_d = \frac{2e\mathcal{E}}{3m_e}\frac{l_{mt}}{\overline{v}_e} \, .$$

In Fig. 2.3b the distribution of electrons in space when they drift in an electric field is shown. From (2.12) we obtain the root-mean-square deviation σ_x of the measured coordinate x for an electron drift length $x = w_d t$:

$$\sigma_x = \sqrt{\frac{2D(\mathcal{E})x}{w_d}} \, , \qquad (2.14)$$

where the diffusion coefficient $D(\mathcal{E})$ corresponds to the enhanced electron temperature in the electric field $\mathcal{E}$.

2.2.4 The Boltzmann Equation and its Solutions

More precise expressions for the drift velocity and the diffusion coefficient can be obtained with the aid of the elaborate theory of electron transport in a gas (Morse et al. 1935; Margenau 1946; Frost, Phelps 1962; Palladino, Sadoulet 1975; Schultz, Gresser 1978; Mathieson, El Hakeem 1979; Kumar 1984; Robson, Ness 1986; Schmidt, Polenz 1988; Loeb 1969; Brown 1959; Lewellyn-Jones 1957; Hasted 1964; Christophorou 1970, 1981; Huxley, Crompton 1974; Rice-Evans 1974; Sauli 1977).

The classical theory of electron motion in a gas is based on the Boltzmann equation reflecting conservation of the number of free electrons in the gas. Consider the Boltzmann equation in the form of a six-dimensional continuity equation for the quantities $n_e f_e = n_e(r, t) \cdot f_e(r, v_e, t)$, where $n_e(r, t)$ is the concentration of free electrons in the gas; $f_e(r, v_e, t)$ is the velocity distribution function of the electrons. The quantity $n_e f_e \, dr \, dv_e \equiv [n_e(r, t) \, dr] \cdot [f_e(r, v_e, t) \, dv_e]$ represents the number of electrons present at the moment of time t inside the elementary volume dr with the coordinate r and exhibiting velocities within the interval from v_e to $v_e + dv_e$.

The terms in the Boltzmann equation reflect the change in population at the points $n_e f_e \, dr \, dv_e$ in the six-dimensional space of coordinates and velocities in the presence of an electric field. Formally, the rate of change of the population can be written as $(\partial/\partial t)(n_e f_e) \, dr \, dt$. The general form of the Boltzmann equation [see the derivation in (Huxley, Crompton 1974)] is the following:

$$\frac{\partial}{\partial t}(n_e f_e) + \mathrm{div}_r(n_e f_e v_e) + \mathrm{div}_{v_e}\left(n_e f_e \frac{e\mathcal{E}}{m_e}\right) + S_{\mathrm{col}} = 0 . \tag{2.15}$$

The first two terms correspond to variations of the population in space due to diffusion and migration, respectively, the third to acceleration in the electric field, while the fourth term S_{col} is the collision integral corresponding to the stepwise change in the electron velocity occurring in collisions involving momentum transfer.

The Boltzmann kinetic equation is an integrodifferential equation, so significant mathematical difficulties arise in the course of its solution; these are overcome with the aid of the following approaches:

1) The analytical approach involving expansion of the electron velocity distribution function into spherical harmonics and subsequent solution of ordinary differential equations. There exist several similar methods of solution, such as the well-known Chapman-Enskog method or the Gred method in the kinetic theory of gases.

In the solution one may restrict oneself to considering the first two terms of the expansion (the two-term approximation), as was done by Frost and Phelps (1962) and Schultz and Gresser (1978); we shall also adopt this method. Such an approach is justified if the inelastic scattering cross section is smaller than the elastic cross section. Otherwise it is necessary to use three (or more) terms of the series (2.16), thus significantly complicating the problem, although providing more precise expressions for the transport coefficients, primarily for the transverse diffusion (Schmidt, Polenz 1988). Accurate estimations of the transport coefficients applying so-called "multiterm Boltzmann calculations" have been performed by Robson and Ness (1986).

2) Straightforward solution of the Boltzmann partial differential equation (2.15). This method is more complicated and requires much computer time.

3) Application of the Monte-Carlo method in which the probability distribution of finding an electron at the "point" $n_e f_e$ is obtained statistically by simulation of many electron collisions with atoms of the gas. Such a method was successfully applied by, for instance Palladino, Sadoulet (1975) and Biagi (1988). A comparison of the methods for solving the Boltzmann equation can be found, for example, in (Christophorou 1970; Huxley, Crompton 1974; Elecki et al. 1975).

For an analytical solution of the Boltzmann equation, a number of simplifications are required. The function f_e is conveniently expressed in terms of spherical coordinates, $f_e(r, v_e, \theta_e, t)$, in the form of the convergent series

$$f_e(r, v_e, \theta_e, t) = f_0(r, v_e, t) + \sum_{k=1}^{\infty} f_k(r, v_e, t) P_k(\cos \theta_e) , \qquad (2.16)$$

where $P_k(\cos \theta_e)$ are Legendre polynomials. Such a spherically symmetric expansion holds in the case of a small electric field strength typical of a proportional detector, when $w_d \ll v_e$.

In the Boltzmann equation the electron energy acquired in the field $\mathcal{E}$ is compared with the energy lost by the electrons in collisions with atoms. Processes of three kinds occur in the gas; these are described by the collision integral S_{col} in the Boltzmann kinetic equation and can be dealt with separately: a) elastic collisions, b) influence of the thermal motion of molecules of the gas on the electrons; c) inelastic collisions related, primarily to excitation of atoms by the electrons.

Elastic Collisions. An electron of kinetic energy $\varepsilon_e = (1/2) m_e v_e^2$ loses a part of its energy $\Delta \varepsilon_e$ in a collision with an atom, such that

$$\frac{\Delta \varepsilon_e}{\varepsilon_e} = \frac{2\Delta v_e}{v_e} = 2 \left(\frac{m_e}{M_a} \right) (1 - \cos \theta_e) ,$$

where θ_e is the angle between the directions of the electron velocities before and after the collision. In the spherically symmetric approach the mean energy loss $< \Delta \varepsilon_e / \varepsilon_e >$ is $\Lambda = 2 m_e / M_a$.

Following calculations by Schultz and Gresser (1978) we shall restrict ourselves to considering the stationary state, when f_e is independent of the time t, and also set $x \parallel \mathcal{E}$. In this case f_e depends only on x and θ_e. In the expansion of f_e in Legendre polynomials we shall take into account only the first two terms (the two-term approximation), since the function f_e is approximately spherically symmetric:

$$f_e(x, v_e, \theta_e) = f_0(x, v_e) + f_1(x, v_e) \cdot \cos\theta_e + \ldots .$$

In the classical approach, according to Schultz and Gresser (1978), v_e can be replaced with ε_e, while the velocity distribution function f_e is replaced with the electron energy distribution function F_e:

$$F_e(\varepsilon_e, \theta_e) = F_0(\varepsilon_e) + F_1(\varepsilon_e)\cos\theta_e + \ldots .$$

It is expedient to substitute two differential equations for F_0 and F_1, respectively, for the general Boltzmann equation (2.15) (Huxley, Crompton 1974) in order to derive from them the transport coefficients: the drift velocity, the diffusion coefficient, etc. Upon simplification these equations assume the form

$$e\mathcal{E}\frac{\partial}{\partial\varepsilon_e}(v_e F_0) - \frac{2e\mathcal{E}}{m_e v_e}F_0 = -\frac{v_e}{l_e}F_1 ; \tag{2.17}$$

$$\frac{e\mathcal{E}}{3}\frac{\partial}{\partial\varepsilon_e}(v_e F_1) = \frac{2m_e}{M_a}\frac{\partial}{\partial\varepsilon_e}\left(\frac{\varepsilon_e v_e F_0}{l_e}\right) . \tag{2.18}$$

The first equation expresses the conservation of momentum; and the second, the energy conservation law. Solving for F_1 we can eliminate it to obtain a new equation:

$$\frac{2}{3}\frac{(e\mathcal{E})^2}{m_e}\frac{\partial}{\partial\varepsilon_e}\left[\varepsilon_e l_e(\varepsilon_e)\frac{\partial(F_0/v_e)}{\partial\varepsilon_e}\right] + \frac{2m_e}{M_a}\frac{\partial}{\partial\varepsilon_e}\left(\frac{\varepsilon_e v_e F_0}{l_e(\varepsilon_e)}\right) = 0 . \tag{2.19}$$

Account of Thermal Motion in the Gas. If the electrons acquire in the electric field an energy comparable to the energy of thermal motion of electrons (about 0.04 eV), then the solution of (2.19) has the form

$$F_0(\varepsilon_e) = C_F\sqrt{\varepsilon_e}\exp\left\{-\int\frac{3\Lambda(\varepsilon_e)\varepsilon_e\,d\varepsilon_e}{[e\mathcal{E}l_e(\varepsilon_e)]^2 + 3\Lambda(\varepsilon_e)\varepsilon_e kT}\right\} . \tag{2.20}$$

where $\Lambda(\varepsilon_e)$ is the mean energy loss occurring in electron collisions. The constant C_F is determined from the normalization condition $\int_0^{\varepsilon_{max}} F_0(\varepsilon_e)\,d\varepsilon_e = 1$. In a weak electric field, the term $[e\mathcal{E}l_e(\varepsilon_e)]^2$ may be neglected, and $F_0(\varepsilon_e)$, then, coincides with the Maxwellian distribution (2.1).

Inelastic Collisions. In the actual transport conditions of electrons in a gas it is necessary to take into account inelastic collisions, in which atoms are excited to energy levels ω_k. The mean free path of an electron between such collisions $l_k = 1/(\sigma_k n_a)$ where σ_k is the cross section of inelastic collisions, accompanied by excitation. In Margenau (1946), Frost and Phelps (1962) and Schultz and

Gresser (1978), it is shown that inelastic collisions can be taken into account by the introduction of an additional term into (2.19), which, then, assumes the form

$$\frac{2}{3}\frac{(e\mathcal{E})^2}{m_e}\frac{\partial}{\partial\varepsilon_e}\left[\varepsilon_e l_e(\varepsilon_e)\frac{\partial}{\partial\varepsilon_e}(F_0/v_e)\right] + \frac{2m_e}{M_a}\frac{\partial}{\partial\varepsilon_e}\left(\frac{\varepsilon_e v_e F_0}{l_e(\varepsilon_e)}\right)$$

$$+ \sum_k\left[\frac{\sqrt{(2/m_e)(\varepsilon_e+\omega_k)}}{l_k(\varepsilon_e+\omega_k)}F_0(\varepsilon_e+\omega_k) - \frac{\sqrt{\varepsilon_e(2/m_e)}}{l_k(\varepsilon_e)}F_0(\varepsilon_e)\right] = 0 .$$

(2.21)

In the case of known elastic and inelastic collision cross sections the electron distribution function F_e can be calculated independently from the electron energy ε_e (Fig. 2.4).

The description of electron transport processes in molecular gases is even more complicated, since it is necessary to take into account the excitation of vibrational and rotational molecular levels, which occurs at extremely low energies (about 0.1 eV) (for example, Hasted 1964). In this case, instead of $\Lambda(\varepsilon_e)$, it is necessary to substitute into (2.20)

$$\Lambda(\varepsilon_e) = \frac{2m_e}{M_a} + \sum_k \frac{\omega_k}{\varepsilon_e}\frac{l_e(\varepsilon_e)}{l_k(\varepsilon_e)} = \frac{2m_e}{M_a} + \sum_k \frac{\omega_k}{\varepsilon_e}\frac{\sigma_k}{\sigma_e} .$$

Transport Coefficients. Solution of the Boltzmann equation results in derivation of the *transport coefficients* describing the motion of electrons in a gas: the drift velocity, the diffusion coefficient and the characteristic energy. In calculations it is necessary to know the collision cross sections of electrons with molecules of the gas as functions of their energy, $\sigma_e(\varepsilon_e)$ and $\sigma_m(\varepsilon_e)$ (Fig. 2.2), as well as the mean energy loss occurring in the collisions of electrons, $\Lambda(\varepsilon_e)$.

The drift velocity in the direction of the field lines of the electric field $\mathcal{E}$ (Schultz, Gresser 1978) is

$$w_d = \int v_e \cos\theta_e F_e(\varepsilon_e, \theta_e)\, d\varepsilon_e$$

and upon transformation (Rice-Evans 1974)

$$w_d = -\frac{2}{3}\frac{e\mathcal{E}}{m_e}\int \varepsilon_e l_e(\varepsilon_e)\frac{\partial}{\partial\varepsilon_e}(F_0(\varepsilon_e)/v_e)\, d\varepsilon_e .$$

If l_e is independent of v_e, and the width of the electron energy distribution $F_e(\varepsilon_e)$ is not large, then

$$w_d \approx \left[\frac{2}{3}\left(\frac{1}{3}\Lambda\right)^{1/2} e\mathcal{E} l_e/m_e\right]^{1/2} .$$

(2.22)

The diffusion coefficient is

$$D(\mathcal{E}) = \frac{1}{3} \int l_e(\varepsilon_e) v_e F_e(\varepsilon_e, \theta_e) \, d\varepsilon_e \, d\theta_e$$

and upon simplification

$$D(\mathcal{E}) = \frac{1}{3} \int l_e(\varepsilon_e) v_e F_0(\varepsilon_e) \, d\varepsilon_e \; . \tag{2.23}$$

If the conditions under which we obtained (2.22) are satisfied, we obtain

$$D(\mathcal{E}) = \left[\frac{2}{9} (3\Lambda)^{-1/2} e\mathcal{E} l_e^3 / m_e \right]^{1/2} . \tag{2.24}$$

The characteristic energy $\varepsilon_c = eD(\mathcal{E})/\mu_e = eD(\mathcal{E}) \cdot \mathcal{E}/w_d$ is approximately

$$\varepsilon_c \approx (3\Lambda)^{-1/2} e\mathcal{E} l_e$$

where $\mu_e = w_d/\mathcal{E}$ is the electron mobility.

The theory of motion of electrons in a gas, presented above, gives dependences of the transport coefficients $w_d(\mathcal{E})$ and $\varepsilon_c(\mathcal{E})$ on the electric field strength, which are in quite good agreement with experimental data (Sauli 1977), under the condition that $\varepsilon_e < \omega_k$.

The cross section and the mean energy loss in gas mixtures are determined by the expressions

$$\sigma_e(\varepsilon_e) = \sum_j a_j \sigma_j(\varepsilon_e); \quad \Lambda(\varepsilon_e) = \frac{1}{\sigma_e(\varepsilon_e)} \sum_j a_j \sigma_j(\varepsilon_e) \Lambda_j(\varepsilon_e) \; .$$

The dependences of drift velocities upon the electric field strength, $w_d(\mathcal{E})$, differ quite significantly for different gases (Fig. 2.5).

The dependences $w_d(\mathcal{E})$ and $D(\mathcal{E})$ have been measured for many gas mixtures utilized in drift and proportional chambers (Sauli 1977; Fehlmann, Viertel 1983; Peisert, Sauli 1984).

2.2.5 Diffusion of Electrons in an Electric Field

The diffusion coefficients along the electric field and across the field differ from each other. In the general case $D(\mathcal{E})$ can be written in the form

$$D(\mathcal{E}) = \begin{vmatrix} D_L & 0 & 0 \\ 0 & D_T & 0 \\ 0 & 0 & D_T \end{vmatrix} , \tag{2.25}$$

where $D_L(\mathcal{E})$ is the longitudinal diffusion coefficient; $D_T(\mathcal{E})$ is the transverse diffusion coefficient, and $\mathcal{E} \parallel x$.

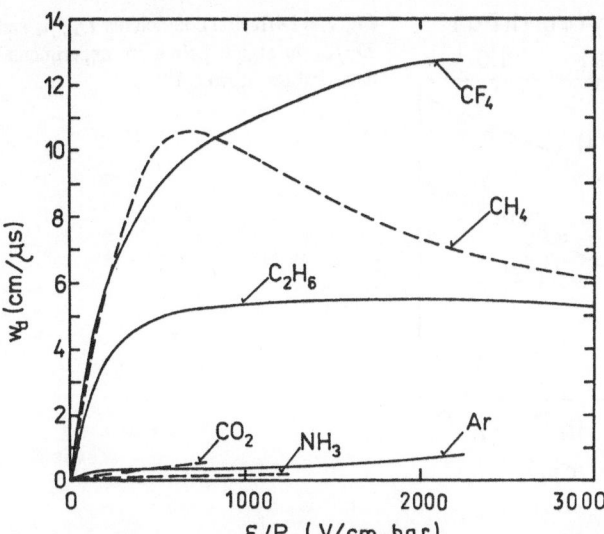

Fig. 2.5. Electron drift velocity in several gases

The diffusion coefficient (2.24) is termed symmetric; it represents the transverse components D_T of the diffusion tensor $D(\mathcal{E})$ (2.25). For many gases $D_L < D_T$ for a fixed $\mathcal{E}$ (Peisert, Sauli 1984).

The decrease of diffusion in the direction parallel to the field $\mathcal{E}$ may be due to the electrons present in the front of the avalanche undergoing diffusion mainly in the direction of motion of the electron swarm, so that they are accelerated to a greater extent by the electric field. Therefore their mean free path l_e decreases and from (2.22) it follows that their drift velocity (mobility) decreases too. On the contrary, the electrons at the rear part of the swarm mainly experience diffusion in a direction opposite to the motion of the cloud, so they are "slowed down" by the electric field and their free path l_e is increased, leading to enhancement of their drift velocity. These processes lead to contraction of the electron swarm in the drift direction, so $D_L < D_T$. The relationship between D_L and D_T depends on the behavior of the function $l_e(\varepsilon_e)$ [or $\sigma_m(\varepsilon_e)$]. The dependence of D_L on $\mathcal{E}/P$ (Fig. 2.6) is determined by the behavior of the cross section $\sigma_m(\varepsilon_e)$ (Fig. 2.2). At low electron energies ε_e, when the cross section σ_m decreases down to the Ramsauer minimum, $D_L > D_T$. As ε_e increases, σ_m starts to rise, while D_L decreases (Fig. 2.6).

The difference between D_L and D_T was first observed by Wagner et al. (1967). Calculations of the longitudinal diffusion coefficient were performed by Parker and Lowke (1969) and by Skullerud (1969).

For calculation of D_L, Parker and Lowke (1969) make use of the Fourier transform of the Boltzmann equation:

$$D_L = D_T + \frac{8\pi e \mathcal{E} kT}{3m_e^2 n_a \sigma_0} \int_0^\infty \frac{\xi_e \sigma_0}{\sigma_m} \frac{\partial F_1}{\partial \xi_e} \, d\xi_e + \frac{4\pi \mathcal{E} kT}{m_e} \left(\frac{2kT}{m_e}\right)^{1/2} \int_0^\infty \xi_e^2 F_1 \, d\xi_e \,,$$

$$(2.26)$$

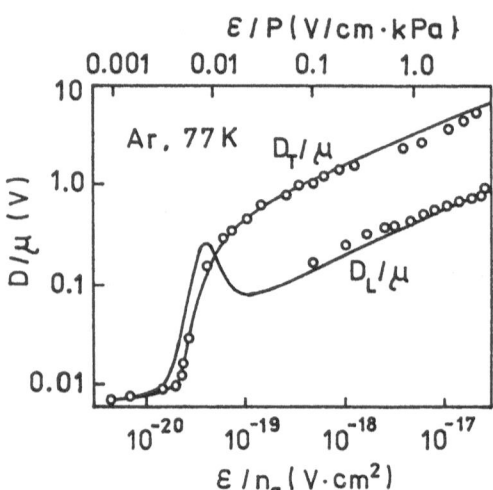

Fig. 2.6. Diffusion coefficients D_L/μ_e and D_T/μ_e in argon; points are experimental data (Parker, Lowke 1969)

where $\xi_e = \varepsilon_e/kT$; n_a is the density of molecules of the gas $(cm^3 \cdot 133\,Pa)^{-1}$; $\sigma_0 = 10^{-16}\,cm^2$; and

$$D_T = \frac{8\pi}{3 n_a \sigma_0} \left(\frac{kT}{m_e}\right)^2 \int\limits_0^\infty (\xi_e F_0 \sigma_0/\sigma_m)\, d\xi_e \ . \tag{2.27}$$

The expression for D_T coincides with (2.23) to within a normalization constant. More details are to be found in Huxley and Crompton (1974).

The Parker-Lowke model agrees well with experimental data (Fig. 2.6). In argon at certain $\mathcal{E}/P$ ratios, the difference between D_L and D_T is quite significant: $D_T \approx 7 D_L$, which is explained by the significant increase of $\sigma_m(\varepsilon_e)$ (Fig. 2.2) and the corresponding decrease of $l_e(\varepsilon_e)$.

In krypton and xenon the functions $(D_T/\mu_e)(\mathcal{E})$ and $(D_L/\mu_e)(\mathcal{E})$ are close to the measured functions for argon. For other gases, however, the ratio D_T/D_L is not so great. Therefore, in estimating the influence of diffusion on the accuracy of coordinate measurements in a drift chamber, it is necessary to take into account the longitudinal component of diffusion; then (2.14) assumes the form

$$l_D = \sigma_x = \sqrt{2 D_L t} = \sqrt{2 D_L x/w_d} \ . \tag{2.28}$$

The results of measurements of the dependences $D_L(\mathcal{E}/P)$ for various gases and gas mixtures may be found in Peisert and Sauli (1984), Fehlmann and Viertel (1983), Piuz (1983), Fulda-Quenzer (1985) and Wong et al. (1988).

In certain so-called cool gases (CO_2, NH_3, isobutane, methylal vapor) the electron energy remains thermal even at relatively high electric field strengths (Christophorou 1970; Bobkov et al. 1984; Va'vra 1986; Sitar 1987; Sosnovtsev 1987). This is due to the large elastic scattering cross section in these gases. For example, in CO_2 and isobutane the electron energy distribution remains maxwellian up to $\mathcal{E} = 200$–$300\,V/cm$, while in NH_3 it remains maxwellian even up to approximately 700 V/cm. In these gases the drift velocity depends linearly on $\mathcal{E}$ and may be expressed in the simple form: $w_{\mathrm{d}} = \mu_{\mathrm{e}}\mathcal{E}/P$, where $\mu_{\mathrm{e}} = \mathrm{const}$. Here, for $P = 10^5\,\mathrm{Pa}$ and $T = 300\,\mathrm{K}$, the Nernst-Townsend equation holds:

$$\frac{D}{\mu_{\mathrm{e}}} = \frac{kT}{e} = \frac{2}{3}\varepsilon_{\mathrm{e}} = \varepsilon_{\mathrm{c}} \approx 0.026\,\mathrm{eV} \ . \tag{2.29}$$

The Nernst-Townsend equation (sometimes incorrectly called the Einstein equation) is valid mainly for the motion of ions in gases, in which case it holds within a wide range of $\mathcal{E}$ values.

Taking account of (2.29) the root-mean-square error may be written

$$\sigma_x = \sqrt{2kTx/(e\mathcal{E})} \ . \tag{2.30}$$

The actual measurement precision cannot exceed this value, which is called the *thermal limit*, due to diffusion. The coordinate resolution of drift chambers filled with cool gases is discussed in Chap. 4.

In other gases the electron energy spectrum $F_{\mathrm{e}}(\varepsilon_{\mathrm{e}})$ differs from the maxwellian distribution even at low electric field strengths. The large difference between electron diffusion coefficients in various gases leads to a significant difference between root-mean-square deviations σ_x. Admixtures with small diffusion coefficients, like cool gases, are preferable for use in drift chambers.

2.2.6 Drift and Diffusion of Electrons in a Magnetic Field

Free electrons in a magnetic field move under the action of the Lorentz force along circular trajectories in between collisions, and so the drift velocity of an electron swarm decreases. In the case of constant electric and magnetic field strengths, $\mathcal{E}$ and H, and if $\mathcal{E} \perp H$, the electron swarm moves at an angle α_{M} to the electric field lines.

From elementary kinetic theory approximate relations follow for the drift velocities w_{M} and for the angle α_{M} in a magnetic field:

$$w_{\mathrm{M}} = w_{\mathrm{d}}/(1 + f_{\mathrm{L}}^2\tau_{\mathrm{M}}^2)^{1/2} \ ; \tag{2.31}$$

$$\tan \alpha_{\mathrm{M}} = f_{\mathrm{L}}\tau_{\mathrm{M}} \ , \tag{2.32}$$

where $f_{\mathrm{L}} = eH/m_{\mathrm{e}}$ is the Larmor frequency; $\tau_{\mathrm{M}}(\mathcal{E}, H)$ is the mean free path time of an electron in crossed electric and magnetic fields. In Fig. 2.7 the functions $w_{\mathrm{M}}(B)$ and $\alpha_{\mathrm{M}}(B)$, where B is the magnetic induction, calculated applying (2.31) and (2.32), are presented. The experimental data are in good agreement with these calculations.

The electron diffusion coefficient in a magnetic field D_{M} is smaller than D_{T}, and

$$D_{\mathrm{M}} \approx D_{\mathrm{T}}(\mathcal{E})/(1 + f_{\mathrm{L}}^2\tau_{\mathrm{M}}^2) \ . \tag{2.33}$$

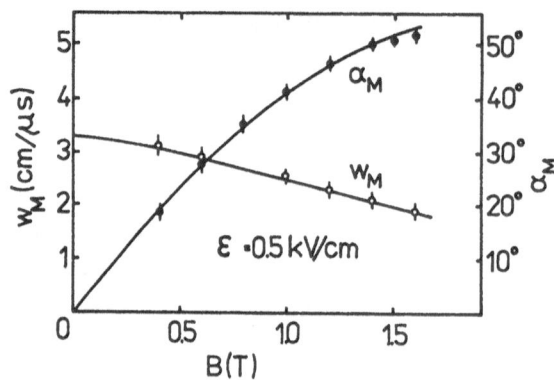

Fig. 2.7. Dependence of w_M and α_M on the magnetic induction in a mixture of 67.2% Ar + 30.3% Iso-C_4H_{10} + 2.5% Methylal at $\mathcal{E} = 500$ V/cm (Breskin et al. 1975)

At low pressures or high H, $D_M \sim H^{-2}$. In this case, in accordance with (2.12), σ_x is inversely proportional to H. Even at $P = 10^5$ Pa this effect is impressive. Thus, in a neon streamer chamber the root-mean-square deviation of streamers from the particle trajectory in a field of $B = 1$ Tesla, occurring during a time delay of the high-voltage pulse $\tau_d = 1.2\,\mu s$, is $\sigma_x = 0.2\,$mm, i.e. nearly two times smaller than for $B = 0$ ($\sigma_x = 0.37$ mm) (Dayon et al. 1970). The suppression of electron diffusion in track detectors, placed in a magnetic field, at the same time leads to enhancement of the measurement precision of momenta and of ionization.

A more correct theory is based on the solution of the Boltzmann equation (2.15) which for a non-zero magnetic field with an induction B has the form

$$\frac{\partial}{\partial t} f_e + v_e \cdot \nabla_r f_e + F_\parallel \cdot \nabla_{v_e} f_e + S_{col} = 0 , \tag{2.34}$$

where $F_\parallel = (e/m_e)(\mathcal{E} + v_e \times B)$. Equation (2.34) is also known as the Fokker-Planck equation. In the case of motion of electrons in a magnetic field (2.18) does not change, while (2.17) assumes the form

$$e\mathcal{E}\frac{\partial}{\partial \varepsilon_e}(v_e F_0) - \frac{2e\mathcal{E}}{m_e v_e}F_0 + G_H(\varepsilon_e)\frac{v_e F_1}{l_e} = 0 , \tag{2.35}$$

where $G_H(\varepsilon_e) = 1 + e^2 H^2 l_e^2(\varepsilon_e)/(2m_e \varepsilon_e)$. The solution of (2.18) and (2.35) gives the drift velocity in the direction of $\mathcal{E}$:

$$w_L = -\frac{2e\mathcal{E}}{3m_e} \int \frac{1}{G_H(\varepsilon_e)} \varepsilon_e l_e(\varepsilon_e) \frac{\partial}{\partial \varepsilon_e}(F_0(\varepsilon_e)/v_e)\,d\varepsilon_e . \tag{2.36}$$

The component of the drift velocity normal to $\mathcal{E}$ and H is

$$w_T = \frac{eH^2}{3m_e} \int \frac{1}{G_H(\varepsilon_e)} v_e l_e^2(\varepsilon_e) \frac{\partial}{\partial \varepsilon_e}(F_0(\varepsilon_e)/v_e)\,d\varepsilon_e . \tag{2.37}$$

Then

$$\tan \alpha_M = w_T/w_L = w_M B/\mathcal{E} ; \quad w_M = (w_L^2 + w_T^2)^{1/2} . \tag{2.38}$$

For small ε_e, when it is sufficient to take into account only the elastic scattering cross section σ_e, the solution of the Boltzmann equation (2.20) for $H \neq 0$ has the form

$$F_0(\varepsilon_e) = C_F \sqrt{\varepsilon_e} \exp \left\{ - \int_0^{\varepsilon_e} \frac{3\varepsilon_e \Lambda(\varepsilon_e) G_H(\varepsilon_e)}{[e\mathcal{E} l_e(\varepsilon_e)]^2 + 3\Lambda(\varepsilon_e) G_H(\varepsilon_e) \varepsilon_e kT} \, d\varepsilon_e \right\} . \quad (2.39)$$

The diffusion coefficient in a magnetic field is

$$D_M = \frac{1}{3} \int \frac{1}{G_H(\varepsilon_e)} l_e(\varepsilon_e) v_e F_0(\varepsilon_e) \, d\varepsilon_e . \quad (2.40)$$

Comparison of (2.23) and (2.40) reveals that $D_M < D_T$. When $B = 0$, (2.40) transforms into (2.23). In Fig. 2.8 it is shown that the electron drift velocity changes significantly in the presence of a magnetic field.

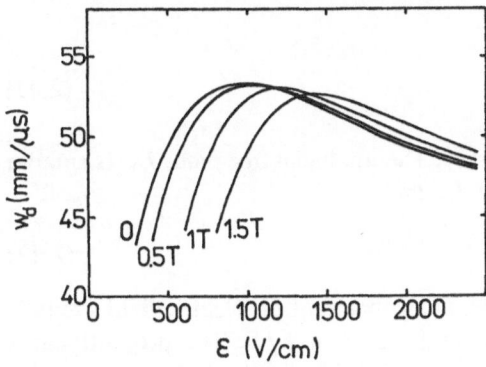

Fig. 2.8. Electron drift velocity in a 50 % Ar + 50 % C_2H_6 mixture versus electric field strength and magnetic induction (Ramanantsizehena et al. 1980)

2.3 Drift and Diffusion of Ions

Sources of Ions in a Proportional Detector. In the ionization process in the gas of the detector, in addition to electrons positive ions are produced. In proportional detectors well-localized clusters of positive ions arise in the vicinity of the signal wires, which then slowly drift into the detector volume. Their presence has a negative influence upon the measurement precision of coordinates and ionization (Sect. 3.4).

Other, not so essential, mechanisms of ion formation in a gas are represented by the capture of electrons by electro-negative admixtures, photo-ionization and photo-dissociation, as well as collisions of excited ions with the atoms and molecules of the gas.

2.3.1 Motion of Ions in a Gas

In the absence of an electric field ions experience diffusion in just the same manner as electrons; thus (2.1) to (2.12) are also suitable (with appropriate constants) for describing the motion of ions. The ion energy distribution retains its maxwellian nature in the presence of an electric field. This follows from the relation between the energy of thermal motion and the energy transferred to the ions under the action of the electric field:

$$\frac{\text{thermal energy}}{\text{drift energy}} \sim \frac{M_a}{M_+} \, ,$$

where M_a is the mass of the atom (molecule) of the gas; M_+ is the mass of the ion. In the case of electrons (where m_e is substituted for M_+) the energy of thermal motion is dominant. Therefore the electron velocity v_e is significantly higher than the thermal velocity of the ions v_+ and from (2.41) it follows that the drift velocity of positive ions w_+ is lower than the electron drift velocity w_d by several orders of magnitude. The relation between w_+ and v_+ for ions has the form

$$w_+ = \frac{eL_+}{M_+ \bar{v}_+} \frac{\mathcal{E}}{P} = \mu_+ \frac{\mathcal{E}}{P} \tag{2.41}$$

where $\bar{v}_+$ is the mean positive ion velocity. The ion mean free path, L_+, is smaller than the mean free path of a molecule L_m and

$$4\sqrt{2}\, L_+ \lesssim 4\sqrt{2}\, L_m = l_e \, . \tag{2.42}$$

The ion energy distribution is almost independent of the electric field strength, so the drift velocity depends linearly on $\mathcal{E}$: $w_+ = \mu_+ \mathcal{E}/P$. The proportionality factor μ_+ is termed the *mobility* of the positive ion. The ion mobility is practically independent of $\mathcal{E}/P$. The values of L_+, $\bar{v}_+$, μ_+ and the diffusion coefficient for positive ions D_+ are given for some gases in Table 2.3. The mobilities of ions in gases utilized in proportional detectors are given in Table 2.4, and the dependence of the positive ion drift velocity on the electric field strength is presented in Fig. 2.9.

The Nernst-Townsend equation (2.29) holds for ions within a wide range of electric field strengths, i. e. the ratio D_+/μ_+ remains constant as $\mathcal{E}$ changes.

Theoretical issues of ion transport in gases are presented in greater detail in Brown (1959), Hasted (1964), Smirnov (1968), Christophorou (1970), and McDaniel (1964).

The ratio D_+/μ_+ for ions is smaller than the same ratio for electrons; thus, for example, in argon $D_+/\mu_+ = 0.068\,\text{V}$, while D_L/μ_e varies within the range 0.1 V to 1.0 V, and D_T/μ_e varies from 0.1 V to 8.0 V, in electric field-strengths typical for drift chambers (Fig. 2.6).

Table 2.3. Mean free path, mean volocity, diffusion coefficient and mobility of positive ions under normal conditions (Sauli 1977)

Gas	L_+ [10^{-5} cm]	$\bar{v}_+$ [10^4 cm/s]	D_+ [cm^2/s]	μ_+ [cm^2/(V $\cdot$ s)]
H_2	1.8	20.0	0.34	13.0
He	2.8	14.0	0.26	10.2
Ar	1.0	4.4	0.04	1.7
O_2	1.0	5.0	0.06	2.2
H_2O	1.0	7.1	0.02	0.7

Table 2.4. Positive ion mobilities in gases [cm^2/V·s], at normal conditions (experimental data) (Smirnov 1968; Elecki et al. 1975; Sauli 1977)

Gas \ Ion	He^+	Ne^+	Ar^+	Kr^+	Xe^+	CO_2^+	CH_4^+	iso-$C_4H_{10}^+$	$CH_2(CH_3O)_2^+$
Same Gas	10.2	4.1	1.7	0.94	0.58	1.09	2.26	0.61	0.26
He		10.2	24.0	19.5	20.2	18.0			
Ne	17.2	4.1							
Ar			1.7	2.3		1.72	1.87	1.56	1.51
Isobutane								0.61	0.55

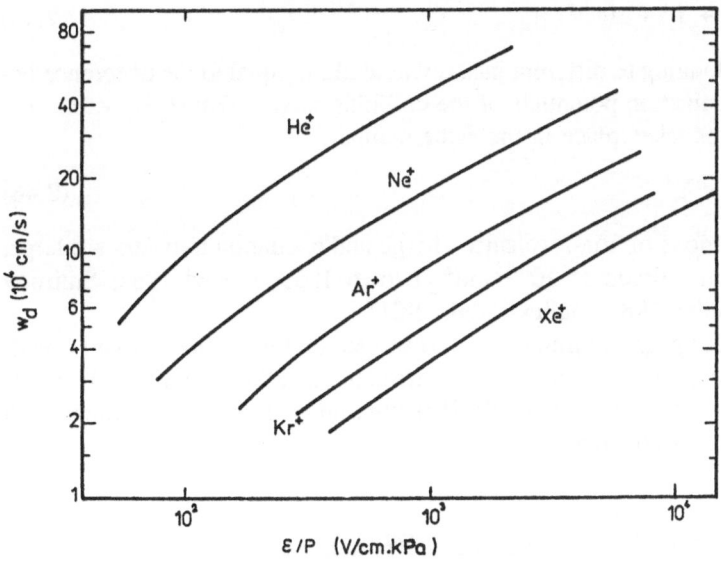

Fig. 2.9. Dependence of the drift velocity of positive ions in noble gases upon $\mathcal{E}/P$ (McDaniel 1964)

The diffusion coefficients D_+ and D_- and mobilities μ_+ and μ_- for positive and negative ions do not differ so strongly, but for most gases $D_+ < D_-$ and $\mu_+ < \mu_-$ (Engel 1965; Huxley, Crompton 1974).

The mobility of ions in a mixture of gases with partial concentrations a_j and mobilities μ_j is determined by the Blank rule,

$$1/\mu_+ = \sum_j a_j/\mu_j \, . \tag{2.43}$$

2.3.2 Charge Transfer During Ion Transport

In ion transport processes in a gas, charge transfer from ions to neutral atoms (molecules) occurs very effectively. A moving ion loses its charge and becomes a neutral atom, but it transfers practically no kinetic energy to the new ion. The new ion proceeds to drift in the electric field with an initial velocity nearly equal to zero.

The process of *symmetric resonance charge transfer*

$$X^+ + X \rightarrow X + X^+ \tag{2.44}$$

occurs with a large cross section: 10^{-15}–$10^{-14}\,\text{cm}^2$ (Galeev, Sudan 1983). *Asymmetric non-resonance charge transfer*

$$X^+ + X \rightarrow X + Y^+ + \Delta\varepsilon \tag{2.45}$$

involves ions belonging to different gases, where $\Delta\varepsilon$ is equal to the difference between the first ionization potentials of the colliding atoms. *Transfer of excitation* between the atoms takes place in the same manner:

$$X^* + Y \rightarrow X + Y^* \, . \tag{2.46}$$

The cross sections of non-resonance charge and excitation transfers are large, of the order of magnitude of $10^{-15}\,\text{cm}^2$ (Brown 1959; Hasted 1964; Smirnov 1968; Galeev, Sudan 1983; Galicki et al. 1981).

Charge transfer plays an important part in gas mixtures. The charge is readily transferred to molecules of lower ionization potentials, so this process is irreversible; after a short time the only ions remaining in the gas are those with the lowest ionization potential.

2.4 Change in the Degree of Ionization Due to Drift of Electrons

For precision ionization measurements in proportional and track detectors it is necessary to take into account effects enhancing or reducing ionization in the electron transport processes. The difference between the true and measured ionizations in a real detector depends, for example, on the charge collection time, the composition of the gas mixture, the field strength etc.

2.4.1 Processes Enhancing the Observed Ionization

Until the high-voltage pulse has been applied in spark or streamer chambers, and with a field strength typical for drift chambers, $\mathcal{E} \leq 1\,\text{kV/cm}$, the electrons do not acquire sufficient energy for ionization or excitation of the gas atoms. In the course of electron diffusion and drift processes, however, reactions occur that do enhance the ionization: the formation of molecular ions and the Penning reaction.

Molecular Ions. The formation of molecular ions leads to the appearance of additional electrons in the gas. In noble gases the molecular ions are produced in the process

$$X^* + X \rightarrow X_2^+ + e^- . \tag{2.47}$$

The threshold energy required for the production of a molecular ion X_2^+ is lower than the ionization potential I_1 of the given gas: for He_2^+ it is 23.18 eV, for Ne_2^+ 20.86 eV, for Ar_2^+ 15.06 eV and for Kr_2^+ 13.23 eV. The formation cross section of molecular ions, $\approx 10^{-15}\,\text{cm}^2$ (Rice-Evans 1974), exhibits a maximum near the threshold. At pressures of $P \geq 10^4\,\text{Pa}$ this reaction is almost instantaneous, so the additional ionization due to it cannot be separated from the primary ionization. The accuracy with which the primary ionization in He is measured, taking account of reaction (2.47), turns out to be about 8 % (Kondratev 1974).

Molecular ions may also be produced in other less probable processes.

The Penning Effect. If the gas mixture contains a gas Y, the ionization potential of which is lower than the excitation level of the other gas (X^*), then the excited atom colliding with the atom of lower ionization potential may ionize it:

$$X^* + Y \rightarrow X + Y^+ + e^- . \tag{2.48}$$

The Penning effect occurs with a large cross section, if of the two gases involved, the atom X is excited to a long-lived metastable state X^{**} (a typical lifetime of a magnetic dipole is of the order of 10^{-3} s, and of a quadrupole of the order of 1 s). An example of such a pair of gases is neon with an admixture of argon. Neon has a metastable state of 16.53 eV, which is higher than the ionization potential of argon, $I_1 = 15.76$ eV. The metastable and resonance levels of the noble gases are presented in Table 2.5.

The cross sections for the Penning process in some gas mixtures are presented in Table 2.6. The contribution of the Penning effect depends on the concentration of the admixture exhibiting the lower ionization potential and on the rate capability of the detector. This effect is utilized, in particular, for memorizing track information in streamer chambers (in the operation mode of a two-pulse supply during 1 ms without loss of spatial resolution).

Table 2.5. Metastable I_m, resonance I_r, and ionization I_1, potentials of inert gases (Engel 1965)

Gas	I_m [eV]	I_r [eV]	I_1 [eV]
He	19.8–20.7	21.2	24.59
Ne	16.6–16.7	16.7–16.8	21.57
Ar	11.5–11.7	11.6–11.8	15.76
Kr	9.9–10.5	10.0–10.6	14.00
Xe	8.3–9.4	8.5–9.6	12.13

Table 2.6. Cross section of the Penning process, σ_P [10^{-16} cm^2], in gases at room temperature (Galeev, Sudan 1983)

Admixture	He*		Ne*	Ar*
	$^3S_2$	$^1S_2$	$^3P_2$	$^3P_2$
Ar	5.5	9.7	2.8	
Kr	7.8	34	1	1
Xe	11.8	50	16	
H_2	2.4	2.4		0.3
N_2	5.3	9	0.6	
O_2	14.5	24	0.9	1.2
CO_2	35	64		

2.4.2 Processes Decreasing the Ionization

The losses of electrons transported in a gas are mainly due to the following three processes: electron-ion recombination, the capture of electrons by electro-negative admixtures and absorption by the walls of the detector. In practice, the latter mechanism can be neglected in ionization detectors of large volume. In the course of electron drift recombination it is almost absent owing to the small number of ions N_+ in the gas volume of the detector. It plays a more significant part in the development of an avalanche in the process of gas amplification.

2.4.3 Electron Attachment

In large drift detectors, spark and streamer chambers the capture of electrons by neutral atoms (molecules) represents a serious problem, since in a gas there are several orders of magnitude more neutral atoms (or molecules), than ions. The probability of electron attachment is high in gases (O_2, H_2O, halogens) in which stable negative ions are formed. The energy of such ions is somewhat lower than the energy of the primary atoms. The binding energy between an electron and ion is called *the electron affinity of the atom*. The electron affinities of certain atoms and molecules, usually present in the form of small admixtures in the gas fillings of proportional and other gas detectors, are presented in Table 2.7. The cross section of electron attachment is correlated with the electron affinity.

Table 2.7. Electron affinity of atoms and molecules forming stable negative ions (Kondratev 1974)

Negative ion	Cl^-	F^-	Br^-	I^-	O^-	C^-	H^-
Electron affinity [eV]	3.614	3.45	3.37	3.08	1.467	1.27	0.754

Molecular ion	Cl_2	F_2	Br_2	I_2	O_2	C_2	OH^-
Electron affinity [eV]	2.38	3.08	2.51	2.58	0.44	3.54	1.83

Various mechanisms are responsible for the capture of electrons:
a) *radiative capture, or photo-attachment,*

$$e^- + Z \rightarrow Z^- + \hbar\nu , \tag{2.49}$$

where $\hbar$ is Planck's constant and ν is the frequency of photon emitted, occurs when the atom Z exhibits positive electron affinity.

The cross section of radiative attachment to an atom exhibiting electron affinity is not large [of the order of $4 \cdot 10^{-23}$ cm^2 in oxygen (Hasted 1964) and 10^{-22} cm^2 in halogens (Galeev, Sudan 1983), since the electron-atom interaction time (10^{-15} s) is small compared with the lifetime of an excited state (10^{-8} s)];

b) *electron attachment involving three body collisions*:

$$e^- + Z + X \rightarrow Z^- + X + \text{energy} ,$$
$$e^- + e^- + Z \rightarrow Z^- + e^- + \text{energy} , \tag{2.50}$$

where the excess energy is transferred to the third body, X or e^-;

c) *electron attachment to a molecule with subsequent transfer of the excitation to a third body*:

$$e^- + YZ \rightarrow (YZ)^{-*} , \quad (YZ)^{-*} + X \rightarrow (YZ)^- + X + \text{energy} . \tag{2.51}$$

If the "catalyst" (the atom X) is capable of taking over all the energy generated in the capture, process (2.51) exhibits a resonant character (Fig. 2.10) and its cross section is large (Table 2.8);

b) *dissociative capture* by polyatomic molecules with positive electron affinity takes place along various channels:

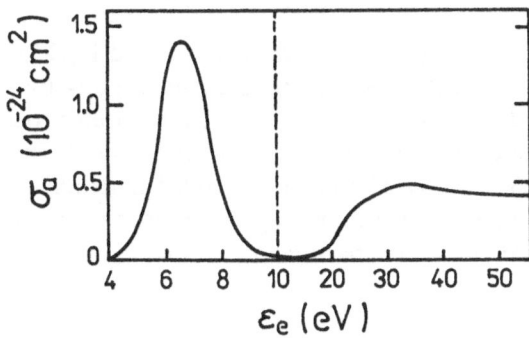

Fig. 2.10. Electron capture cross section in oxygen as a function of the electron energy (Rapp, Englander-Golden 1965)

Table 2.8. Resonance energy, maximum capture cross section σ_{at}^{max} (Christophorou 1970; Hasted 1964), and electron capture cross section σ_{at} at thermal energies (Davidenko et al. 1970)

Gas	O_2	CO	CO_2	SF_6
Resonance energy [eV]	6.2	10.1	7.8	0.0
$\sigma_{at}^{max}[10^{-16} cm^2]$	0.013	0.027	0.005	5.7
σ_{at} [cm^2]	$\sim 10^{-22}$		$< 10^{-20}$	$1.7 \cdot 10^{-14}$

Gas	CCl_4	CCl_2F_2	$C_2Cl_3F_2$	H_2O
Resonance energy [eV]	0.02	0.15	0.16	6.4
$\sigma_{at}^{max}[10^{-16} cm^2]$	1.3	0.54	19.2	0.048
σ_{at} [cm^2]	$2.6 \cdot 10^{-14}$			$7.5 \cdot 10^{-21}$

$$e^-(\varepsilon_e) + (XY \ldots VZ) \rightarrow (XY \ldots VZ)^{-*} \rightarrow$$

$$\rightarrow (XY \ldots VZ)^{(*)} + e(\varepsilon_e') ; \tag{2.52a}$$

$$\rightarrow \left. \begin{array}{l} XY^{(*)} + VZ^- \\ (XY \ldots V)^* + Z^- \end{array} \right\} ; \tag{2.52b}$$

$$\rightarrow (XY)^* + VZ^{-*} \left. \begin{array}{l} \rightarrow VZ^* + e^- \\ \rightarrow V^{(*)} + Z^- \end{array} \right\} ; \tag{2.52c}$$

$$\rightarrow (XY \ldots VZ)^- + energy \tag{2.52d}$$

Channel (a) represents an inelastic collision ($\varepsilon_e' < \varepsilon_e$); (b) represents the dissociative capture of an electron involving the production of stable negative ions; (c) represents dissociative capture accompanied by the creation of metastable negative ions; (d) represents the formation of negative ions.

Dissociative capture by molecules with high electron affinities (mainly by molecules containing halogens; Table 2.7) at low electron energies is characterized by large cross sections σ_{at} of the order of $10^{-15} cm^2$ (Christophorou 1970). These cross sections exhibit resonant behavior (Fig. 2.10). In certain gases (CCl_4, SF_6) the resonance energy is low (Table 2.8), so thermal electrons are captured very effectively. The capture cross sections of thermal electrons are presented in Table 2.8. Small admixtures of these gases are used for reducing memory time of spark and streamer chambers.

The attachment cross sections σ_{at} for thermal electrons in O_2 and H_2O are not high, since the thresholds for molecular dissociation are quite high (4.6 eV for $O_2 \rightarrow O^+ + O^-$, and 5.5 eV for $H_2O \rightarrow H^+ + OH^-$). The capture of electrons in drift and proportional chambers occurs mainly in the admixtures of O_2 and H_2O, through the mechanisms (2.51) and (2.52); this is described in detail, for example, in Huk et al. (1988). Problems related to the capture of electrons in the gas fillings of drift and proportional chambers are discussed in Sect. 3.3.

The number of electrons in a swarm drifting through a distance of x, falls as a result of capture according to the law

$$n_e = n(0) \exp(-\eta_a x) \,, \tag{2.53}$$

where η_a is the attachment coefficient which is related to the mean capture cross section $\overline{\sigma}_{at}$:

$$\eta_a = n_a \overline{\sigma}_{at} \overline{v}_e / w_d \,. \tag{2.54}$$

The experimental data on η_a are contradictory (Huxley, Crompton 1974). Electron attachment is often characterized by the probability of capture in a single collision (Sauli 1977):

$$k_a = l_e \overline{\sigma}_{at} n_a = 1/(\overline{\tau}_{at} \nu_e) \,, \tag{2.55}$$

where $\overline{\tau}_{at}$ is the mean lifetime of the electrons before capture. These quantities are presented for some gases in Table 2.9; it can be seen that in oxygen or water vapor, electrons are captured within several hundreds of nanoseconds and their typical absorption length L_a amounts to several millimeters. On the other hand, the probability of electron capture in pure inert gases is almost equal to zero. Note that electron capture should not only be considered to be a detrimental effect. For instance, small admixtures of electro-negative gases, for example, $10^{-5}\%$ SF_6, are often utilized for reducing the memory time of streamer chambers in accelerator experiments.

Table 2.9. Capture probability k_a, mean capture time $\overline{\tau}_{at}$, and collision frequency ν_e of thermal electrons under normal conditions (Sauli 1977)

Gas	k_a	$\overline{\tau}_{at}$ [s]	ν_e [s^{-1}]
CO_2	$6.2 \cdot 10^{-9}$	$7.1 \cdot 10^{-4}$	$2.2 \cdot 10^{11}$
O_2	$2.5 \cdot 10^{-5}$	$1.9 \cdot 10^{-7}$	$2.1 \cdot 10^{11}$
H_2O	$2.5 \cdot 10^{-5}$	$1.4 \cdot 10^{-7}$	$2.8 \cdot 10^{11}$
Cl_2	$4.8 \cdot 10^{-4}$	$4.7 \cdot 10^{-9}$	$4.5 \cdot 10^{11}$

2.5 Registration of Ionization

The charge arising from the interaction of a charged particle in the gas (about 100 electrons per 1 cm in argon, i.e. $1.6 \cdot 10^{-17}$ C) is too small to be measured directly, so it must be amplified. To this end electron (in ionization chambers), thermodynamic (in condensation chambers), and gas amplification (in gas-discharge chambers) methods are applied. In the strong electric field in the vicinity of the signal wires of a proportional detector, electrons are accelerated, and as a result of collisions, they ionize atoms of the gas. An avalanche ionization process occurs, which results in a high charge density being created within a small distance. The mean free path of an electron undergoing ionizing collisions in the vicinity of the signal wire amounts to 1–$10\,\mu$m at atmospheric pressure; therefore elec-

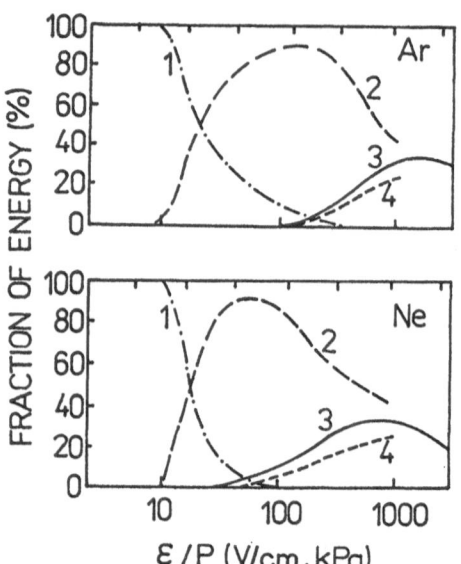

Fig. 2.11. Fraction of energy spent by electrons in the following processes: (*1*) – elastic collisions; (*2*) – excitation by electrons; (*3*) – ionization; (*4*) – enhancement of the electron kinetic energy (Charpak 1970)

tron multiplication occurs along a very short path. This process is termed *gas amplification*.

The probability of inelastic collisions in the gas increases with the field strength. The main inelastic processes taking place in a strong electric field, e. g. excitation of electron levels in an atom, ionization, photo-excitation and photo-ionization, all exhibit a complicated energy dependence (Fig. 2.11). Numerous monographs (Brown 1959; Hasted 1964; Loeb 1969; Marr 1967; Rice-Evans 1974; Sauli 1977; Galeev, Sudan 1983) are devoted to the description of these processes. We list some of their fundamental properties in Sects. 2.5.1–9.

The interesting phenomenon of scintillation in inert gases which also pertains to inelastic collisions, is dealt with in Chap. 6.

2.5.1 Excitation of Atoms by Electron Impact

If an electron in the gas acquires in the electric field an energy higher than the excitation energy ω_k of an atom to the level k, then the following inelastic process may take place:

$$e^- + X \rightarrow X^* + e^- . \tag{2.56}$$

The excitation cross sections σ_k of permissible transitions are characterized by sharp maxima near the thresholds ω_k. In noble gases these cross sections achieve a maximum at electron energies from 10 to 20 eV and amount to $\sigma_k \approx 10^{-17}$ cm^2. Thus, for example, in argon under normal conditions electron collisional excitation occurs effectively when the electrons are accelerated in a field of strength $\mathcal{E} \geq 2.5$ kV/cm (Fig. 2.4). This phenomenon is utilized in electro-luminescence gas chambers (Chap. 5).

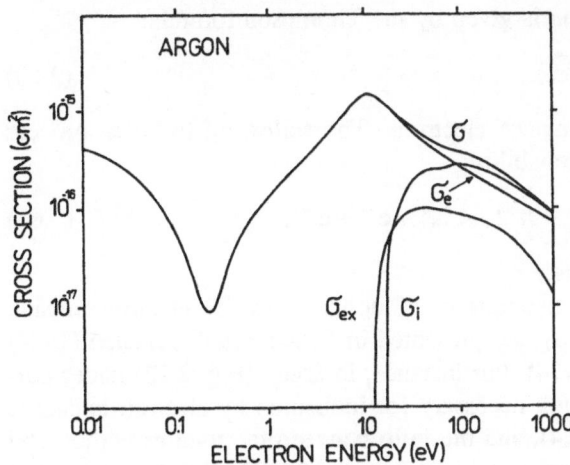

Fig. 2.12. Electron scattering cross sections in argon: σ_{ex} – excitation, σ_i – ionization, σ_e – elastic scattering, σ – total cross section (Biagi 1988)

The total excitation cross section of the atoms of a gas by electron collision is $\sigma_{ex} = \sum_k \sigma_k$. Its maximum value in argon for $\varepsilon_e \approx 20\text{–}30\,eV$ amounts to $\sigma_{ex} \approx 10^{-16}\,cm^{-2}$ (Lapique, Piuz 1980) (Fig. 2.12). Theoretical and experimental data on the excitation of the atoms of inert gases by electron collisions are presented, for example, in Moiseiwitch and Smith (1968).

The excitation of atoms and molecules plays an important part in the process of gas amplification; the emitted resonance photon,

$$X^* \rightarrow X + \hbar\nu \tag{2.57}$$

may ionize a molecule (atom) Y with a lower potential of photo-ionization:

$$\hbar\nu + Y \rightarrow e^- + Y^+ . \tag{2.58}$$

The contribution of photo-ionization to gas amplification is significantly lowered by the Penning effect, which is characterized by a cross section $(10^{-16}\,cm^2)$ close to the photo-ionization cross section $(\sim 10^{-16}\,cm^{-2})$ and thus competes with it.

The excitation of molecules is of a more complex nature owing to the large excitation cross section of rotational and vibrational states. These issues are discussed below.

2.5.2 Ionization by Electron Impact

Electrons accelerated in the electric field to energies higher than the first ionization potential I_1 cause ionization by collision:

$$e^- + X \rightarrow X^{+(*)} + e^- + e^- . \tag{2.59}$$

A positive ion is often produced in an excited state and its transition to the ground state is accompanied by emission of a photon or by excitation transfer to another atom (2.46).

The ionization cross section is given by the Thompson formula:

$$\sigma_i = (n_v \pi e^4 / \varepsilon_e)(1/I_1 - 1/\varepsilon_e) \,, \tag{2.60}$$

where n_v is the number of valence electrons. The following two-step process also takes place with a low probability:

$$e^- + X \rightarrow X^{**} + e^- \,; \quad e^- + X^{**} \rightarrow X^+ + e^- + e^- \,, \tag{2.61}$$

where X^{**} is a metastable state.

The total ionization cross sections σ_i of noble gases, by electron impact, which are possible when $\varepsilon_e > I_1$, are presented in Brown (1952), Hasted (1964) and Huxley and Crompton (1974). For instance, in argon (Fig. 2.12) under normal conditions, the field strength necessary for ionization by electron impact is $\mathcal{E} \geq 10\,\mathrm{kV/cm}$ (see also Fig. 2.4), and the ratio between the total excitation and ionization cross sections in argon $\sigma_{ex}/\sigma_i \approx 1$ for $\varepsilon_e \approx 20$–$35\,\mathrm{eV}$ and $\sigma_{ex}/\sigma_i \approx 0.4$ for $\varepsilon_e > 50\,\mathrm{eV}$ (Lapique, Piuz 1980).

2.5.3 Photo-absorption and Photo-ionization

Photons produced in an electron avalanche are absorbed by atoms or molecules of the gas through various mechanisms [for details see Marr (1967)]:

a) excitation of the atoms or molecules to higher-lying levels by resonance transitions:

$$\hbar\nu + X \rightarrow X^* \,; \quad \hbar\nu + XY \rightarrow XY^* \,; \tag{2.62}$$

b) excitation accompanied by transition of the molecule to states adjacent to the continuous spectrum and pre-dissociation:

$$\hbar\nu + XY \rightarrow XY \rightarrow X + Y \,; \tag{2.63}$$

c) photo-absorption accompanied by molecular dissociation

$$\hbar\nu + XY \rightarrow X^{(*)} + Y^{(*)} \,. \tag{2.64}$$

The latter process may be accompanied not only by excitation, but also by ionization. The cross sections of photo-absorption for some inert gases are presented in Fig. 1.6, and as an example, for ethane in Fig. 2.18. When $\hbar\nu > I_1$, the process of photo-absorption is also accompanied by photo-ionization, which plays an important role in gas amplification;

d) transition of atoms and molecules to highly-excited states adjacent to the continuous ionization spectrum, and auto-ionization of atoms and pre-ionization of molecules:

$$\hbar\nu + X \rightarrow X^* \rightarrow X^+ + e^- \,;$$
$$\hbar\nu + (XY) \rightarrow (XY)^* \rightarrow (XY)^+ + e^- \,; \tag{2.65}$$

e) absorption accompanied by transition to states of the continuous ionization spectrum:

$$\hbar\nu + X \rightarrow X^+ + e^- \; ; \quad \hbar\nu + (XY) \rightarrow (XY)^+ + e^- \; . \tag{2.66}$$

The ionization potentials (energies of the absorption edges) for various electron shells of the noble gases are presented in Fig. 1.6. Photo-ionization cross sections in the inert gases are larger by approximately an order of magnitude than the electron impact ionization cross sections (Fig. 2.11).

2.5.4 Gas Amplification

In an electric field of strength $\mathcal{E} > 10\,\text{kV/cm}$ electrons acquire energy sufficient for ionization. Avalanche multiplication of electrons (gas amplification) arises. The theory of this phenomenon was developed by Townsend. If along a path of $1\,\text{cm}$ in the direction of the electric field an electron creates α_T new electrons, then the increase in the number of electrons N_s along a distance dx equals dN_s

$$dN_s = \alpha_T N_s\, dx \; , \tag{2.67}$$

from which

$$N_s = N_t \exp(\alpha_T x) \; , \tag{2.68}$$

where N_t is the initial number of electrons and α_T is the *first Townsend ionization coefficient*.

The enhancement of the number of electrons along a path x is termed the *gas amplification coefficient*, or *gas gain*, which in the case of a homogeneous electric field is given by the relation

$$M = N_s/N_t = \exp(\alpha_T x) \; . \tag{2.69}$$

In a *cylindrical counter* the field strength at a distance r from the anode is

$$\mathcal{E}(r) = \frac{U_0}{r \ln(r_b/r_a)} \; . \tag{2.70}$$

where r_a is the radius of anode wire; r_b is the radius of the cathode (the distance between the anode and the cathode); U_0 is the voltage applied between the electrodes. The $\mathcal{E}(r)$ dependence is shown in Fig. 2.13. The gas gain in the case of cylindrical geometry is

$$M = \exp\left[\int_{r_a}^{r_c} \alpha_T(r)\, dr\right] \; . \tag{2.71}$$

Here r_c is the critical radius. At this radius the field strength, $\mathcal{E}(r_c)$, is sufficient to accelerate electrons up to energies necessary for ionization by electron impact:

$$r_c = r_a U_0/U_c \; , \tag{2.72}$$

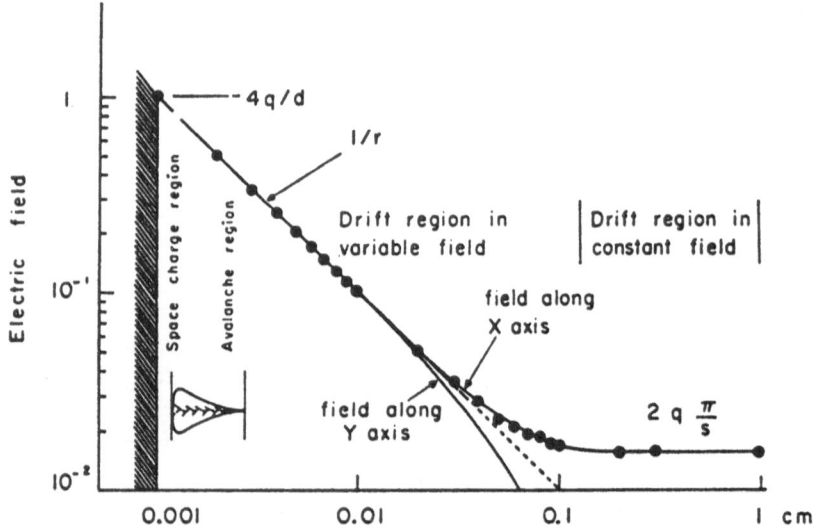

Fig. 2.13. Electric field strength as a function of the distance from the wire in a multiwire proportional chamber (Charpak 1970)

where U_c is the critical potential at which ionization by electron impact starts. In a typical proportional counter, $r_a = 10\,\mu m$, the mean free path before ionization $l_i = 1/\alpha_T \approx 1-10\,\mu m$, and to achieve $M \approx 10^3$ only about 10 generations of electrons are needed; in this case the critical radius $r_c \approx 10(1/\alpha_T) \approx 20-40\,\mu m$ (Fig. 2.13).

It has been demonstrated experimentally that the ratio α_T/P, where P is the pressure of the gas, is a function only of the variable $S = \mathcal{E}/P$, and taking into account that $dr = -(r_a S_a/S^2) \cdot dS$, M can be expressed, with the aid of (2.70) and (2.71), as follows:

$$\frac{\ln M}{P r_a S_a} = \int_{S_{cr}}^{S_a} \frac{\alpha_T}{P} \frac{1}{S^2}\, dS \,, \tag{2.73}$$

where S_a and S_{cr} are the values of S on the surface of the anode and at a distance r_c, respectively. Since S_{cr} is a constant for a given gas mixture, α_T/P depends only on S_a.

Various functions $[\alpha_T/P](S)$ are chosen for different gases, gas mixtures and values of S, and with the aid of (2.73) the dependences of the gas amplification on S_a are obtained (Table 2.10). The Aoyama and Kowalski formulae in Table 2.10 are common: they are valid for many gases and gas mixtures within a wide range of S values (Aoyama 1985; Kowalski 1986). The remaining formulae in this table are obtained as partial cases of the Aoyama and Kowalski formulae, corresponding to certain values of constants $m(0 \le m \le 1)$ and d. In Fig. 2.14 an example of the dependence of α_T/P upon $\mathcal{E}/P$ for inert gases is presented, obtained with the aid of the Rose-Korff formula, defined in Table 2.10.

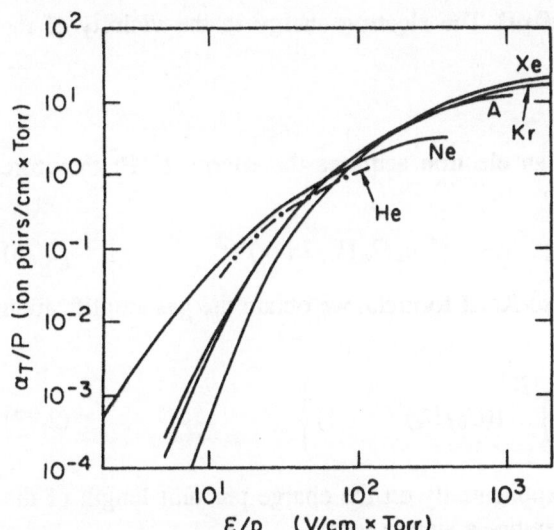

Table 2.10. Formulae for α_T/P and for the gas amplification expressed in the form $\ln M(r_a P S_a)$; C_1-C_{13}, K_1-K_{10}, S_0, m, d are constants (Aoyama 1985; Kowalski 1986)

Authors	α_T/P	$\ln M/(r_a P S_a)$
Rose and Korff (1941)	$\frac{1}{2}C_1\sqrt{S}$	$K_1 - C_1/\sqrt{S_a}$
Khristov (1947)	C_2	$K_2 - C_2/S_a$
Diethorn (1956)	$C_3 S$	$C_3(\ln S_a - \ln K_3)$
Townsend (1915); Williams and Sara (1962)	$C_4 \exp(-C_5/S)$	$(C_4/C_5[\exp(-C_5/S_a) + K_4]$
Ward (1958); Charles (1972)	$C_6 \exp(-C_7/\sqrt{S})$	$K_5 + \dfrac{2C_6}{C_7^2}\left(\dfrac{C_7}{\sqrt{S_a}}+1\right)\exp\left(-\dfrac{C_7}{\sqrt{S_a}}\right)$
Zastawny (1966)	$C_8(S - S_0)$	$K_6 + C_8\left(\ln\dfrac{S_a}{S_0}+\dfrac{S_0}{S_a}-1\right)$
Kowalski (1986)	$\frac{1}{2}C_9 S^{1.5}$	$C_9\sqrt{S_a} + K_7$
Aoyama (1985)	$C_{10}S^m \exp\left(-\dfrac{C_{11}}{S^{1-m}}\right)$	$\dfrac{C_{10}}{C_{11}}\dfrac{1}{1-m}\exp\left(-C_{11}S_a^{m-1}\right) + K_8$
Kowalski (1986)	$C_{12}S^d + C_{13}$	$\dfrac{C_{13}\cdot}{d-1}S_a^{d-1} - \dfrac{S_{13}}{S_a} + K_9$
Kowalski (1985)	$C_{12}S^d$	$\dfrac{C_{13}}{d-1}S_a^{d-1} + K_{10}$

Within a certain range of S values the ionization cross section is proportional to the electron energy, $\sigma_i = k_i \cdot \varepsilon_e$. The mean path made by electrons before undergoing an ionizing collision is $l_i = 1/\alpha_T = 1/\sigma_i n_a$; therefore α_T depends linearly upon the electron energy (Sauli 1977):

$$\alpha_T = k_i n_a \varepsilon_e . \tag{2.74}$$

The capacity per unit length of a cylindrical counter is

$$C_w = 2\pi\varepsilon / \ln(r_b/r_a) \tag{2.75}$$

where ε is the dielectric constant.

A typical value is $C_w \sim 10\,\text{pF}$. The electron energy in the vicinity of the wire varies as

$$\partial \varepsilon_e / \partial r = -C_w U_0 / r .$$

In between ionizing collisions an electron acquires the energy $\mathcal{E}/P$; therefore with the aid of (2.74) we obtain

$$\varepsilon_e = [C_w U_0 / (2\pi\varepsilon k_i n_a r)]^{1/2} ; \quad \alpha_T = (k_i n_a C_w U_0 / 2\pi\varepsilon r)^{1/2} . \tag{2.76}$$

As a result, applying the Rose-Korff formula, we obtain the gas amplification in a cylindrical counter:

$$M = \exp\left\{ 2 \left(\frac{k_i n_a C_w U_0 r_a}{2\pi\varepsilon} \right)^{1/2} [(U_0/U_c)^{1/2} - 1] \right\} . \tag{2.77}$$

When $U_0 \gg U_c$, M depends exponentially on the charge per unit length of the wire, $q_w = C_w U_0$, and (2.73) assumes a simple form:

$$M \sim \exp(C_w U_0) . \tag{2.78}$$

The exponential dependence $M(U_0)$ holds up to $M \sim 10^4$ (Fig. 2.15). The gas amplification doubles as U_0 increases by approximately $150\,\text{V}$.

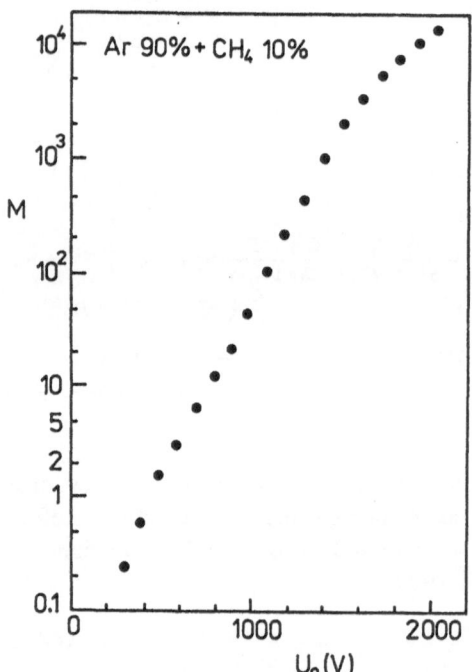

Fig. 2.15. Dependence of the gas amplification M on the voltage applied to the anode of a proportional counter (Sanada 1982)

2.5.5 Quenching of the Discharge

In pure inert gases in high electric fields an essential contribution to the development of an electron avalanche is provided by photo-ionization [(2.65) and (2.66)]: its influence results in a fast development of the avalanche and in increasing dimensions. The interaction of photons with the metal surface of the walls of the detector (of the cathodes) also significantly enhances the number of electrons:

$$\hbar \nu + \text{wall} \rightarrow e^- \,;$$
$$X^{**} + \text{wall} \rightarrow e^- + X^+ \,;$$
$$X^+ + \text{wall} \rightarrow e^- + X^+ \,.$$

These are effective processes, since extraction of an electron from the metal requires an energy of merely 1 eV.

If the number of electrons in an avalanche increases strongly, breakdown in the gas sets in, the beginning of which is determined by the *Raether condition* $\alpha_T \cdot x \approx 20$ (Raether 1964), from which $M \approx 5 \cdot 10^8$. The statistical nature of gas amplification manifests itself in significant fluctuations of M; therefore to avoid breakdown, the average value of M must be limited in gas-discharge detectors to a level of about 10^6. Note that in multilayer proportional detectors it is necessary to achieve a linear dependence between the ionization and the amplitude of the signal, which is possible only when the value of M is not too high. Typical values of the gas amplification in such detectors are of the order of magnitude of 10^3 to 10^4.

In most gas-discharge detectors the development of the electron avalanche has to be restricted. In a streamer chamber such restriction is achieved by the small duration (about 10 ns) of the high-voltage pulse. In proportional detectors and Geiger-Müller counters quenching admixtures are used for this purpose; these are organic gases or vapors containing complex molecules, for instance, methane, propane, isobutane, ethanol, methylal etc. The absorption of a photon by these molecules is highly likely to result in excitation of their vibrational and rotational levels with the energies ω_v and ω_r, respectively. The intrinsic energy of a molecule is $\omega_{in} = \omega_k + \omega_v + \omega_r$, where ω_v and ω_r are lower than the electron excitation energy of the molecule ω_k. As an example, cross sections of the above processes for the isobutane molecule are presented in Fig. 2.16. The rotational and vibrational levels of molecules are effectively excited not only by photo-absorption, but also as a result of electron collisions.

The quenching of a discharge in a strong electric field is based on the fact that collisions with atoms and molecules result in relaxation of the excitation (Christophorou 1970; Hasted 1964; Elecki et al. 1975), i.e. the transformation, in succession, of the energy of rotational and vibrational motions of molecules into their thermal motion. Relaxation represents a multistep process of energy redistribution between a large number of molecules; this is realized in proportional detectors with the aid of *quenching admixtures*.

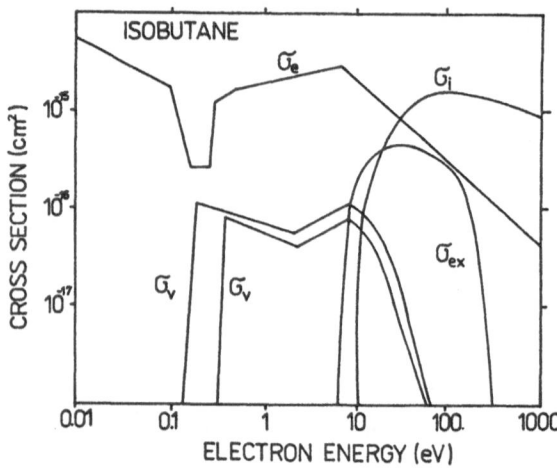

Fig. 2.16. Electron scattering cross sections in isobutane: σ_e – elastic scattering, σ_i – ionization, σ_v – vibration, σ_{ex} – excitation (Biagi 1988)

Another important mechanism limiting the discharge in a gas is the dissociation of molecules that is effectively initiated by electron collisions and by photo-absorption, (2.64). For example, for the dissociation $CH_4 \rightarrow CH_3^- + H^+$ it is necessary to spend 4.45 eV, while for $C_2H_6 \rightarrow C_2H_5^- + H^+$, 4.08 eV are needed. The maximum of the dissociation cross section is situated not far from the threshold and amounts to 10^{-16} cm^2 (Fig. 2.17).

Another mechanism restricting and stabilizing the development of an electron-photon avalanche is the transfer of excitation to a complex molecule (2.46) characterized by a large cross section (10^{-15} cm^2).

On the other hand, the presence of complex molecules also contributes to the enhancement of the number of electrons, owing to the high cross section of their photo-ionization, to the large ionization cross section through electron collisions, and to the Penning effect occurring intensively on molecules with low ionization potentials.

The photo-absorption cross section in organic gases is of the order of magnitude of 10^{-17}–10^{-16} cm^2; however it substantially exceeds the photo-ionization cross section (Fig. 2.18), so that on the whole quenching properties prevail.

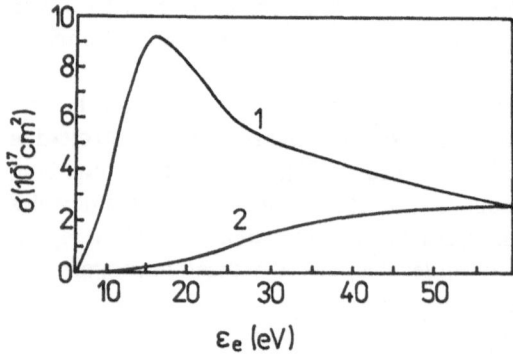

Fig. 2.17. Cross section for dissociation of a hydrogen molecule by electron impact: (*1*) – with the formation of two atoms in the ground state; (*2*) – when one of the atoms is produced in the excited state H(2*P*) (Slovetzki 1974)

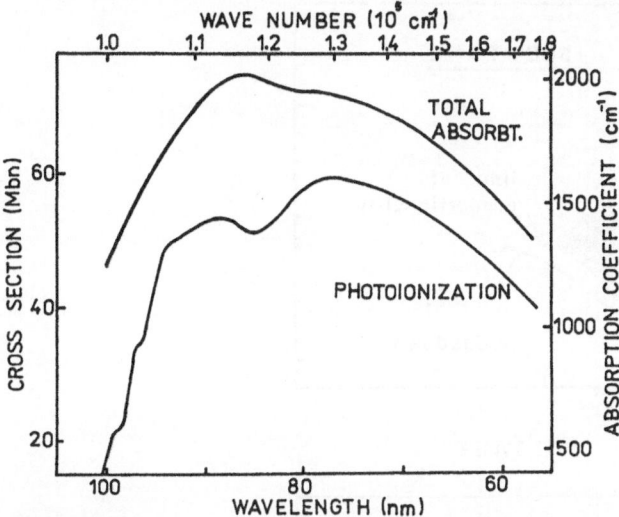

Fig. 2.18. Photo-ionization cross sections and total photo-absorption cross sections in C_2H_6 (Marr 1967)

As a result, the addition of an organic gas leads to the stabilization of the gas amplification process within a wide range of voltages, although the voltage itself necessary for the required gas amplification increases (Fig. 2.19).

Recently the self-quenching streamer (SQS), or limited streamer (LS), mode of operation of gas detectors has become widely applied. It arises in a gas mixture with a high concentration of the quenching admixture. The SQS mode is described in detail, for example, in the review article by Alekseev et al. (1982). We note that, as in the Geiger mode, ionization measurements cannot be performed in a detector operating in the SQS mode.

2.5.6 Amplitude Resolution of a Proportional Counter

Amplitude resolution σ_A depends on the dispersion of the number of electrons $\sigma_{N_t}/\overline{N_t}$ on the measured section of the particle track and on the fluctuations σ_M of the gas amplification $\overline{M}$. If the average number of electrons produced by a particle in the gas (including transport processes) is $\overline{N_t}$, then the relative dispersion of the amplitude $\overline{A}$ may be described by

$$\left(\frac{\sigma_A}{\overline{A}}\right)^2 = \left(\frac{\sigma_{N_t}}{\overline{N_t}}\right)^2 + \frac{1}{\overline{N_t}}\left(\frac{\sigma_M}{\overline{M}}\right)^2 = \frac{F}{\overline{N_t}} + \frac{1}{\overline{N_t}k_1} = \frac{w}{\overline{\Delta}}\left(F + \frac{1}{k_1}\right) , \qquad (2.79)$$

where F is the Fano factor and $1/k_1$ is the relative variance of the single electron spectrum (2.81), w is the mean energy loss for creation of an electron-ion pair and $\overline{\Delta}$ is the mean energy loss. The second term gives the contribution due to fluctuations of the gas amplification; these are substantially smaller than the ionization fluctuations described by the first term.

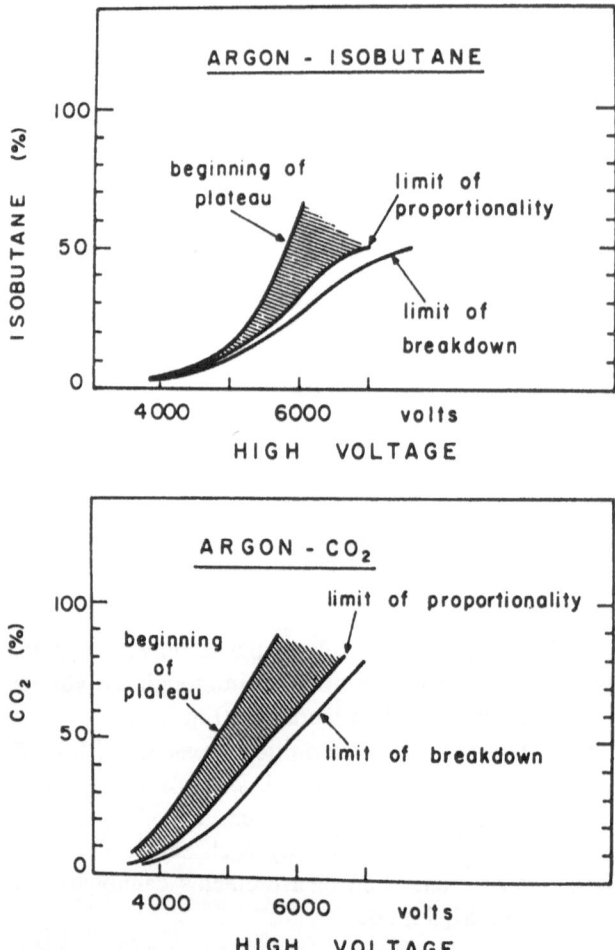

Fig. 2.19. The operating range of a proportional chamber with an anode to cathode distance of $l = 8\,\text{mm}$, and an anode wire pitch of $s = 2\,\text{mm}$ as a function of the content of isobutane or CO_2 in the gas mixture (Bouclier et al. 1970)

We mentioned in Sect. 1.5 that for a fixed energy loss ($\Delta = \text{const}$) the relative variance of the number of secondary electrons, N_Δ, can be expressed as

$$\left(\sigma_{N_\Delta}/\overline{N}_\Delta\right)^2 = F/N_\Delta \; ,$$

where σ_{N_Δ} is the standard deviation of N_Δ and F is the Fano factor, which ranges between 0.2 and 0.4 (for a typical argon-methane mixture used in a proportional detector $F \approx 0.2$).

The theoretical resolution of a proportional detector R_{th}, is, in accordance with (2.79)

$$R_{\text{th}} = 2.36 \left[\frac{w}{\overline{\Delta}} \left(F + \frac{1}{k_1}\right)\right]^{1/2} \; . \tag{2.80}$$

The number of electrons in the avalanche and the gas gain M exhibit large fluctuations which are characterized by the root-mean-square deviation σ_M. An essential part, here is played by the first electron. It may travel several lengths, $l_i = 1/\alpha_T$, without ionizing or, on the contrary, ionize several times over the length l_i.

In a low electric field an electron must travel a relatively long distance and undergo numerous elastic collisions before it acquires an energy $\varepsilon_e > I_1$ necessary for ionization. In such conditions equilibrium between elastic and ionizing collisions arises, while the probability of ionization is independent of time. If the Snyder (1947) model is adopted, the probability $\mathcal{P}(M)$ of there being M electrons in an avalanche, when $M \gg 1$, will be:

$$\mathcal{P}(M) = (1/\overline{M}) \exp(-M/\overline{M}) .$$

The relative variance due to gas multiplication is

$$f_M = (\sigma_M/\overline{M})^2 = 1/\overline{N_t} k_1 \tag{2.81}$$

where k_1 is the relative dispersion of the spectrum for a single electron. When $f_M = 1$ the probability $\mathcal{P}(M)$ falls exponentially (Fig. 2.20a). As $\mathcal{E}/P$ increases, the assumption of equilibrium between elastic and ionizing collisions is no longer valid, and the dependence $\mathcal{P}(M)$ will no longer be exponential (Fig. 2.20). Several models have been proposed to describe the "statistics" of gas multiplication, a list of which can be found, for example, in Sephton et al. (1984) and Obelic (1985).

The efficiency of the avalanche development is characterized by the quantity $\chi_i = \alpha_T I_1/\varepsilon_e$ known as the *ionization efficiency*. In the case of large $\mathcal{E}/P$ values, $\chi_i \approx 1$, i.e. $I_1/\varepsilon_e \approx 1/\alpha_T$, and the electron free path before it undergoes an ionizing collision is small. The probability of ionization will now depend on the distance travelled by the electron from the last ionizing collision in the direction towards the anode. The model developed by Alkhazov (1970) takes into account the relationship between the frequencies of collision, excitation and ionization in the gas amplification process. In this theory the probability $\mathcal{P}(M)$ in the inhomogeneous field of a cylindrical gas counter corresponds approximately to the Poya distribution:

$$\mathcal{P}(M) = M^{\Theta} \exp(-(1 + \Theta)M) \quad \text{when} \quad f_M = 1/(1 + \Theta) , \tag{2.82}$$

where Θ is a parameter depending on the choice of the gas and on the conditions of electron collection. In calculations it is assumed that $\Theta = 0.5\text{--}1.0$. For example, in argon the distribution $\mathcal{P}(M)$ is in agreement with the experimental data for $\Theta = 0.5$ ($f_M = 0.66$) (Alkhazov 1970; Lapique, Piuz 1980).

The best amplitude (ionization) resolution, i. e. the minimum fluctuation of the gas amplification, is achieved in proportional counters with low gas amplification coefficients ($M \approx 10^2$) (Fig. 2.21). Deterioration of the resolution when $M > 10^2$, is due to the creation of secondary avalanches, owing to photo-ionization,

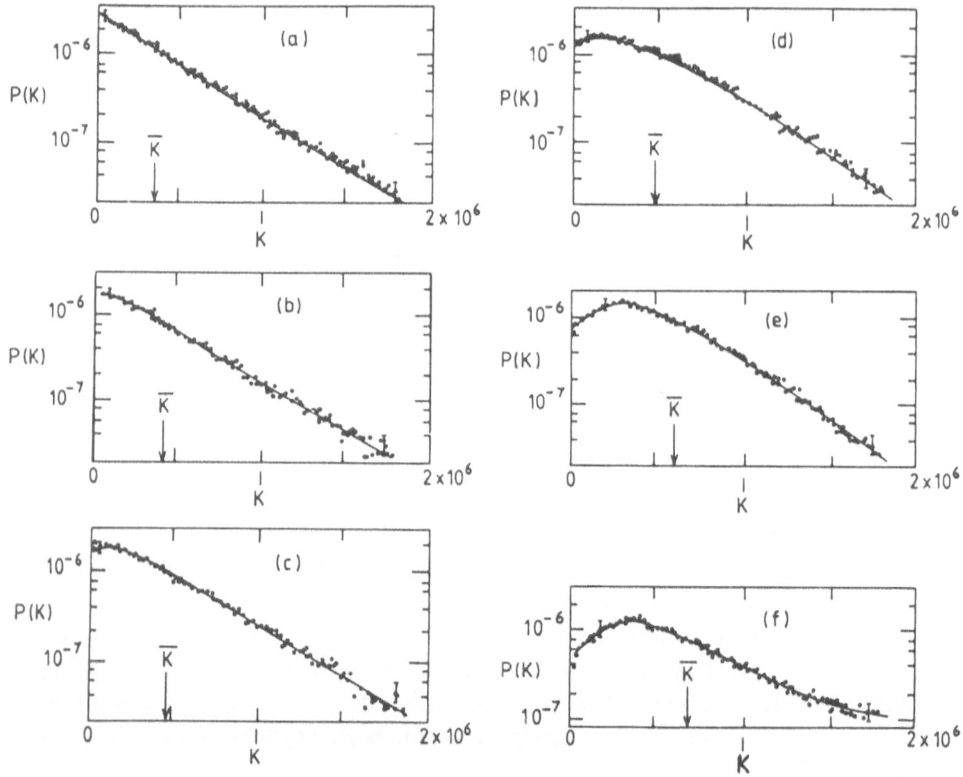

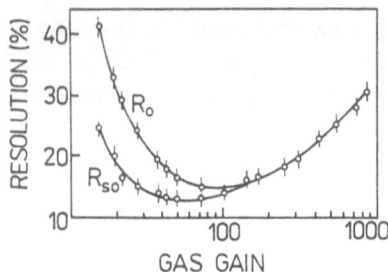

Fig. 2.20. The distribution of electron avalanche sizes in methane as a function of ionization efficiency χ_i, $\mathcal{E}/P$ in units of V/cm $\cdot$ 133 Pa: (a) $\mathcal{E}/P = 48.2$, $\chi_i = 0.026$; (b) $\mathcal{E}/P = 51.3$, $\chi_i = 0.034$; (c) $\mathcal{E}/P = 78.9$, $\chi_i = 0.088$; (d) $\mathcal{E}/P = 120$, $\chi_i = 0.13$; (e) $\mathcal{E}/P = 156.0$, $\chi_i = 0.19$; (f) $\mathcal{E}/P = 218$ V/(cm $\cdot$ 133 Pa), $\chi_i = 0.26$ (Cookson, Lewis 1966)

Fig. 2.21. Amplitude resolution of a proportional counter R_0 [full width at half maximum (fwhm) of a peak of energy 5.9 keV] as a function of gas gain; R_{s0} – upon subtraction of the noise (Sephton et al. 1984)

and to the saturation of χ_i as $\mathcal{E}$ increases. [We recall that $\chi_i(\mathcal{E}/P)$ depends on the behavior of the function $\alpha_T(\mathcal{E}/P)$ (Fig. 2.14).]

To enhance the amplitude resolution in ionization measurements the Penning effect can be utilized (Sephton et al. 1984; Sauli 1984). In the gas amplification process in a Penning mixture (for example, Ne + 0.1 % Ar) the ionization efficiency (i.e. the number of electrons present in the avalanche) is increased (Fig. 2.11) resulting in the enhancement of the amplitude resolution. For a number of reasons experiments reveal that improvement of the resolution by the Penning effect is insignificant (Sephton et al. 1984).

2.5.7 The Proportionality Condition

The amplitude of the output signal of a proportional counter is

$$A = eN_t M/C_w ,$$ (2.83)

and under typical conditions amounts to $A \approx 1\,\text{mV}$ per electron for $M = 10^5$.

The proportionality between the ionization and the amplitude of the output signal, which is achieved for values of M between 10 and 10^5 (Fig. 2.22), is of paramount importance for ionization measurements (for particle identification). The proportionality is strongly affected by the form of the electron avalanche in the vicinity of the wire. Figure 2.23 illustrates the development of an avalanche from a source of α-radiation for various gas amplifications in a proportional detector. The amplitude of the induced signal is conventionally represented by the distance from the anode. Up to $M \approx 10^3$ an avalanche develops only from the side from which the electrons drift. As M increases, the charge distribution acquires a heartlike shape, and when $M > 10^4$ it totally envelopes the anode.

Likewise important is the development of an avalanche along the wire. Straightforward measurements of this process were carried out by Sanada and Matsuda (1982). The distribution of charge along the wire in the case of high ionization from α-particles is shown in Fig. 2.24. The decay length of the avalanche

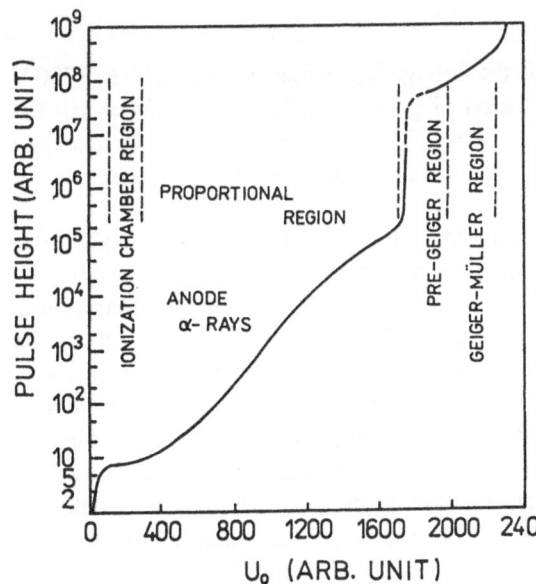

Fig. 2.22. Dependence on the anode voltage U_0 of the single amplitude from a proportional counter (Sanada, Matsuda 1982)

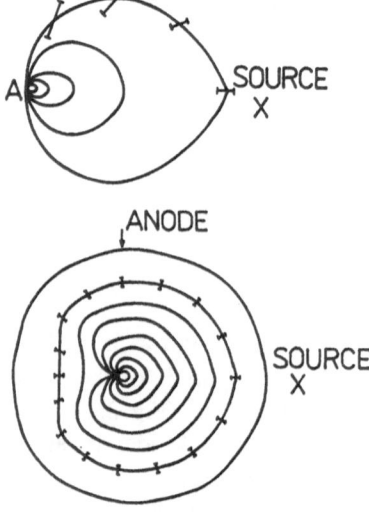

Fig. 2.23. Charge created as an avalanche from an alpha ray develops around the signal wire in a 90 % Ar + 10 % CH$_4$ mixture (polar coordinates): The lines connect the regions of identical charge for given values of gas gain M, which increase from the outer to the periphery *upper part* M: 16.4, 33.3, 76.5, 169, 269; *lower part* M: 772, 1580, 2360, 4550, 6670, 9170, 12600, 15200, 20000. A is the anode wire (Sanada 1982)

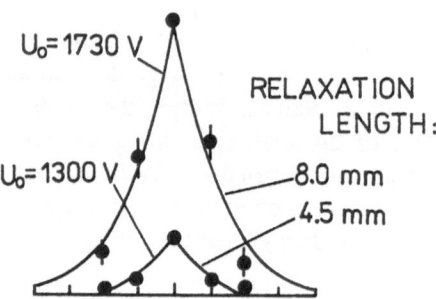

Fig. 2.24. Charge distribution along the signal wire in the region of avalanche development at the voltages U_0 = 1730 and 1300 V (Sanada, Matsuda 1982)

l_{dec} increases from 4.5 to 8.0 mm as the voltage U_0 increases from 1300 to 1730 V. It is noteworthy that within the region of proportionality, where $M = 10–10^4$, the decay length does not change. When $M > 10^4$ a sharp rise is observed, which may be related to the enhancement of photo-ionization processes in the avalanche development.

The development of an avalanche around the anode is interrelated with its development along the anode. When the avalanche's size along the anode is constant ($M < 10^4$), it has a heartlike shape (Fig. 2.23), but when its longitudinal dimensions start to grow ($M > 10^4$) the avalanche gradually envelopes the anode.

The relatively large dimensions of an avalanche along the wire in the proportional region can probably be explained by positive ions, which appear as a result of avalanche development by the primary electrons arriving at the wire and which start to screen it. Owing to the field strength falling in the vicinity of the wire because of space charge, electrons moving near the edges of the avalanche undergo more effective multiplication. This process usually manifests itself when the ionization density is high (for example, with ionization from α-particles).

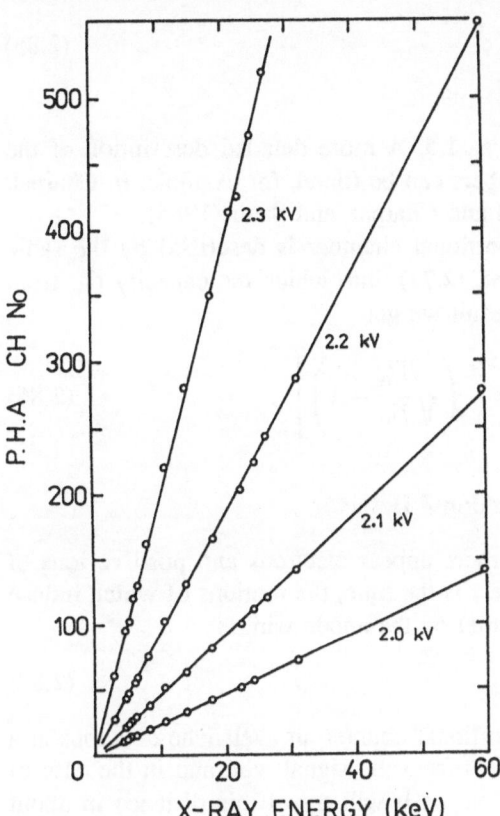

Fig. 2.25. Dependence of the signal amplitude on X-ray energy for various voltages applied to the proportional chamber (Lehraus et al. 1978)

The figure axes are labelled P.H.A CH No (vertical) with values 100, 200, 300, 400, 500, and X-RAY ENERGY (keV) (horizontal) with values 0, 20, 40, 60. The curves are labelled 2.3 kV, 2.2 kV, 2.1 kV, and 2.0 kV.

Note that in spite of the relatively large dimensions of the avalanche along the wire, its center of mass can be measured quite accurately (Charpak, Sauli 1984) (see also Sect. 4.3).

The main condition for reliable ionization measurements in proportional detectors is the linear dependence between the measured ionization and the amplitude of the output signal. Figure 2.25, obtained with an multilayer proportional identifier of relativistic charged particles EPI (External Particle Identifier), demonstrates that this condition is satisfied very well for various values of the gas amplification coefficient.

2.5.8 Gas Amplification in a Proportional Chamber

The distribution of the potential in a symmetric proportional chamber where the distance between the wires is s (in the x direction) and the anode-cathode gap is l (in the y direction), is approximately given by the expression (Erskine 1972)

$$U(x,y) = \frac{q_w}{4\pi\varepsilon}\left[\frac{2\pi l}{s} - \ln\left(4\sin^2\frac{\pi x}{s} + 4\sinh^2\frac{\pi y}{s}\right)\right] , \qquad (2.84)$$

where the charge per unit length of the wire is

$$q_w = U_0 C_w = \frac{2\pi\varepsilon U_0}{\pi l/s - \ln(2\pi r_a/s)} \; ; \tag{2.85}$$

ε is the dielectric constant.

Expression (2.85) holds for $l/s > 1.5$. A more detailed description of the field in multiwire proportional chambers can be found, for example, in Charpak (1970), Erskine (1972), Sauli (1977) and Charpak and Sauli (1984).

The gas amplification in a proportional chamber is described by the same expression as in a cylindrical counter, (2.77), into which the capacity C_w from formula (2.85) is substituted. As a result we get

$$M = \exp\left[\left(\frac{2k_i n_a U_0}{\pi l/s - \ln(2\pi r_a/s)}\right)^{1/2}\left(\sqrt{\frac{U_0}{U_c}} - 1\right)\right] . \tag{2.86}$$

2.5.9 Signal Formation in a Proportional Detector

In the process of gas amplification there appear electrons and positive ions of total charges $Q^-(t)$ and $Q^+(t)$, where t is the time, the motions of which induce a signal on the electrodes. The potential on the anode wire is

$$U(t) = [Q^-(t) + Q^+(t)]/C_w . \tag{2.87}$$

As already mentioned, in a proportional counter an avalanche develops at a distance of several tens of micrometers from the signal wire and in the case of a typical electron drift velocity of $5\,\mathrm{cm}/\mu s$ it will travel this distance in about $1\,\mathrm{ns}$. Such a short pulse due to the electrons gives almost no contribution to the observed signal, which mainly is induced by the motion of positive ions (of total charge Q^+). Wilkinson (1950) has shown that in absence of gas amplification, the voltage induced on the anode of a cylindrical counter is

$$U_1(t) = -\frac{Q^+}{C_w L_w}\frac{\ln r_0 - \ln\left\{r_0^{3/2} - \frac{3}{2}\mu_+\left[\frac{U_0}{\ln(r_b/r_a)}\right]^{1/2}t\right\}^{2/3}}{\ln(r_b/r_a)} , \tag{2.88}$$

where L_w is the length of the wire. The electron-ion pairs appear at the time $t = 0$ at a distance r_0 from the anode. The signal $U_1(t)$ achieves its maximum at the time

$$t_{s1} = \frac{r_0^{3/2} - r_a^{3/2}}{\frac{3}{2}\mu_+\left[\frac{U_0}{\ln(r_b/r_a)}\right]^{1/2}} . \tag{2.89}$$

The voltage $U(t)$ induced in a proportional counter in the case of a gas amplification M (Sanada 1982) is

$$U(t) = U_1(t) - \frac{Q^+\ln\left[1 + \frac{2\mu_+ U_0(t - t_{s1})}{r_a^2 \ln(r_b/r_a)}\right]}{C_w L_w 2\ln(r_b/r_a)} , \tag{2.90}$$

and $U(t)$ achieves saturation during a time

$$t_s = t_{s1} + \frac{(r_b^2 - r_a^2)\ln(r_b/r_a)}{2U_0\mu_+} . \qquad (2.91)$$

Using (2.75), setting $U_1(t) = 0$ and $t_{s1} = 0$, and taking account of the pressure P, we obtain for a proportional counter the following expression

$$U(t) = -\frac{Q^+}{4\pi\varepsilon L_w}\ln\left(1 + \frac{\mu_+ U_0 C_w t}{\pi\varepsilon P r_a^2}\right) = -\frac{Q^+}{4\pi\varepsilon L_w}\ln\left(1 + \frac{t}{t_c}\right) . \qquad (2.92)$$

where t_c is a *characteristic time*:

$$t_c = \pi\varepsilon r_a^2 P/(\mu_+ C_w U_0) . \qquad (2.93)$$

The signal achieves its maximum value $U_s = Q^+/L_w C_w$ at $t_s = P(r_b^2/r_a^2 - 1)t_c$, when all the positive ions arrive at the cathode. In a typical proportional chamber $t_c \approx 0.5\,\mathrm{ns}$ and $t_s \approx 1\,\mathrm{ms}$. In Fig. 2.26 the time dependence of the signal formation is presented. The signal has a sharp front; within several nanoseconds it achieves 20 % of its maximum value U_s, after which it slowly rises logarithmically. Signal formation is dealt with by Wilkinson (1950), Sauli (1977), Jaros (1980) and Sanada (1982).

For further processing by electronics it is useful to have a short signal (of the order $\leq 100\,\mathrm{ns}$); this is achieved by differentiation of the signal at the amplifier input. Typically the input impedance R of an amplifier is about $1\,\mathrm{k}\Omega$, while its time constant $\tau_c = RC_i \sim 10\text{--}100\,\mathrm{ns}$, where C_i is the input capacity. The differentiated signal achieves a maximum at $t \approx \tau_c$, and its amplitude

$$U = U_s\frac{\tau_c}{t_s}\frac{r_b^2/r_a^2}{2\ln(r_b/r_a)} \qquad (2.94)$$

achieves 10–40 % of the undifferentiated maximum U_s.

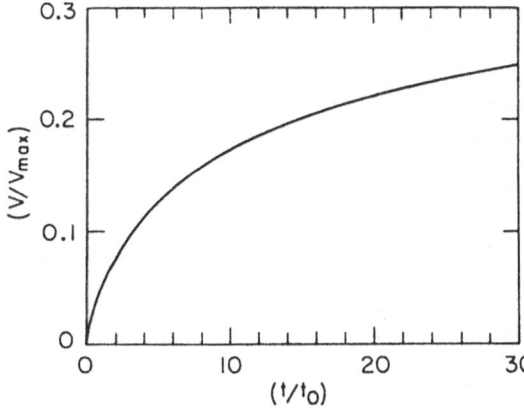

Fig. 2.26. Time dependence of the signal amplitude from a proportional counter (Jaros 1980)

References

Alekseev G.D., Kruglov V.V., Khazins D.M. 1982: Fiz. Elem. Chastits At. Yadra **13**, 703 (English transl. Sov. J. Part. Nucl. **13**, 293). Sect. 2.5

Alkhazov G.D. 1970: Nucl. Instr. Meth. **89**, 155. Sect. 2.5

Allkofer O.C. 1969: *Spark Chambers* (Thiemig, München). Sect. 2.2

Aoyama T. 1985: Nucl. Instr. Meth. A **234**, 125. Sect. 2.5

Biagi S.F. 1988: Nucl. Instr. Meth. A **273**, 533. Sects. 2.2, 5

Bobkov S., Cherniatin V., Dolgoshein B., Evgrafov G., Kalinovsky A., Kantserov V., Nevsky P., Sosnovtsev V., Sumarokov A., Zelenov A. 1984: Nucl. Instr. Meth. **226**, 376. Sect. 2.2

Bouclier R., Charpak G., Dimcovski Z., Fischer G., Sauli F., Coignet G., Flügge G. 1970: Nucl. Instr. Meth. **88**, 149. Sect. 2.5

Breskin A., Charpak G., Sauli F., Atkinson M., Schultz G. 1975: Nucl. Instr. Meth. **124**, 189. Sect. 2.2

Brown S.C. 1959: *Basic Data of Plasma Physics* (Wiley, New York). Sects. 2.2, 3, 5

Charles M.W. 1972: J. Phys. E **5**, 95. Sect. 2.5

Charpak G. 1970: Ann. Rev. Nucl. Part. Sci. **20**, 195. Sect. 2.5

Charpak G., Sauli F. 1984: Ann. Rev. Nucl. Part. Sci. **34**, 285. Sect. 2.5

Christophorou L.G. 1970: *Atomic and Molecular Radiation Physics* (Wiley, New York). Sects. 2.2–5

Christophorou L.G. 1981: *Electron and Ion Swarms* (Pergamon, New York). Sect. 2.2

Cookson A.H., Lewis T.J. 1966: Brit. J. Appl. Phys. **17**, 1437. Sect. 2.5

Davidenko V.A., Dolgoshein B.A., Somov S.V. 1970: Zh. Exp. Teor. Phys. **58**, 130. Sects. 2.2, 4

Dayon M.I., Egorov O.K., Krylov S.A., Pozharova E.A., Smirnitsky V.A., Chechin V.A., 1970: Prib. Tech. Exp. No. **5**, 64. Sect. 2.2

Diethorn W. 1956: Report No. NYO–6628, New York. Sect. 2.5

Dolgoshein B.A. 1970: XII Int. Conf. on Instrumentation in High Energy Phys., Dubna. Sect. 2.2

Elecki A.V., Palkina L.A., Smirnov B.M. 1975: *Transport Phenomena in a Weakly Ionized Plasma.* (Atomizdat, Moscow) (in Russian)

Engel von A. 1965: *Ionized Gases* 2nd ed. (Clarendon, Oxford). Sects. 2.3, 4

Erskine G.A. 1972: Nucl. Instr. Meth. **105**, 565. Sect. 2.5

Farr W., Heintze J., Hellenbrand K.H., Walenta A.H. 1978: Nucl. Instr. Meth. **154**, 175. Sect. 2.2

Fehlmann J., Viertel G. 1983: *Compilation of Data for Drift Chamber Operation*, Preprint ETH Zürich IHP. Sect. 2.2

Frost L.S., Phelps A.V. 1962: Phys. Rev. **127**, 1621. Sect. 2.2

Frost L.S., Phelps A.V. 1964: Phys. Rev. A **136**, 1538. Sect. 2.2

Fulda-Quenzer F., Haissinski J., Jean-Marie B., Pagot J. 1985: Nucl. Instr. Meth. A **235**, 517. Sect. 2.2

Galeev A.A., Sudan R. (Eds.) 1983: *Foundations of Plasma Physics* Vol. 1 (Energoatomizdat, Moscow) (in Russian). Sects. 2.3–5

Galicki V.M., Nikitin E.E., Smirnov B.M. 1981: *The Theory of Atomic Collisions* (Nauka, Moscow) (in Russian). Sect. 2.3

Hasted J.B. 1964: *Physics of Atomic Collisions* (Butterworths, London). Sects. 2.2–5

Huk M., Igo-Kemenes, Wagner A. 1988: Nucl. Instr. Meth. A **267**, 107. Sect. 2.4

Huxley L.G., Crompton R.W. 1974: *The Diffusion and Drift of Electrons in Gases* (Wiley, New-York). Sects. 2.2–5

Jaros J.A. 1980: Stanford Linear Accelerator Report SLAC–PUB–2647. Sect. 2.5

Khristov L.G. 1947: Dokl. Bulg. Akad. Nauk **10**, 453. Sect. 2.5

Kondratev V.N. (Ed.) 1974: *Energies of Chemical Bonds, Ionization Potentials, Electron Affinities* (Nauka, Moscow) (in Russian). Sects. 2.3, 4

Kowalski T.Z. 1985: Nucl. Instr. Meth. A **234**, 521. Sect. 2.5

Kowalski T.Z. 1986: Nucl. Instr. Meth. A **243**, 501. Sect. 2.5

Kumar K. 1984: Phys. Rep. **112**, 319. Sect. 2.2

Lapique F., Piuz F. 1980: Nucl. Instr. Meth. **175**, 297. Sect. 2.5

Lehraus I., Matthewson R., Tejessy W., Aderholz M. 1978: Nucl. Instr. Meth. **153**, 347. Sect. 2.5

Lenard P. 1903: Ann. Phys. (Germ.) **12**, 714. Sect. 2.2

Llewellyn-Jones 1957: *Ionization and Breakdown in Gases* (Methuen, Oxford). Sect. 2.2

Loeb L. B. 1969: Basic Processes of Gaseous Electronics (University of California Press, Berkeley). Sects. 2.2, 5

Margenau H. 1946: Phys. Rev. **69**, 508. Sect. 2.2

Marr G. F. 1967: *Photoionization Processes in Gases* (Academic, New York). Sect. 2.5

Mathieson E., El Hakeem N. 1979: Nucl. Instr. Meth. **159**, 489. Sect. 2.2

McDaniel E. W. 1964: *Collision Phenomena in Ionized Gases* (Wiley, New York). Sect. 2.3

Merson G. I., Sitar B., Budagov Yu. A. 1983: Fiz. Elem. Chastits At. Yadra **14**, 648 (English transl. Sov. J. Part. Nucl. **14**, 270). Sect. 2.2

Moiseiwitch B. L., Smith S. J. 1968: Rev. Mod. Phys. **40**, 238. Sect. 2.5

Morse P. M., Allis W. P., Lamar E. S. 1935: Phys. Rev. **48**, 412. Sect. 2.2

Obelic B. 1985: Nucl. Instr. Meth. A **241**, 515. Sect. 2.5

Palladino V., Sadoulet B. 1975: Nucl. Instr. Meth. **128**, 323; Prep. Lawrence Berkeley Laboratory LBL–3013 (1974). Sect. 2.2

Parker J. H., Lowke J. J. 1969: Phys. Rev. **181**, 290. Sect. 2.2

Peisert A., Sauli F. 1984: Prep. CERN 84–08. Sect. 2.2

Piuz F. 1983: Nucl. Instr. Meth. **205**, 425. Sect. 2.2

Raether H. 1964: *Electron Avalanches and Breakdown in Gases* (Butterworths, London). Sect. 2.5

Ramanantsizehena P., Gresser J., Schultz G. 1980: Nucl. Instr. Meth. **178**, 253. Sect. 2.2

Rapp D., Englander-Golden P. 1965: J. Chem. Phys. **43**, 1464. Sect. 2.4

Rice-Evans P. 1974: *Spark, Streamer, Proportional and Drift Chambers* (Richelieu, London). Sects. 2.2–5

Robson R. E., Ness K. F. 1986: Phys. Rev. A **33**, 2068; A **34**, 2185. Sect. 2.2

Rose H. E., Korff S. A. 1941: Phys. Rev. **59**, 850. Sect. 2.5

Sadoulet B. 1982: Proc. Int. Conf. on Instrumentation for Colliding Beam Physics, Stanford Linear Accelerator Report SLAC, p. 1. Sect. 2.2

Sanada J. 1982: Nucl. Instr. Meth. **196**, 23. Sect. 2.5

Sanada J., Matsuda K. 1982: Nucl. Instr. Meth. **203**, 605. Sect. 2.5

Sauli F. 1977: Prep. CERN 77–09. Sects. 2.2–5

Sauli F. 1984: in *The Time Projection Chamber*, ed. by. J. A. McDonald (Amer. Inst. Phys., New York) p. 171. Sect. 2.5

Schmidt B., Polenz S. 1988: Nucl. Instr. Meth. A **273**, 488. Sect. 2.2

Schultz G., Gresser J. 1978: Nucl. Instr. Meth. **151**, 413. Sect. 2.2

Sephton J. P., Turner M. J. L., Leake J. W. 1984: Nucl. Instr. Meth. **219**, 534. Sect. 2.5

Sitar B. 1987: Fiz. Elem. Chastits At. Yadra **18**, 1080 (English transl. Sov. J. Part. Nucl. **18**, 460). Sect. 2.2

Skullerud H. R. 1969: J. Phys. B **2**, 696. Sect. 2.2

Slovetzki D. I. 1974: in *Chemistry of Plasma*, ed. by B. M. Smirnov (Atomizdat, Moscow) p. 156 (in Russian). Sect. 2.5

Smirnov B. M. 1968: *Atomic Collisions and Elementary Processes in Plasma* (Atomizdat, Moscow) (in Russian). Sect. 2.3

Snyder H. S. 1947: Phys. Rev. **72**, 181. Sect. 2.5

Sosnovtsev V. V. 1987: Thesis, Institute of Physics and Engineering (MIFI), Moscow. Sect. 2.2

Townsend J. S. 1915: *Electricity of Gases* (Clarendon, Oxford). Sect. 2.5

Va'vra J. 1986: Nucl. Instr. Meth. A **244**, 391. Sect. 2.2

Wagner E. B., Davis F. J. Hurst G. S. 1967: J. Chem. Phys. **47**, 3138. Sect. 2.2

Ward A. L. 1958: Phys. Rev. **112**, 1852. Sect. 2.5

Wilkinson D. M. 1950: *Ionization Chambers and Counters* (The University Press, Cambridge). Sect. 2.5

Williams A., Sara R. I. 1962: Int. J. Appl. Radiation Isotopes **13**, 229. Sect. 2.5

Wong L., Armitage J., Waterhouse J. 1988: Nucl. Instr. Meth. A **273**, 476. Sect. 2.2

Zastawny A. 1966: J. Sci. Instrum. **43**, 179. Sect. 2.5

3. Ionization Measurements in Proportional Detectors

3.1 Multilayer Proportional Detectors

The distribution of ionization energy loss in a gas detector narrows very slowly, as the length of the detector l and the gas pressure P increase. Even when $Pl \gtrsim 10^7 \, \mathrm{Pa \cdot cm}$ ($\xi/l \gtrsim 50$), its relative half-width W amounts to 20–30 % (Fig. 1.16). So the separation of relativistic hadrons requiring W = 5–6 % is impossible in a one-layer detector with reasonable values of P and l. The solution consists in multiple measurements in several detectors of the ionization produced by a particle, or, more conveniently, in a single multilayer detector. If the number of layers is sufficiently large, then the most probable ionization, corresponding to the distribution obtained by measurements in each layer, turns out to be a much more precise measure of the ionizing power of a particle than can be obtained from measurement in a single layer of the same total width (Sect. 3.6).

The price of a multilayer ionization detector depends mainly on the number of electronics channels, so it is approximately proportional to the number of layers, n. On the other hand, its length (along the beam in a fixed target experiment, or radius in a collider detector) $L = nl$ is usually restricted by the admissible size of the experimental apparatus. It is therefore important to choose the optimum number of layers of the identifier, such that there is maximum accuracy in determining the most probable ionization for L = const. The solution of this problem also depends on the efficiency of the applied method for a statistical analysis of the data (Sect. 3.6). Under the right conditions, the precision W_n in determining the most probable ionization in a multilayer detector is significantly higher than the precision W for a one-layer detector.

It was realized many years ago that multiple ionization measurements were necessary..A two-layer proportional counter and the corresponding procedure for statistical data processing were first applied by Alikhanian et al. (1945) in studies of the composition of cosmic radiation at mountain level. In subsequent years proportional detectors with 4 and 5 layers (Alikhanov et al. 1956) were used, and effective methods were developed for the statistical analysis of multiple ionization measurements. In 1956 a 50-layer proportional detector, made for the identification of relativistic hadrons (Alikhanov et al. 1956), was tested at the Institute of Theoretical and Experimental Physics (ITEP). However, in the 1970s cheap semiconductor electronics appeared; this made possible the application of multilayer proportional detectors for the separation of high energy particles in experiments at accelerators and with cosmic radiation (Table 3.1). The main con-

Table 3.1. Multilayer proportional detectors for relativistic charged particles identification (based on data available by 1990)[1]

Class of detector	Detector	Type (name) of chamber	Length of MDC; inner-outer radii [cm]	number of samplings n	Sample thickness [cm]	Number of sense wires	Pressure [10^5 Pa]	Calculated dE/dz resolution σ_n [%]
MPC	BEBC	EPI	768	128	6	4096	1	2.6
	PION	PION	100	24	3	792	0.7	7.0
MDC	EHS	ISIS	510	320	1.6	320	1	2.8
	FHS	CRISIS	300	192	1.6	192	1	3.3
	ASTRON	ASTRON	450	250	1.6	250	1	2.8
CDC	PEP-4	TPC	20-100	183	0.4	2196	8.5	2.5
	DELPHI	TPC	30-120	188	0.4	2400	1	5.5
	ALEPH	TPC	31-180	338	0.4	6336	1	5.0
	TOPAZ	TPC	36-118	173	0.4	1384	4	4.6
	JADE	Jet	21-79	48	1.0	768	4	4.7
	AFS	Jet	20-80	42	1.0	1722	1	9.6
	HRS	Jet	21-102	47	1.5		2	4.5
	OPAL	Jet	25-185	159	1.0	3816	4	3.5
	H1	Jet	20-79.5	64	1.0	2560	1	6.0
	MARK III	SSC	14-114	33	1.0	2000	1	15
	MARK II/SLC	SSC	19-145	72	0.83	5832	1	6.9
	SLD	SSC	20-100	80	1.0	5120	1	6.2
	CDF	SSC	27-138	84	1.0	6156	1	
	ZEUS	SSC	16-79	72	0.45	4608	1	5.0
	UCD	SSC	18-140	100	0.6	8000	1	5.0
	BES	SSC	15.5-115	40	1.0	2808	1	8.1
	KEDR	SSC	13-50	42	0.5		1	12
	ARGUS	CDC	15-86	36	1.8	5940	1	4.2
	MAC	CDC	12-45	20	1.6		1	15
	CLEO II	CDC	17-95	51	1.4	12240	1	6.0
	AMY	CDC	15-64	40	0.44-0.57	9796	1	
	UA-1	PDC	10-122	100-180	0.8	6100	1	6.0
CPC	CLEO	CPC	110-200	117	0.7	112320	2.4	5.8

[1] Tables 4.5 and 4.6 are supplements to this table; these give parameters for the devices operating as three-dimensional coordinate detectors

tribution to the development of ionization measurement methods in multilayer proportional detectors has been made by the Oxford ISIS (Identification of Secondaries by Ionization Sampling) group (Allison 1981, 1982; Allison et al. 1979, 1984; Allison, Wright 1984) and by the CERN EPI (External Particle Identifier) group (Lehraus et al. 1978, 1982a, 1982b, 1982c, 1983a, 1983b).

3.1.1 Types of Multilayer Proportional Detectors

Multilayer proportional detectors provide a three-dimensional reconstruction of events of high multiplicity, and permit identification of charged secondary particles. This new class of devices represents the apex of modern experimental tech-

107

nique developed for registration of relativistic charged particles (Nygren 1981; Sauli 1984, 1988a, 1988b; Wagner A. 1981; Walenta 1981; Merson et al. 1983; Sitar 1985, 1987; Saxon 1988).

Multilayer proportional detectors can be identified according to their geometry into four types: (1) multilayer proportional chambers (MPC); (2) multilayer planar drift chambers (MDC); (3) cylindrical drift chambers (CDC); (4) cylindrical proportional chambers (CPC).

Devices of the first and second types are usually applied as external identifiers of relativistic charged particles in magnetic spectrometers in fixed target experiments (Lehraus et al. 1978; Allison et al. 1974; Goloskie et al. 1985; Babaev et al. 1978).

Cylindrical drift chambers are widely used within the large-scale detectors in collider experiments (Fig. 3.1). Multilayer drift chambers situated in the central parts of devices are usually termed *central detectors*. They are operated inside cylindrical magnets and are, therefore, extremely compact (TPC 1976; Delpierre 1984; Blum 1984; Wagner 1982; Calvetti et al. 1982; and others). Cylindrical drift chambers are also used in experiments with extracted particle beams, if particles are to be recorded within large solid angles (Bryman et al. 1985; Alard et al. 1987; Chapin et al. 1984; Etkin et al. 1989; Garabatos 1989).

The most important parameters characterizing multilayer proportional detectors under operation and/or development are presented in Table 3.1, with special attention drawn to ionization measurements for particle identification. We note that the principal aim of these detectors is the spatial reconstruction of events namely the determination of track coordinates and particle momenta in

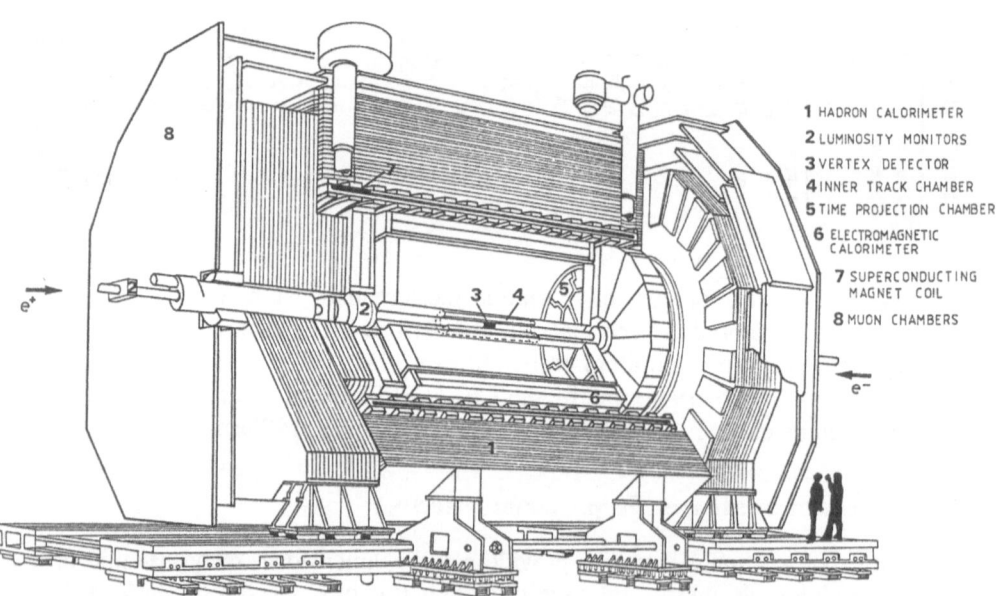

1 HADRON CALORIMETER
2 LUMINOSITY MONITORS
3 VERTEX DETECTOR
4 INNER TRACK CHAMBER
5 TIME PROJECTION CHAMBER
6 ELECTROMAGNETIC CALORIMETER
7 SUPERCONDUCTING MAGNET COIL
8 MUON CHAMBERS

Fig. 3.1. Layout of typical experimental apparatus in colliding beams (ALEPH)

the magnetic field. Cylindrical drift chambers are usually used for this, applied in detectors operating in colliders. Certain parameters of multilayer chambers as coordinate detectors are presented in Chap. 4. Only a few chambers (EPI, ISIS, CRISIS) have served solely for particle identification, the coordinate information from them being of minor importance.

The computed ionization resolution is presented in the last column of Table 3.1. As we show below, for the identification of hadrons (π, K, p), an ionization resolution σ_n better than 3–3.5 % is required, which is achieved only in a certain number of detectors (EPI, ISIS, CRISIS, TPC, OPAL, ASTRON). The number of layers in these chambers approaches 150–400; the thickness of a layer is $Pl = 1.6$–$6.0 \cdot 10^5$ Pa $\cdot$ cm; the length L of plane multilayer chambers amounts to 3–7.5 m; while the diameter of cylindrical chambers is 1.5–3 m.

Chambers in which $\sigma_n > 4$ % (for example, JADE, UA–1, ARGUS, DELPHI, ALEPH, SLD and others) only permit separation of electrons from relativistic hadrons or, at best, of relativistic pions from protons. Problems related to particle identification are analyzed below.

Multilayer chambers are widely used in detectors at the LEP storage rings of CERN, at the Fermilab p$\bar{\text{p}}$–collider, at the SLAC linear collider, and at the e^+e^-–collider TRISTAN of KEK; such chambers are planned for LHC at CERN, for SSC in Texas and for UNK at Serpukhov. We shall now consider some features characteristic of several types of multilayer chambers.

3.1.2 Multilayer Proportional Chambers (MPCs)

Two devices of this type are known: the multilayer proportional chamber EPI (External Particle Identifier) (Jeanne et al. 1973), which was in operation from 1978 up to 1982 behind the bubble chamber BEBC at the SPS accelerator of CERN; and the MPC of the PION detector (Bashindzhagyan et al. 1977), developed for the identification of cosmic particles (Table 3.1).

We shall present the main parameters of the EPI chamber with a sensitive volume of length $L = 7.7$ m. It has 128 layers each consisting of 32 proportional counters of an area of 6×6 cm^2 (Fig. 3.2). The electronics involves 4096 spectrometric channels: each channel consists of a preamplifier, an eight-bit analog-to-digital converter and a buffer memory.

The advantages of MPCs over multilayer drift chambers are the following: simpler (and cheaper) electronics; higher speed of operation; and less influence of various infavorable factors (to be considered below) upon the measured signal. The disadvantages of MPCs include a large number of electronics channels and bad spatial and double-track resolution, which prevent using such chambers in the environment of high multiplicities of charged particles.

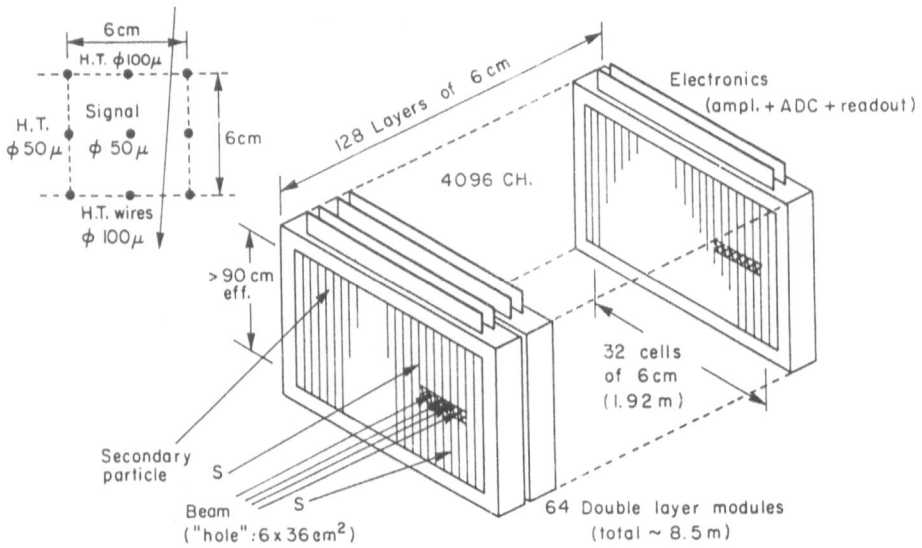

Fig. 3.2. Structure of the EPI multilayer proportional chamber (Baruzzi et al. 1983)

3.1.3 Multilayer Drift Chambers (MDCs)

In several large fixed-target spectrometers multilayer drift chambers are employed as identifiers of relativistic particles; for example, ISIS in the European Hybrid Spectrometer (EHS) at CERN (Allison et al. 1974, 1984; Allison 1981, 1982), CRISIS in the Fermilab Hybrid Spectrometer (FHS) at FNAL (Goloskie et al. 1985; Toothacker et al. 1988); while ASTRON at ITEP is under construction (Babaev et al. 1978) (Table 3.1).

We shall give, as an example, the parameters of the ISIS detector (Fig. 3.3) which was under operation in the CERN European hybrid spectrometer during the period from 1981 to 1985. The dimensions of the fiducial volume are the following: length 5 m, width 2 m and height 4 m; the gas container is about $60 \, m^3$ in volume. The container has entrance and exit windows along the beam which are covered with two layers of mylar. In the middle of the sensitive volume, 640 sense wires are stretched in the direction normal to the beam. They are combined in pairs, thus producing a single elementary layer 16 mm thick, in which the ionization energy loss is measured. On both sides of the 2 meter-long wire plane there are large drift volumes bounded by metal cathodes, to which a constant high voltage is supplied. Within the drift volumes a uniform electric field is created. The field homogeneity is provided by annual field-shaping electrodes to which a linearly falling potential is applied. In the electric field, the electrons produced along tracks of relativistic charged particles drift toward the central plane. The electron drift time determines the x coordinate of a track. The z coordinate along the beam is given by the number of wires hit. The y coordinate along the sense wire can be determined in various ways, which are described in Chap. 4.

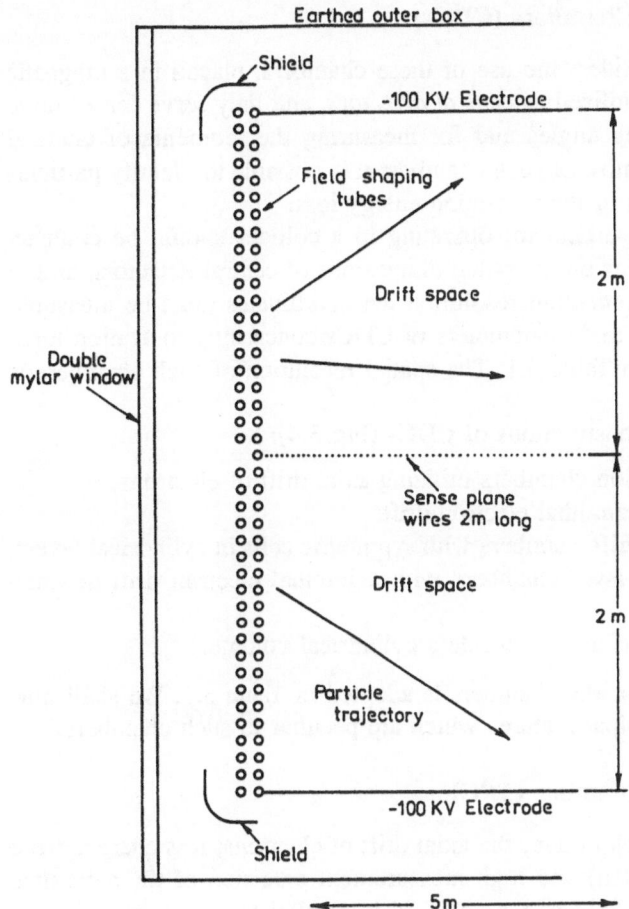

Earthed outer box

Shield

−100 KV Electrode

Field shaping tubes

Drift space

2 m

Double mylar window

Sense plane wires 2m long

Drift space

2 m

Particle trajectory

−100 KV Electrode

Shield

5 m

Fig. 3.3. Schematic view of the multilayer drift chamber ISIS (Allison et al. 1984)

In many fixed-target spectrometers, MDCs of smaller size are used as vertex detectors. In these detectors the trajectories of secondary particles are determined accurately, and by multiple measurements of energy loss it is also possible to identify particles in the non-relativistic region, or to separate electrons and hadrons of momenta $< 10 \, \mathrm{GeV}/c$. Typical dimensions of such detectors are: length ≤ 2 m, width ≤ 1 m, height ≤ 0.5 m; the number of layers is about 100. A more detailed description of such detectors is given in Chap. 4.

Advantages of multilayer drift detectors over MPCs are good spatial resolution and the capability to distinguish between adjacent tracks. The number of channels required for an MDC is much lower than for an MPC, although they are more complex and consequently more expensive (Sect. 4.4). The drawbacks of MDCs are their low speed and large memory time (for instance, the collection time in ISIS is about 50 μs). Serious technical problems arise from the usage of high constant voltage in such chambers (up to 120 kV).

111

3.1.4 Cylindrical Drift Chambers (CDCs)

In many detectors in colliders the use of these chambers, placed in a magnetic field, is basic. They are utilized as *central detectors*, and they serve for accurate determination of outgoing angles and for measuring the momenta of charged particles. In addition, in most of such chambers it is possible to identify particles by multiple measurement of the ionization energy loss.

The requirement that a detector operating in a collider should be compact imposes serious restrictions on the radial dimensions of central detectors, and to achieve the necessary momentum resolution the coordinates must be measured with high precision. The main parameters of CDCs concerning ionization measurement are presented in Table 3.1. The spatial resolution of such chambers is considered in Sect. 4.3.

There exist various constructions of CDCs (Fig. 3.4):

1) TPCs – time projection chambers utilizing axial drift of electrons;
2) Jet chambers with azimuthal electron drift;
3) CDCs – cylindrical drift chambers with symmetric cells in cylindrical layers;
4) SSCs – stereo superlayer chambers with azimuthal electron drift at small distances;
5) PDCs – planar drift chambers inside a cylindrical volume.

The above notation for the chambers is adopted in Table 3.1. We shall now proceed to consider principal features which are peculiar to such chambers.

3.1.5 Time Projection Chamber (TPC)

A TPC is a chamber which utilizes the axial drift of electrons; it is characterized by large drift length (1–2 m) and high measurement precision of the azimuthal coordinate; the magnetic field of the solenoid is parallel to the electric field in the chamber, i.e. $B \parallel \mathcal{E}$ (Fig. 3.5).

The first chamber of this type was the TPC (TPC 1976; Barbaro-Galtieri 1982; Lynch, Hadley 1982; Aihara et al. 1983) working at the PEP electron-positron collider of SLAC. In Fig. 3.5 the TOPAZ TPC is depicted as an example; other TPCs differ from it only in certain details. TPCs provide registration of charged particles within nearly a 4π geometry. A TPC is a cylindrical vessel several cubic meters in volume, filled with a working gas. In the bases of the cylinder there are multiwire proportional chambers, composed of 8 sectors in which sense wires are stretched in the form of concentric octagons (Fig. 3.5). Thus, each sector consists of 175 radial layers, in each of which particle track coordinates and the value of $(-dE/dx)$ are measured.

The high-voltage electrode system creates a uniform electric field $\mathcal{E}$, parallel to the axis of the chamber, in which "tracks" drift towards the endcaps to be "projected" onto them. This gave rise to the name of the device: "Time Projection Chamber". Information is recorded in short time intervals (10–50 ns) by flash ADCs or multihit electronics. A TPC is placed inside a solenoid, so that within its sensitive volume $\mathcal{E} \parallel B$.

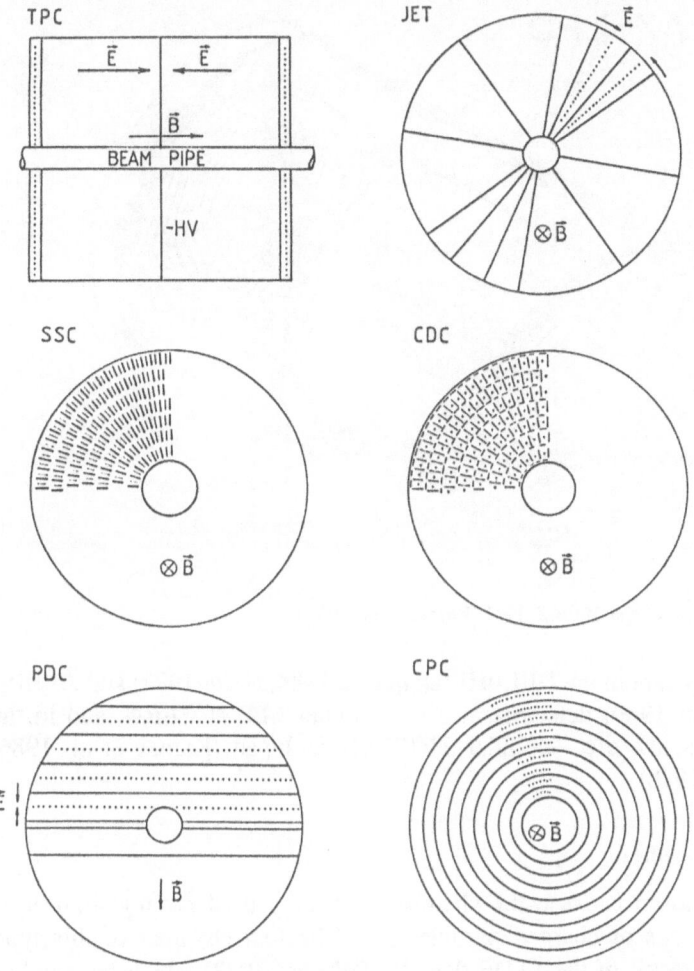

Fig. 3.4. Schematic views of different types of cylindrical drift and proportional chambers

The z coordinates of tracks (along the chamber axis) are determined from the electron drift time with an uncertainty equal to about ± 0.3 mm, on the average. The azimuthal coordinate along the track is determined at 10 points by measuring the charge induced on the cathode pads of the proportional chambers. This method permits determination of the value of $r\varphi$, with a root-mean-square deviation equal to $\sigma_{r\varphi} \approx \pm 0.1$ mm, where r is the distance from the measured point on the track to the chamber axis and φ is the azimuthal angle.

TPCs provide good three-dimensional track images and good particle identification on the basis of multiple measurements of $(-dE/dx)$. Its disadvantages include a large drift length (1–2 m), from which follows a large memory time (20–50 μs), preventing their application in a high-rate environment.

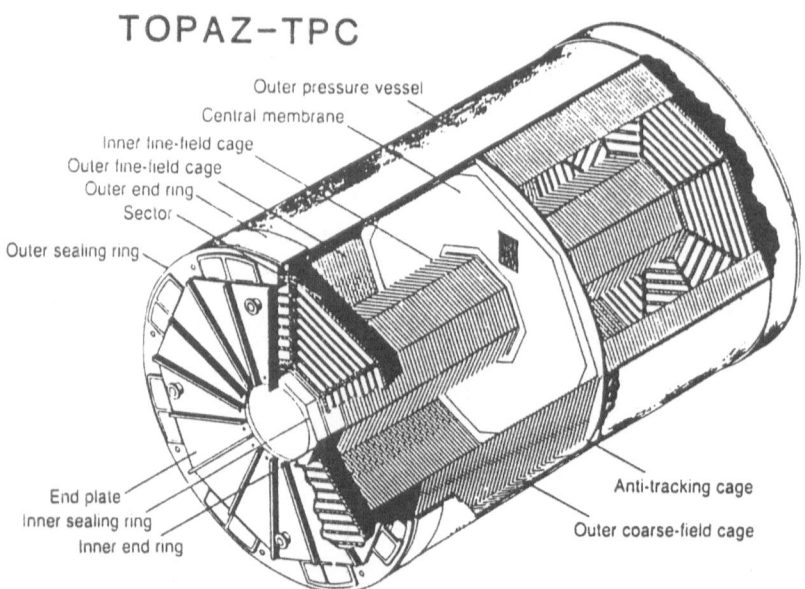

TOPAZ-TPC

Outer pressure vessel
Central membrane
Inner fine-field cage
Outer fine-field cage
Outer end ring
Sector
Outer sealing ring
End plate
Inner sealing ring
Inner end ring
Anti-tracking cage
Outer coarse-field cage

Fig. 3.5. Schematic view of the TOPAZ TPC (Kamae et al. 1986)

Large TPCs operate in the DELPHI (Delpierre 1984, Hilke 1987) and ALEPH (Blum 1984; Blum 1989) detectors in e^+e^-–collider LEP at CERN, and in the detector TOPAZ at the e^+e^-–collider TRISTAN in Japan (Kamae et al. 1986; Shirahashi et al. 1988).

3.1.6 The Jet Chamber

This name was chosen because the chamber is mainly used for registration of densely packed tracks of charged particles-jets. The first chamber of this type was the central detector of the JADE detector (Wagner 1982) which operated at the storage ring PETRA of DESY.

The sensitive volume of a Jet chamber forms a cylinder surrounding the vacuum pipe of a collider. The wire electrodes are stretched parallel to the axis of the chamber which is partitioned into segments. The layout of two of the 24 sections of the OPAL central detector is shown in Fig. 3.6. The electric field produced by the field-shaping electrodes, which are under a radially increasing potential ($B \parallel \mathcal{E}$), compels the electrons to drift at the angle α_M to the plane of the sense wires (thus, for example, in the chamber JADE, α_M amounted to 18.5°). The drift length does usually not exceed 10 cm (although in the chamber OPAL it equals 24 cm), which allows achievement of high spatial resolution (100–200 μm) and good double-track resolution (5–10 mm) (Table 4.6). Sense and field wires alternate in the central plane of each segment (Fig. 3.6). Even and odd wires are staggered, thus permitting resolution of the right-left ambiguity. The coordinate along the wires is usually determined by the current division, with a precision of 10–40 mm.

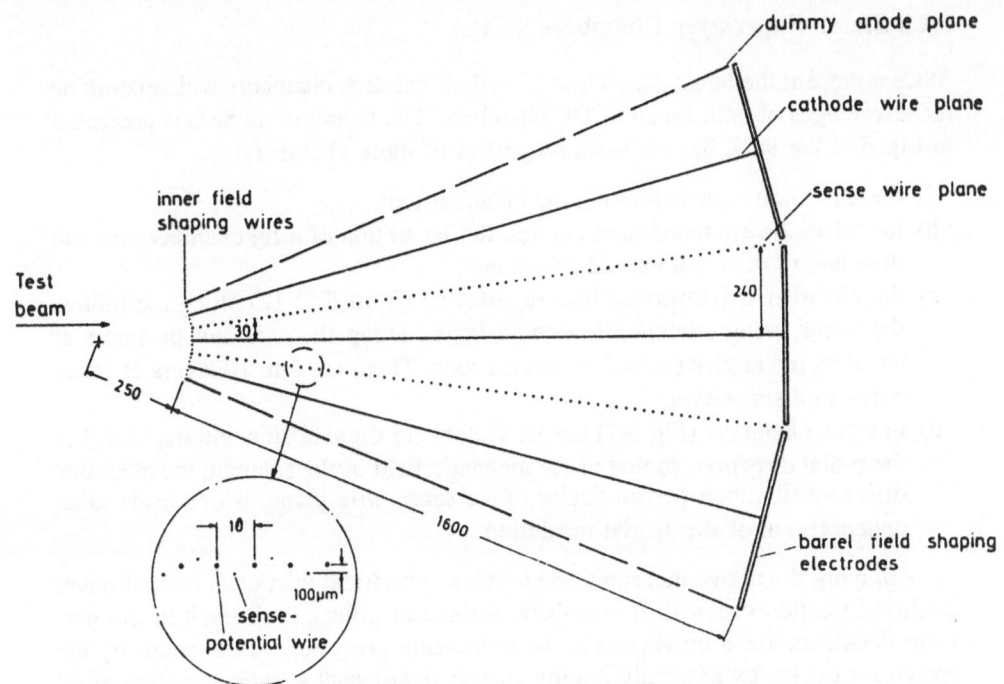

Fig. 3.6. Schematic view of two modules of a cylindrical MDC of the Jet type (OPAL) (OPAL Detector 1983)

Jet chambers were used in the AFS detector (Cockerill 1981) at the ISR at CERN, in the HRS detector (Va'vra 1982) in the e^+e^- beams at SLAC, in the OPAL detector (OPAL Detector 1983; Heuer, Wagner 1988; Breuker et al. 1987) at the LEP of CERN, and in the H1 detector (H1 Collaboration 1986) at the ep–collider HERA at DESY.

3.1.7 CDCs with Symmetric Cells

In these chambers the sense wires are arranged in concentric circular layers (Fig. 3.4). Electrons drift radially from all sides towards the sense wires.

Such chambers operate in the ARGUS detector (ARGUS Collaboration 1989) at the DORIS II ring at DESY, in the MAC at SLAC (Ash et al. 1987), in CLEO II (Cassel et al. 1986) at the e^+e^-–collider CESR of the Cornell University, and in AMY (Ueno 1988) at the e^+e^-–collider at KEK. The advantages of CDCs are that they are simply constructed, that the number of field-shaping wires is small, and that all the drift cells are identical, which is convenient for the measurement of ionization energy losses.

3.1.8 Stereo Superlayer Chambers (SSCs)

SSCs represent the latest generation of cylindrical drift chambers which combine the advantages of both Jet and CDC chambers. The layout of an SSC is presented in Fig. 3.7. We shall list the main properties of these chambers:

a) the wires are stretched along the chamber axis:
b) the wires are stretched in structures similar to that of a Jet chamber, and the drift length does not exceed 25–35 mm;
c) the chamber is partitioned into superlayers (from 8 to 12) along the radius, the wires being stretched in certain layers along the axis and in others at small stereo angles ($\alpha_s = 3$–$5°$) to the axis. There are 6 to 10 layers of sense wires in a superlayer;
d) in some chambers (Fig. 3.7) the Jet sectors are tilted at an angle α_M (2.32) to the radial direction, so that in the magnetic field of the solenoid the electrons drift in a direction perpendicular to the sense wire plane, which leads to an enhancement of the spatial resolution.

Applying the above measures one obtains cylindrical drift chambers of novel quality, so-called vector drift chambers. Instead of points, each track in a superlayer is characterized by vectors in the processing procedure. This speeds up the search for tracks, by gradually joining vectors in adjacent superlayers. The speed with which complex multitrack events are reconstructed is enhanced to such an extent that the procedure can be included in the second-level trigger.

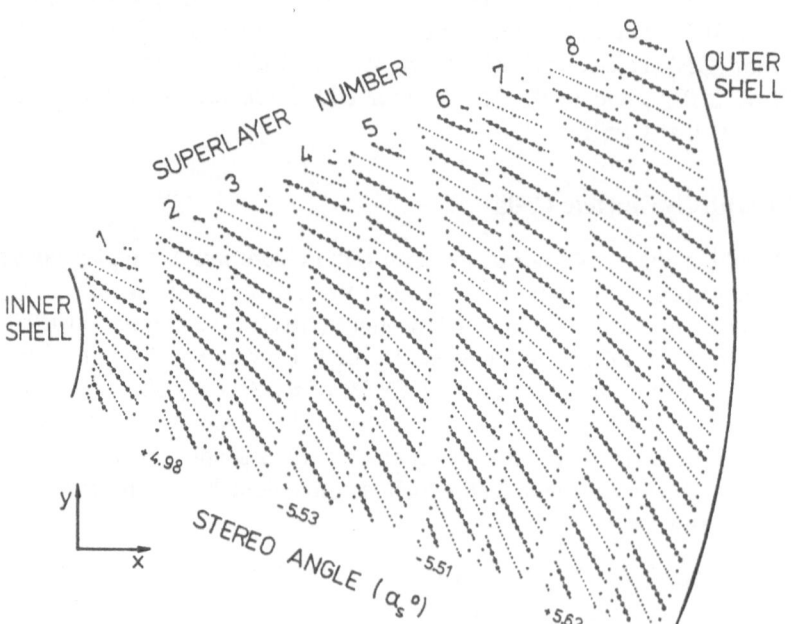

Fig. 3.7. Layout of the octant of the ZEUS Central Tracking Detector (Brooks et al. 1989)

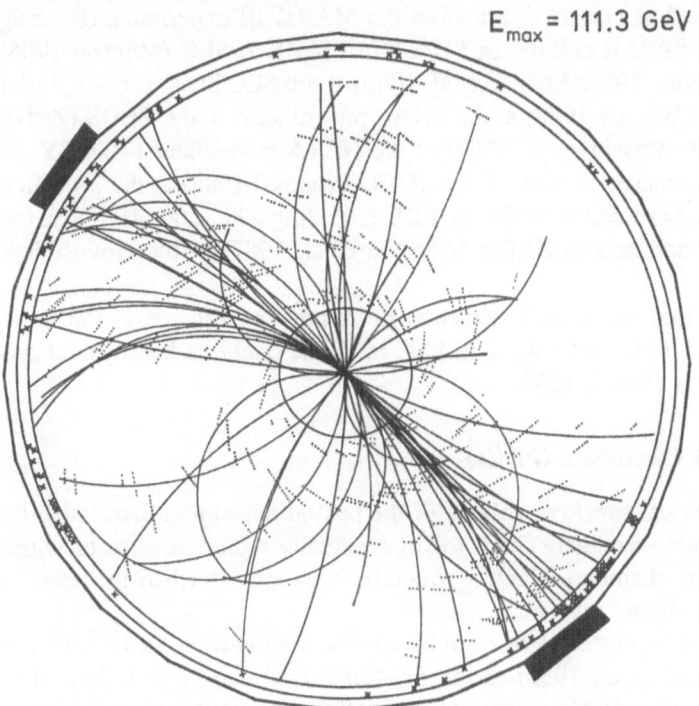

$E_{max} = 111.3$ GeV

Fig. 3.8. Event display showing two jets in the central tracking chamber of CDF (Wagner 1988)

The quality of the image of a complex multitrack event registered by the central detector of the Collider Detector of Fermilab (CDF) is illustrated in Fig. 3.8.

SSC chambers exhibit the following advantages:

- fast track finding;
- resolution of right-left ambiguity;
- calibration on the basis of one's own data;
- rejection of out-of-time tracks;
- low transverse momentum p_T track rejection;
- possibility of second-order triggering.

A detailed description of an SSC can be found, for example, in the review by Saxon (1988).

SSCs are excellent three-dimensional coordinate detectors for high-background environments, but are less suitable for ionization measurements than, for example, TPC's. This is due to the existence of empty gaps in between the superlayers, which leads to a significant reduction of the number of layers and, also, of the track length required for measuring ($-dE/dx$). The number of layers in an SSC (from 70 to 90) is insufficient for the identification of relativistic hadrons (Sect. 3.2).

SSCs are applied as central detectors in the MARK III experiment (Roehrig et al. 1984) at the SPEAR collider of SLAC; in MARK II/SLC (Manson 1986) and SLD (SLD Design 1984; Atwood et al. 1986) at the SLC linear e^+e^-–collider of SLAC; in CDF (Wagner 1988) at the FNAL $p\bar{p}$–collider; in the ZEUS (ZEUS Collaboration 1986; Brooks et al. 1989) at the HERA ep–collider of DESY; in the Universal Calorimetric Detector UCD (UCD Proposal 1988) at the $3+3$ TeV $p\bar{p}$–collider of the Serpukhov UNK; in BES (Bai Jing-Zhi et al. 1987) at the Beijing e^+e^-–collider, and in KEDR (Anashin et al. 1987) at the Novosibirsk e^+e^-–collider.

It is probable that the central detectors of future experiments at the new-generation colliders LHC at CERN and SSC in Texas could be based on stereo superlayer chambers (Saxon 1988).

3.1.9 Planar Drift Chambers (PDCs)

In PDCs the cylindrical sensitive volume of the central detector is partitioned by wires into rectangular sections. These sections actually represent a set of wide-gap multilayer drift chambers. Such a structure is expedient when included in the detector with a dipole magnet.

A representative of this class of chambers is the central detector of the UA–1 apparatus (Calvetti et al. 1982) at the $p\bar{p}$–collider of CERN (Fig. 3.9). It is situated in the field of a dipole magnet of magnetic induction equal to 0.7 Tesla. The detector consists of six modules. The wire planes in the two central modules are orthogonal to the beam, while in the four front ones the wire planes are parallel to the beam. Such partitioning provides good measurement precision of the momenta of particles travelling within a wide range of angles.

In each wide-gap drift chamber there is a central plane in which the sense and field-shaping wires alternate in steps of 5 mm. On both sides of this plane there are drift volumes with a maximum drift length of 18 cm. Up to 180 points per track are measured for particles travelling at a small angle to the beam and up to 100 points per track for particles with trajectories nearly perpendicular to the beam.

Modification of the principal types of cylindrical drift chambers differ mainly in how the layers are arranged along the radius of the chamber.

3.1.10 Cylindrical Proportional Chambers (CPCs)

CPCs have sense wires arranged concentrically around the beam. Electron drift time is not measured in this type of chamber. The advantages of such chambers are a better ionization resolution than in CDCs, and, first of all, more simple electronics channels. Drawbacks are the large amount of electronics and bad spatial resolution; because of these problems, such chambers are practically out of use.

A chamber of this type operated in the CLEO detector (Ehrlich 1982), which was installed in the colliding e^+e^- beams of the CESR accelerator at Cornell

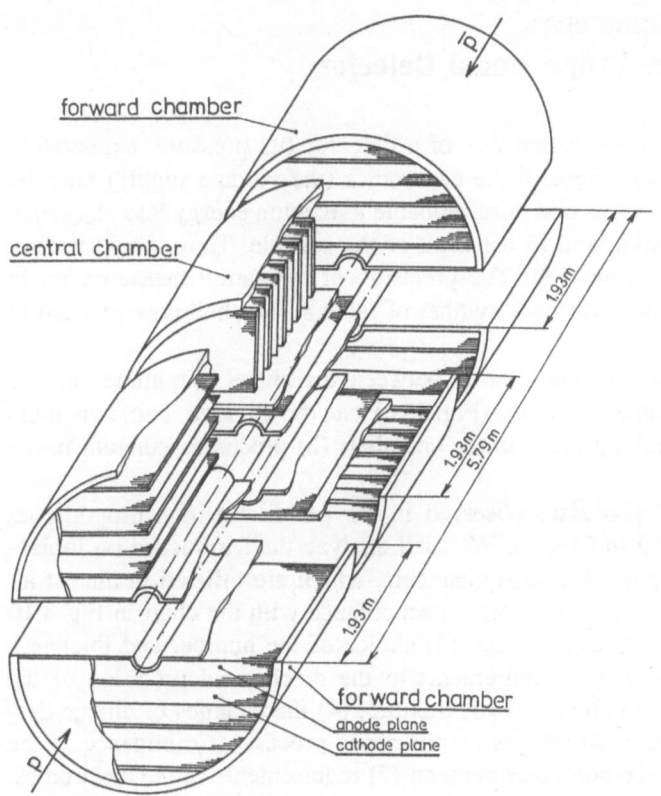

Fig. 3.9. Layout of the cylindrical MDC operating in the UA–1 apparatus (Calvetti et al. 1982)

University. In this detector the regions where the particle momenta and their ionization losses are measured are separated in space. The momenta are determined with the aid of drift chambers in the inner part of the apparatus. The ionization losses are measured, for particle identification, in a multilayer proportional chamber. It is divided into eight modules, each of which contains 117 layers. The number of sense wires is quite impressive: 112 320.

CDC + CPC. Almost all experiments in colliding beams make use of both drift and proportional chambers. Moreover, in many experiments, precision drift chambers (or other precision coordinate detectors) are arranged directly around the vacuum pipe of the accelerator for accurate determination of the angles of outgoing particles (Sect. 4.2).

3.2 Choice of Parameters
of a Multilayer Proportional Detector

Experiments reveal that the parameters of a detector (its pressure, temperature and gas composition) and those of the electronics (the voltage supply) must be extremely stable for the measured most probable ionization energy loss $\Delta_0(p, m)$, where p is the momentum, and m the mass of the particle. They should remain constant with time to within 1 %. The precision of ionization measurement in proportional detectors depends on a number of factors. We shall now proceed to analyze them.

The factors determining the relation between the signal amplitude and the ionization in proportional detectors (Fig. 3.10) can be divided into two main groups: (a) processes taking place in the chamber; (b) processes inherent in the electronics.

The main physical processes observed in the gas filling of a proportional chamber were described in Chap. 2. We shall analyze their influence on ionization measurement, emphasizing drift chambers, which are utilized in almost all modern identifiers of charged particles. In accordance with the chart in Fig. 3.10 we shall consider the following issues: (1) choice of the number and thickness of layers in an identifier; (2) requirements to the mechanical precision of the chamber construction; (3) choice of gas mixture; (4) the changes of charge during the drift of electrons; (5) the gas amplification process; (6) influence of the space charge on the ionization measurement; (7) requirements of the electronics; (8) long-term stabilization of the identifier operation; (9) data handling in the conditions of a physical experiment.

On the basis of such an analysis it is possible to determine the range of particle momenta within which the identification method under discussion can be applied and, also, to compare the expected and experimental confidence levels of identification.

3.2.1 Thickness and Number of Layers

The ionization resolution W_n (fwhm) of a proportional identifier with a fixed length L is primarily determined by the number of layers n, their width l, the pressure P of the gas mixture and its composition. The length of a multilayer identifier is usually limited by the dimensions permitted by the design of the experimental set-up as a whole, and by economic factors. In choosing the optimal number of layers in a chamber of fixed length L the following circumstances must be taken into consideration: (1) the value of W_n decreases relatively slowly as the number of layers increases; (2) the most expensive part (60–75 %) of the apparatus is the electronics, so that enhancement of the number of channels is not desirable; (3) the amplitude of the signal and the signal-to-noise ratio depends on the thickness of the gas layer and on the arrangement of the wires; (4) the effective width lP of a layer increases linearly with the pressure, but

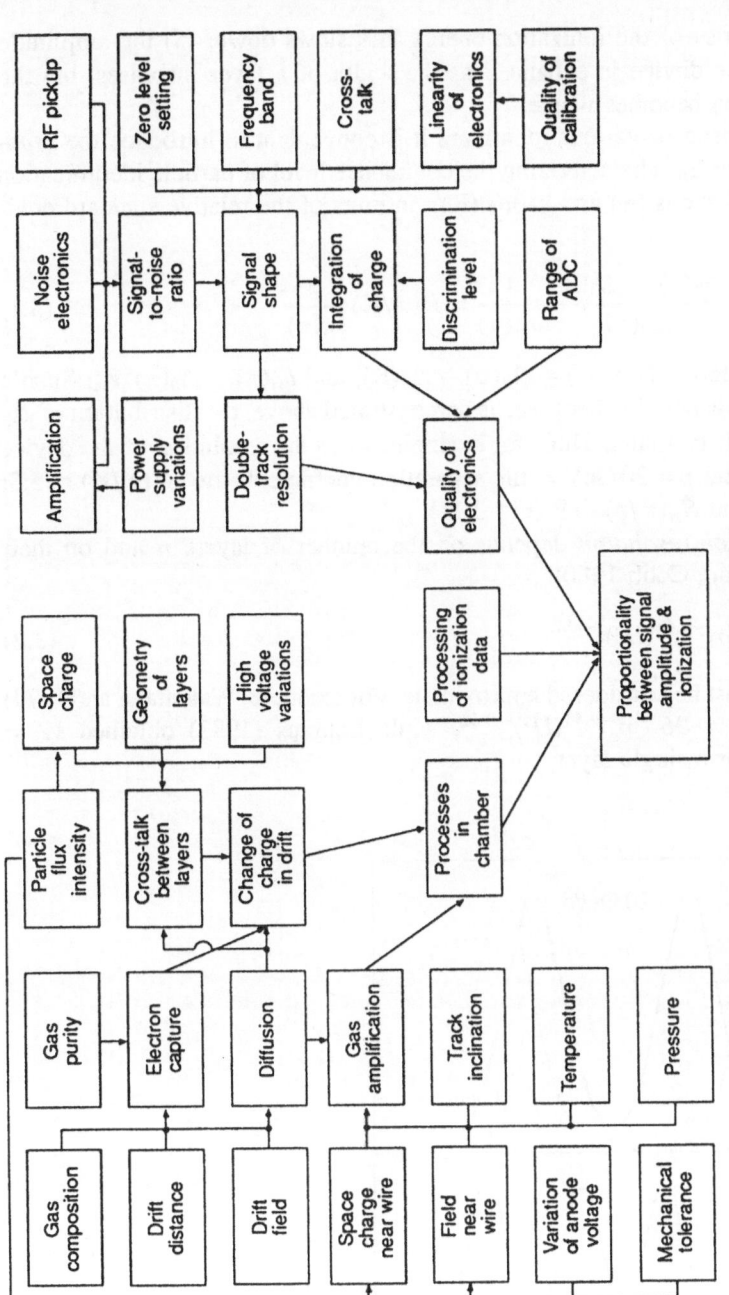

Fig. 3.10. Factors determining the relation between the signal amplitude and the measured ionization

the relativistic rise of the ionization energy loss slows down; (5) the amplitude resolution of the device is enhanced as the width of a layer increases, but the spatial resolution becomes worse.

For comparison of various identifiers it is convenient to introduce the *separation coefficient* S_n, characterizing the confidence level of particle identification [for instance, of pions (π) and kaons (K)], in units of the relative standard deviation σ_n:

$$S_n(\pi/K) = \frac{\Delta_0(\pi) - \Delta_0(K)}{\Delta_0(\pi)} \frac{1}{\sigma_n(\pi)} = D(\pi/K)\frac{2.36}{\delta_n(\pi)} , \tag{3.1}$$

where $W_n = 2.36\sigma_n$: $D(\pi/K) = \Delta_0(\pi) - \Delta_0(K)$; and $\delta_n(\pi) = \Delta_0(\pi)W_n$. Such a definition of S_n is possible because, as demonstrated above, the distribution of Δ_0 [or Δ_α (3.24)] is gaussian. Thus, for example, when the resolution of the device is $\sigma_n = 2.5\%$ and $p = 20\,\text{GeV}/c$, the separation coefficients are $S_n(\pi/K) \approx 5.5$; $S_n(p/K) \approx 3$ and $S_n(\pi/p) \approx 9$ (Fig. 3.11).

The *ionization resolution* depends on the number of layers n and on their thickness (Allison, Cobb 1980):

$$W_n(\%) = 96n^{-0.46}(lP)^{-0.32} . \tag{3.2}$$

This relation must be considered approximate. For example, Walenta et al. (1979) found that $W_n = 96 \cdot n^{-0.43}(lP)^{-0.32}$, while Lehraus (1983) obtained $W = 85 \cdot (lP)^{-0.26}$ for a single layer.

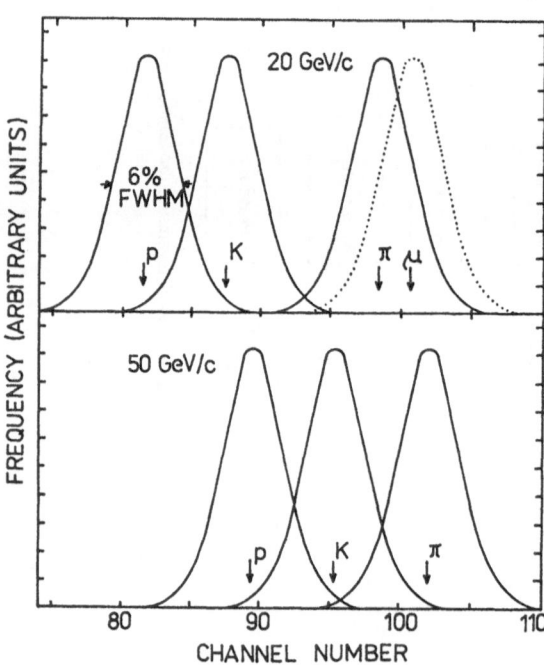

Fig. 3.11. Computed distribution of Δ_α for π, K, p, μ in a 95 % Ar + 5 % CH$_4$ mixture in the 128-layer EPI detector with $\sigma_n = 2.5\%$ (Aderholz et al. 1974)

A more general formula can be obtained by relating the thickness of a single layer to the quantity ξ/I where $\xi = A_1 x/\beta^2$ and I is the mean ionization potential (Sect. 1.5). In this case the ionization resolution is

$$W_n(\%) = 81 n^{-0.46} (\xi/I)^{-0.32} \ . \tag{3.3}$$

This formula holds with an uncertainty between 10 and 20 % in the interval $0.5 < \xi/I < 10$.

Calculations by Allison (1982) show that for a fixed length of the identifier the ionization resolution W_n deteriorates and the relativistic rise of ionization increases with the increase of the number of layers (Fig. 3.12).

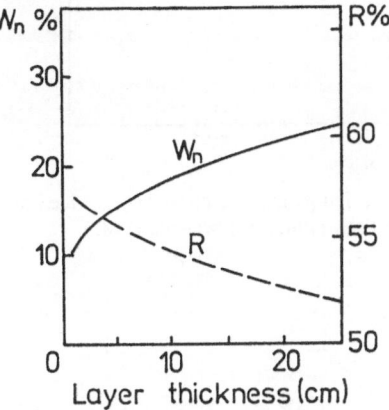

Fig. 3.12. Computed ionization resolution W_n in a proportional chamber 1 m long at $p/(mc) = 10$, and relativistic rise of ionization R plotted against the layer thickness in argon under normal conditions (Allison 1982)

The ionization resolution W_n calculated by Allison (1981), Aderholz et al. (1974) and Walenta (1981) is presented in Fig. 3.13. With the aid of this plot one can determine the identifier length and the number of layers to be used, in order to achieve the required ionization resolution (in Fig. 3.14 $W_n = 5\%$ and 10 %, respectively). From this figure it follows that starting from a certain (optimum) layer thickness a further decrease does not result in an enhancement of the ionization resolution.

In handling experimental results involving very thin layers (Arai et al. 1983) the models proposed in Lehraus et al. (1982a) and Walenta (1981) are preferred. In measurements with thin gas layers, certain peculiarities must be taken into account (Sect. 3.9).

In accordance with calculations performed by Lehraus (1982a) and Walenta (1981), the resolution W_n in argon remains nearly unaltered for most identifiers, when $1 < lP < 4 \cdot 10^5$ cm $\cdot$ Pa (Fig. 3.14). Therefore the main parameter applied for the determination of W_n is the total length L of the identifier. The values of $W_n(L)$ obtained in various experiments are presented in Fig. 3.15. Although the data are taken from devices involving different numbers of layers n of varied thicknesses and gas pressures, almost all experimental data are in good agreement with the universal dependence $W_n = 13.5 \cdot L^{-0.37}$ (Lehraus 1983).

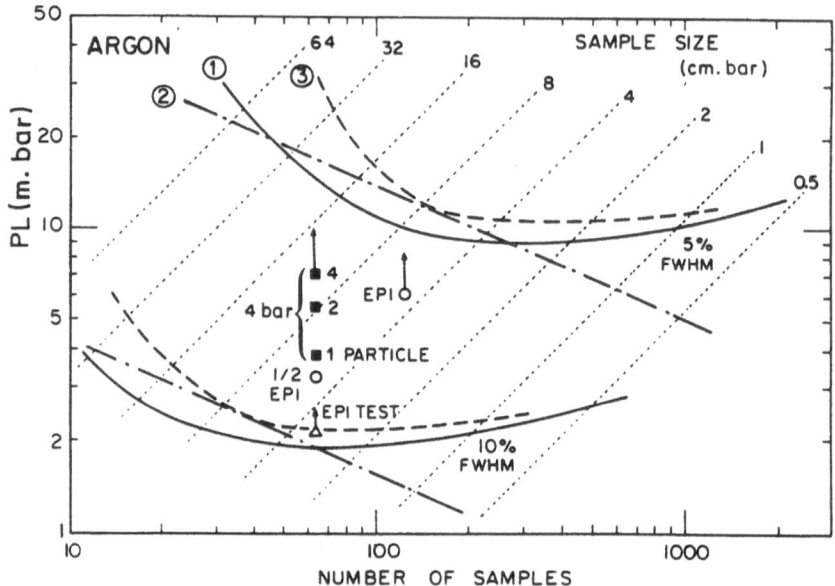

Fig. 3.13. Comparison of three methods of determining design parameters of $(-dE/dx)$ identifiers (Lehraus et al. 1982a). (*1*) Aderholz et al. (1974); (*2*) Allison (1981); (*3*) Walenta (1981)

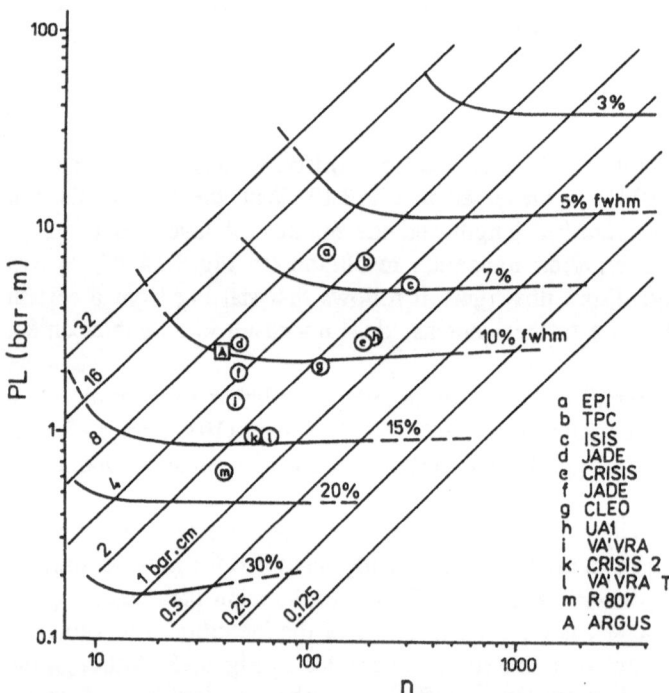

Fig. 3.14. Computed resolution W_n in Ar under normal conditions, plotted against detector length and number of layers involved. Circles indicate resolutions calculated for various identifiers (Hasemann 1982)

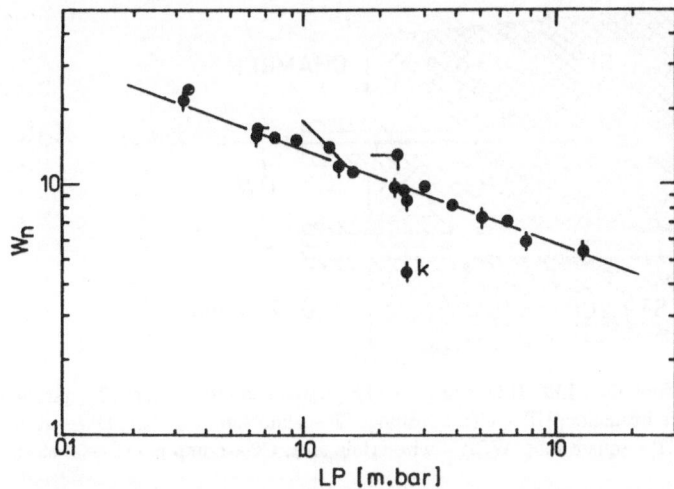

Fig. 3.15. Dependence of the resolution W_n on the detector length LP; the experimental points are obtained in different gases. The point k corresponds to C_3H_8 (Lehraus 1983)

Usually the layer thickness lP is chosen to be within the limits $(1-4)\cdot10^5$ cm·Pa; the length of an identifier must be at least $4\cdot10^5$ m · Pa, since separation of relativistic pions and kaons requires $W_n \lesssim 8\,\%$.

3.2.2 Mechanical Tolerances in the Construction of a Proportional Detector

Variation of the wire diameter, the precision with which it is installed in the chamber, and its sag due to gravitational or electrostatic forces, influence significantly the distribution of the electric field in the chamber and, consequently, the gas gain M (Sect. 2.5).

In proportional chambers with small wire spacing, a small displacement of a wire gives a large change in the multiplication factor. These issues are analyzed, for example, by Sauli (1977) and Erskine (1972). In chambers used for ionization measurements the wires are mounted so that the field surrounding them exhibits axial symmetry (Fig. 3.4). In this case M is less dependent upon the precision with which the wires are fixed.

Precise mounting of wires in large chambers represents quite a difficult task. Various methods are applied in order to overcome such difficulties. In large cylindrical chambers the wires are stretched through crimp pins inserted into precisely machined holes (made in the supporting constructions), and fixed by soldering or by crimping the pins. This technique provides the following mounting precisions of the wires: $100\,\mu m$ – MARK II, $120\,\mu m$ – TASSO, $20\,\mu m$ – AFS. Another technique, involving utilization of a support with a precise comb for the wires, provides an even higher precision: for example, 10–$20\,\mu m$ in the JADE chamber. In Fig. 3.16 an example of the feedthrough of the CLEO II chamber is presented.

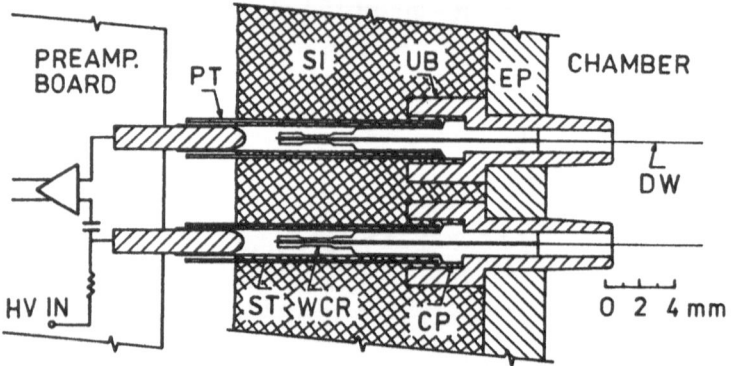

Fig. 3.16. Cross section through the CLEO II feedthrough with ground and sense wires; PT – plastic tube insulation, SI – sylastic insulation, UB – ultem bushing, EP – aluminum end plate, DW – drift wire, HV – high voltage, ST – square tube, WCR – wire crimp area, CP – crimp pin (Cassel et al. 1986)

High precision of wire positioning is necessary both for accurate measurement of coordinates on particle tracks and for ionization measurements.

A change in the wire diameter may give rise to large variations in the gas gain. The gas gain factor depends on the charge on the wire q_w:

$$\Delta M/M = \ln M (\Delta q_w/q_w) ; \qquad (3.4)$$

while variation of the charge is related to a change of the wire radius r_a:

$$\Delta q_w/q_w = (C_w/2\varepsilon_g)(\Delta r_a/r_a) , \qquad (3.5)$$

where ε_g is the dielectric constant of the utilized gas and C_w is the capacity per unit length of the wire. Typical non-uniformity in the diameter of a gold-plated tungsten wire, with $2r_a = 20\,\mu$m, amounts to $\Delta r_a/r_a \approx 0.01$; this leads to variations of 2–4 % in M along the wire (Sauli 1977).

A charged wire in a chamber is in unstable equilibrium; the tension in the wire prevents deviation from its equilibrium position. Calculations of the electric forces and of the required wire tension have been performed, for example, by Erskine (1972) and Sauli (1977). This effect is particularly important in large chambers where thicker sense wires and significant tensions have to be used. Sauli (1977) presents admissible wire tensions as a function of the wire diameter.

An unpleasant effect in large chambers is the gravitational sagging of wires, which is significant when the wires are more than 1 m long. For example, in the JADE chamber, which had wires 234 cm long, this sag amounted to 70 μm; the electrostatic displacement in this chamber was 50 μm. These displacements are usually taken into account in the data processing procedure, but they must always be minimized.

Computed variations of the gas gain on individual wires in a chamber are presented in Table 3.2.

Table 3.2. Sources of gas gain variation (Allison et al. 1976)

Cause	Variation	$\Delta M/M$ [%]
Instability and ripple of drift voltage	$< 10^{-3}$	< 0.7
Instability and ripple of anode voltage	$< 10^{-4}$	< 0.1
Instability of the electronics power supply	$< 10^{-2}$	< 1
Time variations of M related to changes in temperature, pressure and composition of gas mixture	$< 1.6 \cdot 10^{-2}$	< 2
Variation of M along the wire	$< 2 \cdot 10^{-2}$	< 2
Differences between values of M on individual wires: these include		
a) wire diameter variation	$< 0.2\,\mu m$	
b) sense wire position	$< 8.0\,\mu m$	
c) potential wire position	$30\,\mu m$	≈ 2
d) drift electrode position	$40\,\mu m$	≈ 2
Random electronic noise	$120\,eV/channel$	1

In constructing large chambers one must take into account that temperature variations may cause deformations of the chamber elements which lead to undesirable changes of the gas gain, and to a deterioration of the spatial resolution.

3.3 Choice of Gas Mixture

Although gas amplification is possible in any gas, the gas mixture for proportional or drift chambers must fulfill many, often contradictory, requirements. The principal requirement of a gas mixture is a minimum electron capture probability, which excludes application of gases exhibiting electron affinity. The gas mixture in a proportional chamber must provide for high gas amplification at a relatively low voltage: for the signal being proportional to the measured ionization, for a good amplitude resolution, for the possibility of operating at high beam intensities, and it must be inexpensive. A good signal-to-noise ratio is achieved in gases with a large specific ionization, for example, in Xe or Kr (Table 1.3). Unfortunately the cost of these gases is high (the approximate relationship between the costs of the gases Ar : Ne : Kr : Xe is roughly 1 : 10 : 35 : 170).

The base component of practically all gas mixtures used in proportional detectors is Ar. In Sect. 2.5 it was shown that to quench the discharge, a molecular gas must be introduced into the chamber. The utilization of pure molecular gases is inconvenient owing to the necessity of applying high voltages to achieve the required multiplication. Therefore, most frequently used in proportional detectors are binary or triple mixtures, with Ar as the base gas and a 10–20 % admixture of molecular gases.

An ideal gas mixture for coordinate measurements in proportional chambers is the so-called "magic gas" consisting of 75 % argon + 24.5 % isobutane + 0.5 %

freon. With such a mixture a gas amplification of up to 10^8 is achieved (Sauli 1977).

The gas mixture for drift chambers must fulfill additional requirements. This mixture, in addition to everything else, must allow the electron drift velocity to remain constant over a wide range of voltages, and minimum electron diffusion and capture. Here the "magic gas" involving electro-negative freon is no longer suitable.

A constant electron drift velocity is achieved in many gas mixtures. In Fig. 3.17 mixtures of inert and molecular gases are presented, which exhibit constant drift velocities within a wide range of $\mathcal{E}/P$ values. Data for a large number of other mixtures can be found in Peisert and Sauli (1984).

A minimum electron diffusion is observed in so-called *cool gases*, in which the electron energy differs little from the thermal energy up to extremely high

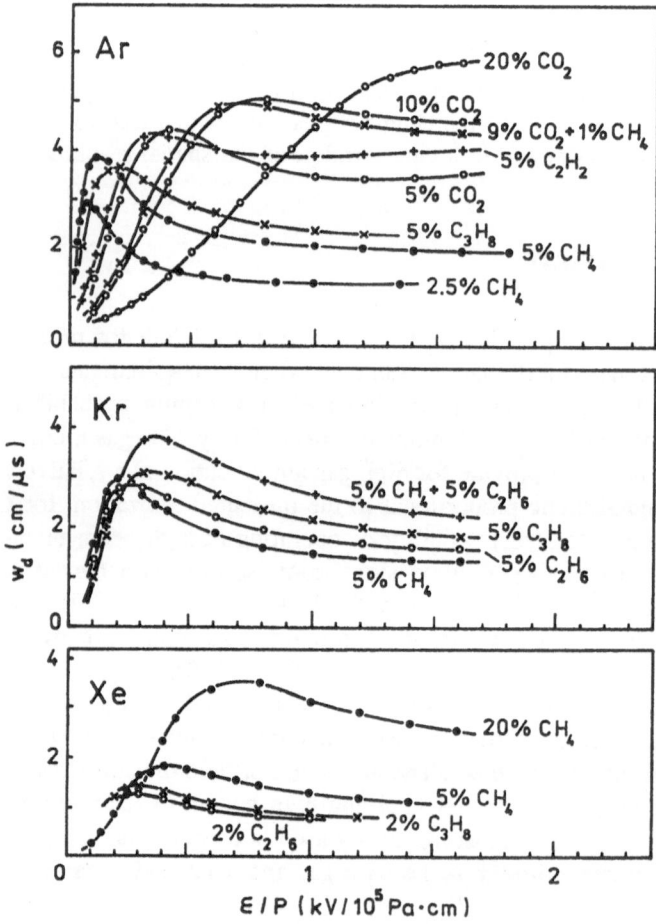

Fig. 3.17. Electron drift velocity in Ar, Kr, and Xe with admixtures of various hydrocarbons and CO_2 (Lehraus et al. 1982b)

electric field strengths. We noted above that such gases are CO_2, NH_3, isobutane, methylal, dimethylether.

In addition, the requirements of the properties of a gas mixture depend on the sort of chamber. Thus, for instance, for large projection drift chambers, "fast" mixtures are recommended. From Fig. 2.5 one can see that such mixtures should contain CH_4 or CF_4. Fast mixtures are also used in chambers operating in particle beams of high density.

In chambers with longitudinal electron drift (Sect. 3.8) it is expedient to utilize *slow gases*, i.e., gases exhibiting a low drift velocity even at relatively high electric field strengths. Gases with this property are CO_2, dimethylether, isobutane, and He–C_2H_6 mixtures. Special slow mixtures based on CO_2 with a small admixture (5–10 %) of argon or isobutane were proposed in Becker et al. (1983).

For chambers operating in a magnetic field, gases are chosen which have a small deviation angle α_M of the electron drift direction. Such gases include CO_2 or slow mixtures of CO_2 with argon or isobutane (Becker et al. 1983), and also other cool gases.

3.3.1 Gas Mixture for Particle Identification

Gas mixtures for identifiers must also provide for a significant relativistic rise of ionization and for minimal fluctuations of ionization energy loss. The maximum relativistic rise is observed in heavy inert gases (Xe, Kr) (Fig. 3.18). These gases are also attractive because their Fermi plateau starts at high $\beta\gamma$ values. It must be pointed out that the relativistic rise in the other inert gases is noticeably lower than in Xe, in molecular gases even smaller (Fig. 3.18), and they saturate at low $\beta\gamma$.

On the contrary, the ionization resolution W is much better in molecular gases than in inert gases (Fig. 3.19). The ionization resolution in all inert gases

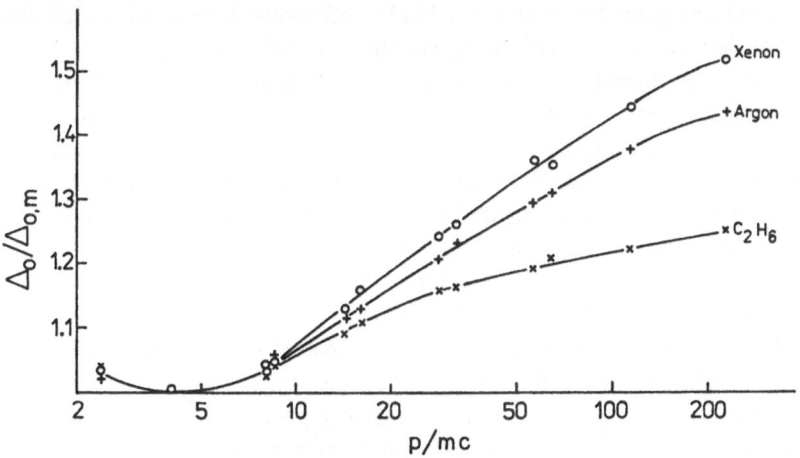

Fig. 3.18. Relativistic rise of probable ionization versus p/mc in some gases at atmospheric pressure (Sauli 1984)

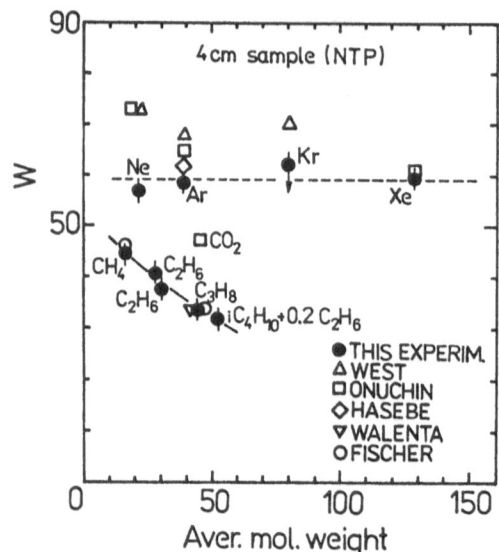

4 cm sample (NTP)

W

• THIS EXPERIM.
△ WEST
□ ONUCHIN
◇ HASEBE
▽ WALENTA
○ FISCHER

Aver. mol. weight

Fig. 3.19. Computed ionization resolution $W(\%)$ in a layer of gas 4 cm thick under normal conditions as a function of an average molecular mass; • Lehraus, △ West, □ Onuchin, ◇ Hasebe, ▽ Walenta, ○ Fischer (Lehraus et al. 1982b)

is approximately the same, while in heavy molecular gases it is higher than in light ones.

Thus it is clear that in inert gases the advantage due to the large relativistic rise is reduced by insufficient ionization resolution; on the contrary, in molecular gases gains due to a good ionization resolution are reduced because of the small relativistic rise. In Table 3.3 computed values are presented for the relativistic rise of probable energy loss (the ratio between the values of Δ_0 for protons and electrons of momentum 4 GeV/c) and for the ionization resolution W_n. The separation coefficients S'_n are also presented in Table 3.3 in various gases, for momenta 4 GeV/c and 20 GeV/c. The values of $S'_n = [\Delta_0(e) - \Delta_0(\pi)]/\Delta_0(\pi)W_n$ are given in units of relative ionization resolution $W_n(\%)/\sqrt{L(m)}$. Calculations were performed applying the maximum likelihood method. It is clear that the separation coefficients are virtually independent of the choice of gas.

In Fig. 3.20 experimental S_n values are presented plotted against the gas pressure. The differences between the separation coefficients in different gas mixtures are insignificant; however, in separation of hadrons preference should be given to mixtures based on Ne, which at atmospheric pressure in the region of 15 GeV/c provides the same separation as observed in Xe or Kr. At higher momenta it is more expedient to utilize Xe or Kr.

The most effective mixtures for separating electrons from hadrons are mixtures containing Xe or Kr (Fig. 3.20). However, because these gases are very expensive the most frequently used mixtures are based on Ar. Pure molecular gases are not suitable for the separation of electrons and hadrons.

Practical reasons (cost, risk of explosion, etc.) prevail that Ar is used in large identifiers, with a molecular gas admixture, such as the $Ar + CH_4$ mixture in EPI and TPC; $Ar + 20\% CO_2$ in ISIS and CRISIS; and $Ar + C_2H_6$ in UA–1. Molecular gases – $75\% C_3H_8 + 25\% C_2H_4$ – were used in the JADE chamber.

Table 3.3. Calculated relativistic relativistic rise, ionization resolution and separation coefficients S'_n in various gases for $l = 1.5$ cm (Allison 1982)

Gas	Relativistic rise $\Delta_0(e)/\Delta_0(p)$ [%]	Ionization resolution $W_n/L^{1/2}$ [$\%/m^{1/2}$]	Separation coefficient $S'_n/\sqrt{L}$ in units of $[W_n(\%)/m^{1/2}]$					
			$p = 4$ GeV/c			$p = 20$ GeV/c		
			e/π	π/K	K/p	e/π	π/K	K/p
He	58	15	1.5	0.9	0.3	0.3	0.9	0.5
Ne	57	13	1.8	1.1	0.4	0.6	1.1	0.6
Ar	57	12	1.6	1.1	0.4	0.6	0.9	0.6
Kr	63	11	2.2	1.3	0.5	0.9	1.1	0.7
Xe	67	13	2.2	1.2	0.4	1.0	1.0	0.6
N_2	56	11	1.8	1.3	0.4	0.7	1.0	0.7
O_2	54	9	2.0	1.5	0.5	0.7	1.1	0.8
CO	55	11	2.0	1.3	0.5	0.7	1.1	0.8
NO	54	10	2.0	1.5	0.5	0.7	1.1	0.8
CO_2	48	8	2.0	1.6	0.5	0.6	1.1	0.9
N_2O	48	8	2.0	1.6	0.5	0.7	1.1	0.8
CH_4	43	9	1.6	1.5	0.5	0.5	1.0	0.8
C_2H_2	42	8	1.7	1.5	0.5	0.6	1.0	0.8
C_2H_6	38	8	1.6	1.6	0.5			
iso-C_4H_{10}	23	6	1.8	1.3	0.3	0.6	1.0	0.8
Ar + 20 % CO_2	55	12	1.9	1.2	0.4			

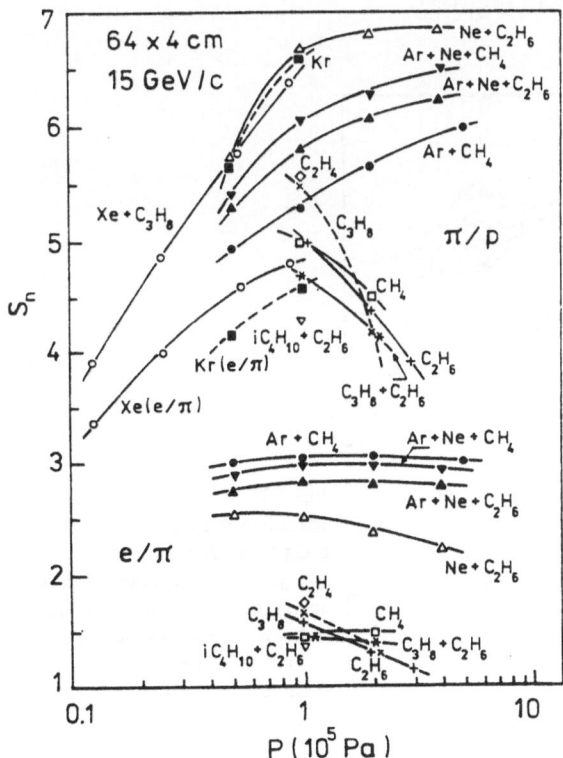

Fig. 3.20. Separation coefficients $S_n(\pi/p)$ and $S_n(e/\pi)$ in terms of σ_n in various gases as functions of the gas pressure (Lehraus et al. 1982b)

3.3.2 Purity of a Gas Mixture

The capture of electrons by admixtures (O_2, H_2O, halogens and others) in a gas significantly influences the collection of charge, especially in the case of a large electron drift length. The mechanism of electron capture and its cross sections are presented in Sect. 2.4. Oxygen and water are the principal gas "contaminants" in proportional and drift chambers. Usually several parts per million (ppm) of oxygen are present, and its concentration increases, due to its diffusion through the mylar windows of the chamber and to degassing of the chamber constructions. The amount of water in a gas mixture is sometimes even greater. These problems are considered in detail, for example by Huk et al. (1988).

The capture of electrons by admixtures is enhanced if there are present in the admixture such gases as CO_2 or isobutane, which act as catalysts for the reactions (2.50) and (2.51). A particularly significant role is played by carbon dioxide in the enhancement of electron capture. When the concentration of this gas in the mixture is a_k, and the concentration of a gas exhibiting electron affinity (for instance, O_2, H_2O) is a_a, the frequency of electron capture is $\nu_a = (3.1\pm0.3)\cdot10^{-30}a_a a_k \, s^{-1}$, where concentrations are in molecules per cm^3 (Allison et al. 1974). Thus, for example, when the volume concentration of oxygen is 10^{-6} and the amount of CO_2 is 20%, the lifetime of free electrons $\tau_{at} = 1/\nu_a \approx 2.8\,ms$.

The electron capture rate in various mixtures is presented in Fig. 3.21. In mixtures containing methane the lowest capture rate is observed. The change of the signal amplitude in an argon-methane-isobutane mixture is shown for different oxygen concentrations in Fig. 3.22.

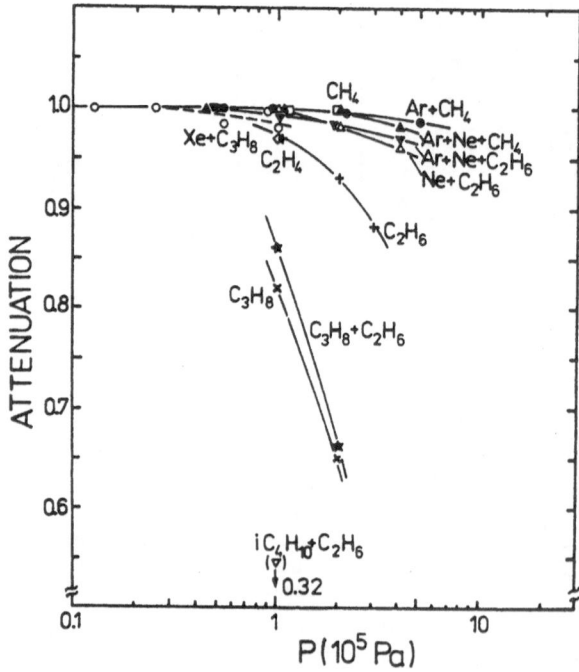

Fig. 3.21. Attenuation of the signal due to electron attachment for a drift length of 41 cm in various gases as a function of pressure. The admixture of oxygen was (5–10) ppm in inert gases and (15–20) ppm in molecular gases (Lehraus et al. 1982b)

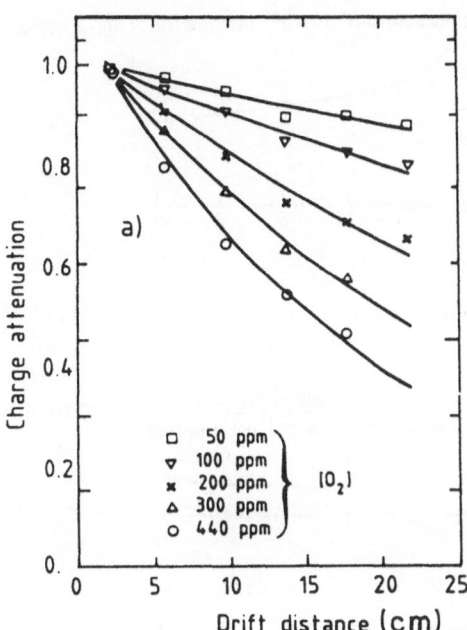

Fig. 3.22. Charge attenuation coefficient for various O_2 concentrations in the gas mixture 88 % Ar + 10 % CH_4 + 2 % isoC_4H_{10} at $4 \cdot 10^5$ Pa. $\mathcal{E}/P = 200$ V/(cm $\cdot 10^5$ Pa) (Huk et al. 1988)

3.3.3 Gas Pressure and Range of Momenta for Reliable Particle Identification

The ionization resolution W_n and the relativistic rise R_E of ionization depend on the gas pressure (Fig. 3.23a, b). The ionization resolution is enhanced as the pressure increases, but the relativistic rise is reduced and, as a result, the separation coefficient remains almost unchanged (Fig. 3.23c, d). These calculations are confirmed experimentally.

The dependence of the ionization resolution W_n upon the pressure differs significantly in inert and molecular gases (Fig. 3.24). Figure 3.25 presents the dependence of the relativistic rise of ionization upon pressure, in various gases and mixtures. The dependence observed in Ar and Ne is in good agreement with the computed dependence $R_E \sim P^{-0.065}$, represented in Fig. 3.25 by the dashed line. The relativistic rise in molecular gases is significantly lower than in inert gases. As a result of the behavior of $W_n(P)$ and $R_E(P)$, the separation coefficient is nearly independent of the pressure in molecular gases (Fig. 3.20), although it does improve somewhat as the pressure increases in mixtures containing large amounts of inert gases.

The range of momenta within which particle identification can be performed by ionization measurements in a gas depends on its relativistic rise and on the ionization resolution of the device. As an example, in Fig. 3.26 the points indicate the probable ionization losses Δ_0 of the particles kinematically identified in the EHS spectrometer, and measured in the ISIS chamber, with $W_n = 6$ % (Allison et al. 1984). Also presented are the dependences $\overline{\Delta}_0(p/mc)$ for particles π, K, p and e obtained from calibration measurements.

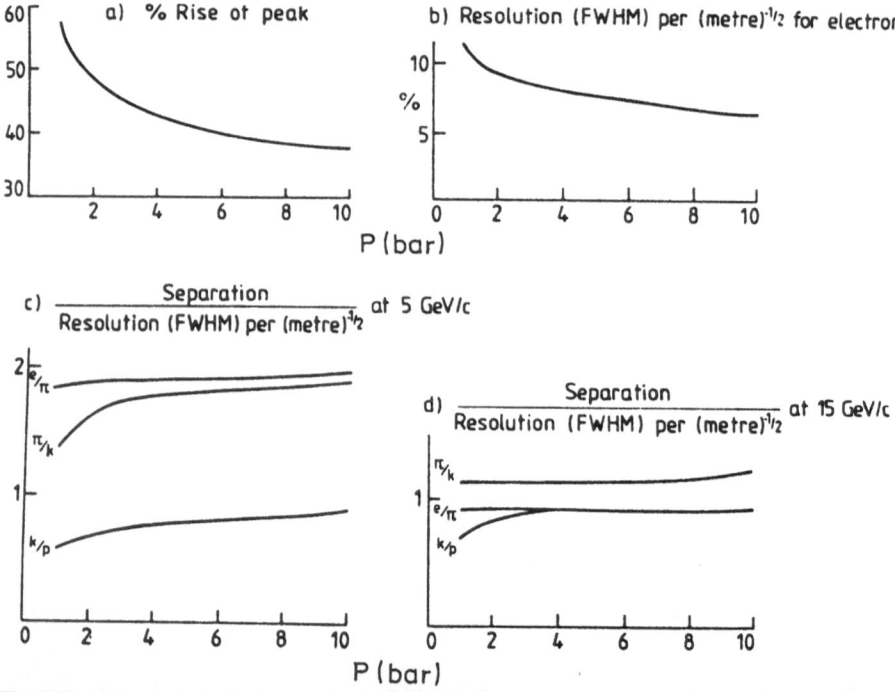

Fig. 3.23. Computed dependences of relativistic rise (**a**), ionization resolution (**b**) and separation coefficient S'_n (**c**), (**d**) in argon as functions of gas pressure for $L = 1$ m and $l = 1.5$ cm (Allison 1982)

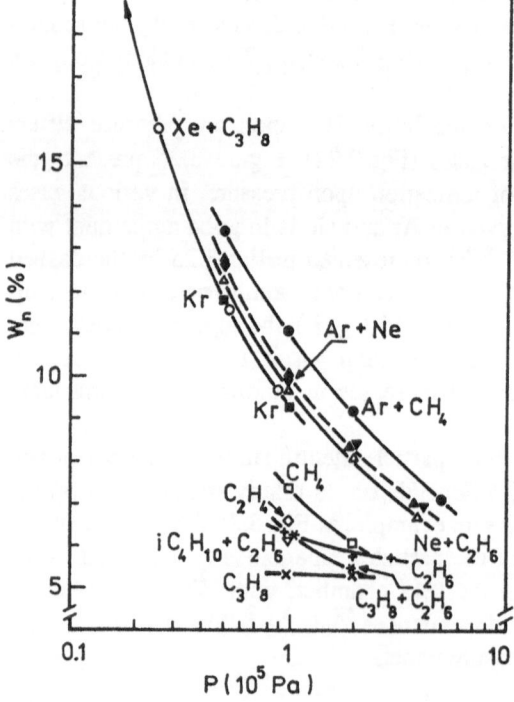

Fig. 3.24. Ionization resolution W_n as a function of gas pressure in an identifier, with $n = 64$, $l = 4$ cm, $\alpha = 0.4$ for pions with $p = 15$ GeV/c (Lehraus et al. 1982b)

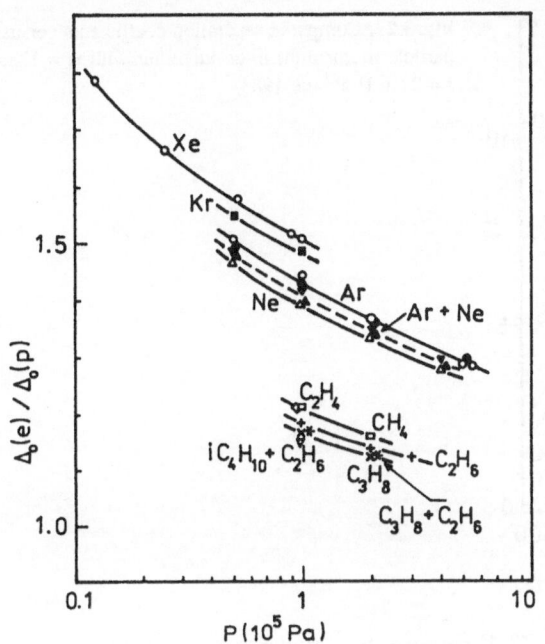

Fig. 3.25. Relativistic rise [ratio $\Delta_0(e)/\Delta_0(p)$] at $15\,\text{GeV}/c$ as a function of gas pressure; $n = 64$, $l = 4\,\text{cm}$, $\alpha = 0.4$ (Lehraus et al. 1982b)

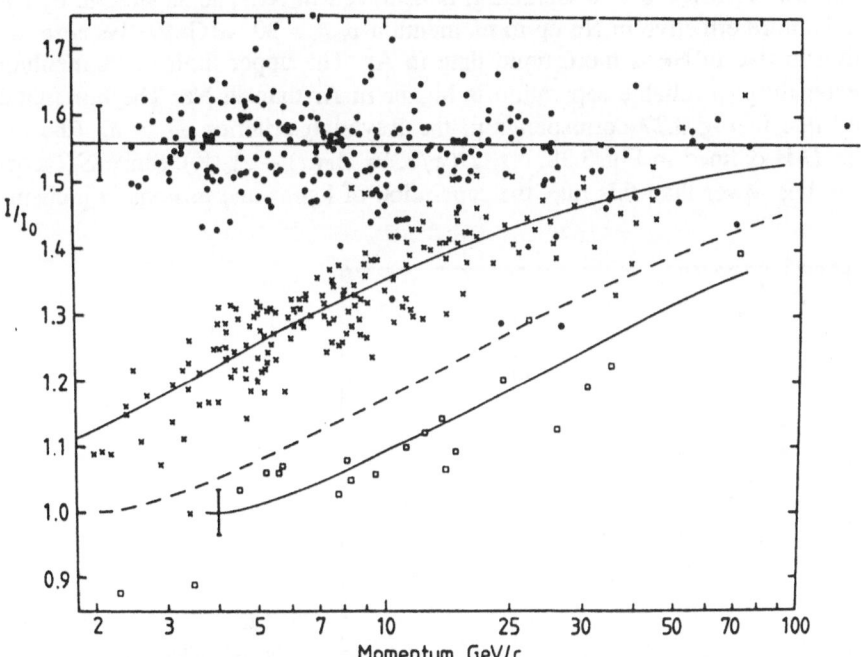

Fig. 3.26. A scatter plot of the measured ionization versus momentum for kinematically identified tracks. Full circle – electron; cross – pion; open square – proton; the full lines are calculated curves for electrons, pions and protons; dashed line – kaons; the error bars – root-mean-square (r.m.s.) for tracks with 250 points (Allison et al. 1984)

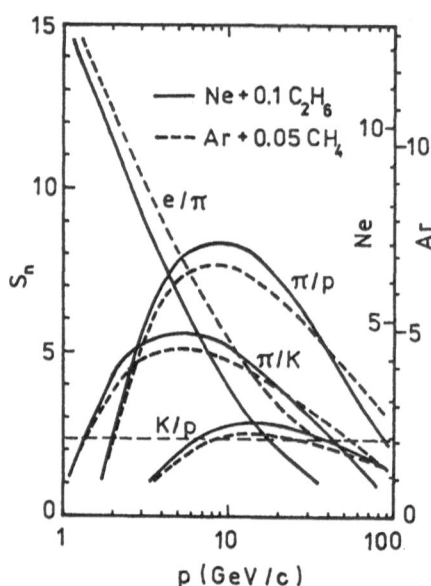

Fig. 3.27. Computed separation coefficients versus particle momentum in an identifier with $n = 128$, $l = 2$ cm (Lehraus 1983)

The separation coefficient S_n (3.1) represents a measure of the particle identification reliability. The computed dependence $S_n(p)$ is given for Ar and Ne in Fig. 3.27. A better $e - \pi$ separation is achieved in Ar. The separation of hadrons is more effective in Ne up to momentum $p_{max} = 30$–40 GeV/c because the relativistic rise in Ne is more rapid than in Ar. The upper limit of momentum corresponding to reliable separation is higher in Ar than in Ne. The horizontal dotted line in Fig. 3.27 corresponds to the Rayleigh criterion $D > \delta_n$ (the distance D is defined in Fig. 3.38, e.g. $D(\pi/K) = \Delta_0(\pi) - \Delta_0(K)$. Since $S_n(K/p)$ usually lies lower than this line, the separation of kaons and protons in a cham-

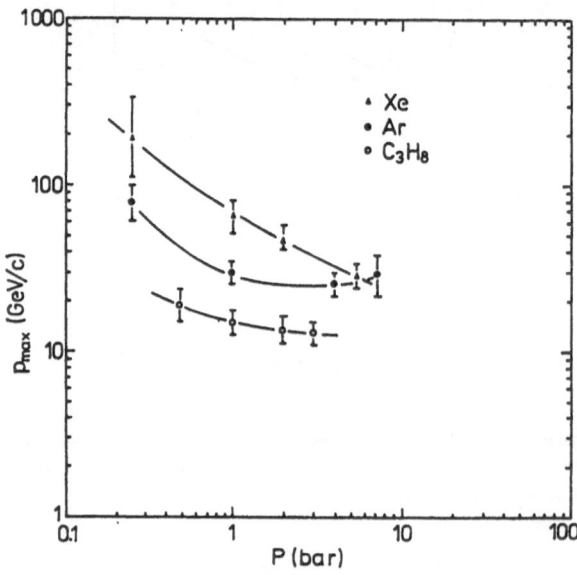

Fig. 3.28. Dependence of the maximum momentum p_{max} corresponding to reliable separation of pions and kaons, upon the gas pressure for $l = 2.3$ cm (Walenta et al. 1979)

Table 3.4. Range of momenta for reliable separation of particles in accordance with the Rayleigh criterion in the ISIS chamber filled with $Ar + 20\%$ CO_2 (Allison, Wright 1984)

Particles separated	e/π	π/K	K/p	π/p
Range of momenta [GeV/c]	< 25	1.5–60	5–40	2–150

ber of such dimensions ($L = 256$ cm) is almost impossible. Table 3.4 gives the momentum ranges corresponding to the Rayleigh criterion $D > \delta_n$, applied in the ISIS chamber as the condition for sufficiently reliable particle separation.

Since R_E decreases as the pressure goes up, a decrease also occurs in the maximum particle momentum p_{max} at which reliable identification is still feasible (Fig. 3.28). In computing this dependence, instead of the Rayleigh criterion, the requirement that $S_n \geq S_{n0}/\sqrt{2}$ was imposed, where S_{n0} is the maximum separation coefficient for the given pair of particles. It is clear that to identify particles of high energies ($p > 100$ GeV/c) Xe at a reduced pressure must be used.

The maximum momenta at which it is possible to separate pions and kaons at $P = 10^5$ Pa in various gases are presented in Table 3.5. It can be seen that molecular gases can only be used for identification of particles of low energies (up to 30 GeV), while the heavy inert gases (Xe in particular) are suitable for identifying high-energy particles.

Table 3.5. The computed maximum momentum for reliable separation of pions and kaons in a drift chamber with $L = 5$ m, $l = 1.5$ cm and $P = 1 \cdot 10^5$ Pa in various gases (Allison, Cobb 1980)

Gas	Relativistic rise R_E [%]	Ionization resolution W_n [%]	$p_{max}(\pi/K)$ [GeV/c]
He	1.41	5.2	45
Ne	1.54	5.4	50
Ar	1.58	5.3	55
Kr	1.60	5.2	55
Xe	1.70	5.5	95
CH_4	1.36	3.9	30
NH_3	1.39	4.2	45
N_2	1.48	4.6	45
CO_2	1.45	3.9	50
$Ar + 20\%$ CO_2	1.55	4.9	55

3.4 Physical Processes in a Gas and Precision of Ionization Measurements

In the process of gas amplification in the vicinity of a sense wire, N_+ positive ions are produced per unit time:

$$N_+ = \Phi P M N_t l / \cos\theta , \tag{3.6}$$

where Φ is the number of charged particles traversing the gas layer of width l per unit time, P the pressure, M a gas gain, N_t the number of electrons; θ is the angle between the track and the plane of sense wires. The ions move slowly into the drift volume with a velocity w_+ (Sect. 2.3). Accumulation of the positive space charge leads: 1) to reduction of the gas gain in the vicinity of the sense wire; and 2) to a change in the velocity and direction of motion of the electrons in the drift volume.

3.4.1 Space Charge Near the Wire

This issue has been raised in Becker et al. (1983), Mori et al. (1982), Hendricks (1969), Sauli (1977) and Sipilä and Vanha-Honko (1978). Let us write out the equations of motion of positive ions in an electric field $\mathcal{E}$:

$$\nabla\mathcal{E} = \varrho_+/\varepsilon ; \quad J_+ = \varrho_+ w_+ ;$$
$$\nabla J_+ = 0 ; \quad w_+ = \mu_+\mathcal{E} ; \tag{3.7}$$

where J_+ is the current density of the positive ions; μ_+ is the ion mobility; ε is the dielectric constant; and ϱ_+ is the density of positive ions. From (3.7) it follows that

$$\nabla[\mathcal{E}(\nabla\mathcal{E})] = 0 . \tag{3.8}$$

In the vicinity of the wire, where the field configuration is cylindrical, (3.8) assumes the form

$$\frac{d}{dr}\left(\mathcal{E}\frac{d\mathcal{E}}{dr}\right) + \frac{1}{r}\left(\mathcal{E}\frac{d\mathcal{E}}{dr}\right) = 0 . \tag{3.9}$$

Upon integrating we obtain the following expressions for the field strength, the space charge density and the potential $U(r)$:

$$\mathcal{E} = (C_1/r)(1 + r^2 C_0/C_1^2)^{1/2} ;$$
$$\varrho_+(r) = \varepsilon C_0(C_1 + C_0 r^2)^{-1/2} ; \tag{3.10}$$
$$U(r) = (C_1^2 + C_0 r^2)^{1/2} - C_1 \ln\{[C_1 + (C_1^2 + C_0 r^2)^{1/2}]/(C_0^{1/2} r)\} .$$

The constants of integration C_0 and C_1 are determined taking account of the wire geometry, of the voltage U_0 and of the current I in the chamber:

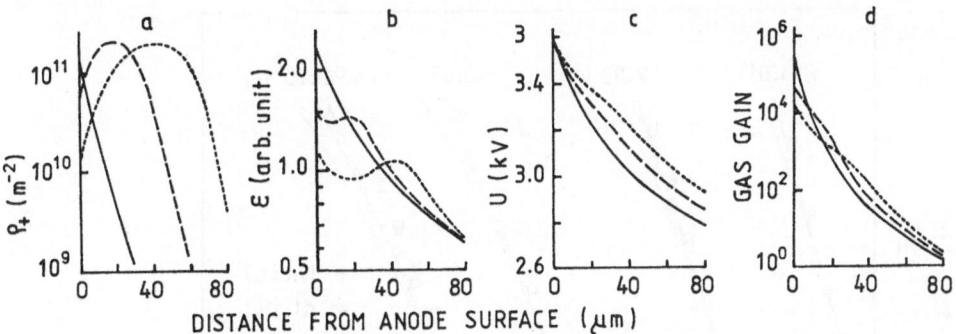

Fig. 3.29. Computed dependences of the positive charge density $\varrho_+(x, t)$ (**a**); of the electric field strength $\mathcal{E}(x, t)$ (**b**); of the potential $U(x, t)$ (**c**); and of the gas amplification (**d**); upon the distance from the surface of the sense wire (x) for electrons hitting it at times $t = 0$ (———), $t = 8\,\mathrm{ns}$ (- - -) and $t = 16\,\mathrm{ns}$ (·····) after the arrival of the first electron; $U_0 = 3.6\,\mathrm{kV}$, $P = 10^5\,\mathrm{Pa}$, $r_a = 25\,\mu\mathrm{m}$ (Mori et al. 1982)

$$C_0 = \frac{I/L_w}{2\pi\varepsilon\mu_+}, \quad C_1 = \frac{U_0}{\ln(r_b/r_a)}\left[1 + \frac{C_0 r_b^2 \ln(r_b/r_a)}{4U_0^2}\right]. \tag{3.11}$$

where L_w is the wire length; r_a is the anode wire radius; and r_b is the distance between the anode and cathode.

The decrease in the electric field strength resulting from the accumulation of charge around the wire can be expressed by a fictitious decrease of the voltage U_0 by ΔU_0:

$$\Delta U_0 \approx -\frac{r_b^2 I/[L_w \ln(r_b/r_a)]}{8\pi\varepsilon\mu_+ U_0}. \tag{3.12}$$

Mori et al. (1982) computed the distributions for the density of positive charge, the electric field strength, the electric potential and the gas multiplication factor, in the vicinity of the wire for times $t = 0, 8, 16\,\mathrm{ns}$ after the arrival of the first electrons at the wire. These calculations clearly reveal (Fig. 3.29) noticeable changes in time of the electric field strength and potential, which result in a decrease of the gas amplification and, consequently, in a decrease of the signal amplitude. The space charge produced by the primary electrons has the shape of an elongated droplet (Fig. 2.23). This charge screens the wire as if enhancing its diameter. Thus, the field gradient is reduced, which results in reduction of the efficiency of ionization by electron impact.

The influence of space charge in the vicinity of the wire manifests itself in the deviation of the dependence $M(U_0)$ from an exponential (Figs. 2.15 and 3.30). As expected, this influence is more important in a chamber with thin sense wires at elevated pressures (Mori et al. 1982). Experimental data are in good agreement with computed curves.

Electrons produced by particles travelling in a direction perpendicular to a sense wire ($\theta = 0$) are collected on it at almost a single "point", where a high

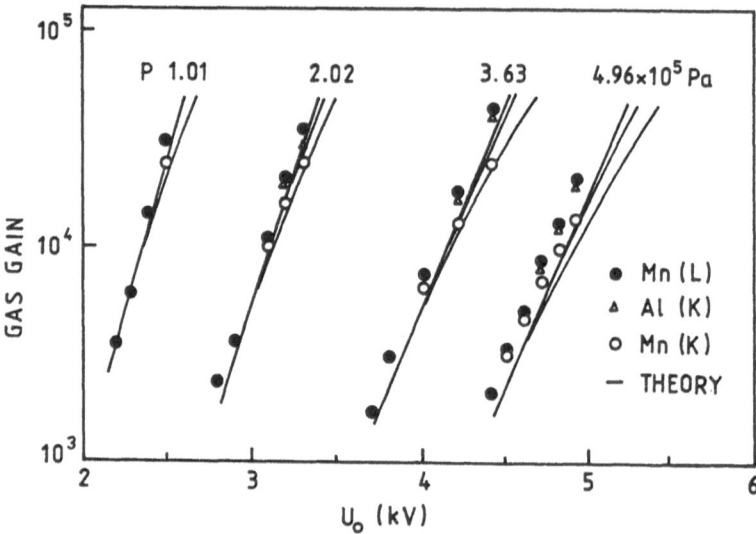

Fig. 3.30. Dependence of the gas gain on the anode voltage U_0 for various X-ray energies and gas pressures (Mori et al. 1982)

charge density forms in the process of gas multiplication; this effectively reduces the field strength in the vicinity of this point. The charge density along the wire, when the ionization density is $dN_t/dx \approx 10$ electrons per mm in argon, is

$$\varrho_+ = 1.6 \cdot 10^{-18} M P l \cos \theta . \tag{3.13}$$

The space charge changes the signal amplitude if its density $\varrho_+ \gtrsim 10^{-2} q_w$ where the charge on the wire q_w usually amounts to 10 pC/mm. In a typical chamber of thickness $l = 10$ mm filled with Ar ($P = 10^5$ Pa), such conditions are achieved when $M \sim 10^3$; but even when $M \approx 10^4$ the signal amplitude depends explicitly upon the track angle θ (Fig. 3.31). The angular dependence of the signal amplitude is noticeable only at large $M(U_0)$. In the case of low gas amplification it is nearly absent (Fig. 3.32). Thus, when it is necessary to work with a high gas amplification ($M \geq 10^3$) one must first perform precise measurements of the angular dependence of the signal amplitude, and then introduce the appropriate corrections in the course of data processing. The angle θ is usually well known in experiments.

Gas amplification in a chamber also depends on the drift length of electrons. A cluster of electrons spreads out along its drift path, because of diffusion, and the wider cluster is amplified more effectively near the wire. Breskin et al. (1977) observed that the signal amplitude increases by 10 %, if the source (^{55}Fe) situated close to the sense wires is displaced by a distance of 15 cm. The space charge diffuses very slowly, so that it affects the gas amplification and, consequently, the signal amplitude of subsequent events. An integral characteristic of this phenomenon, the dependence of the signal amplitude on the particle flux,

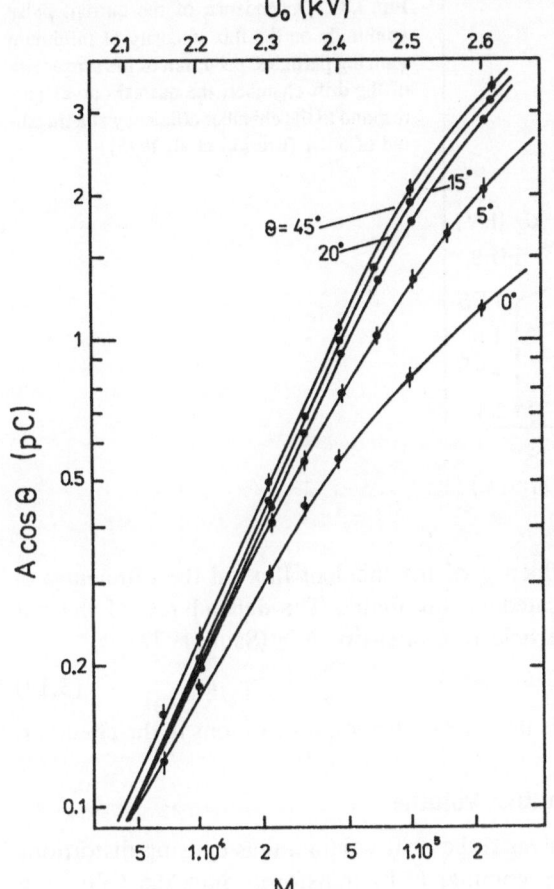

Fig. 3.31. Signal amplitude versus gas gain for particle tracks inclined at different angles (Sauli 1984)

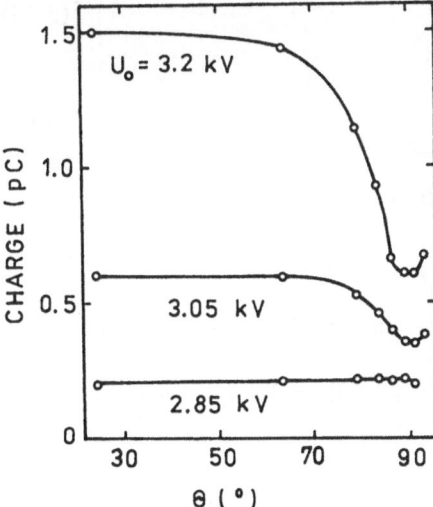

Fig. 3.32. Dependence of the charge collected in the chamber upon the angle $(90° - \theta)$ for various voltages U_0. The track length is normalized to a gas layer thickness of 18 mm (Hasemann 1982)

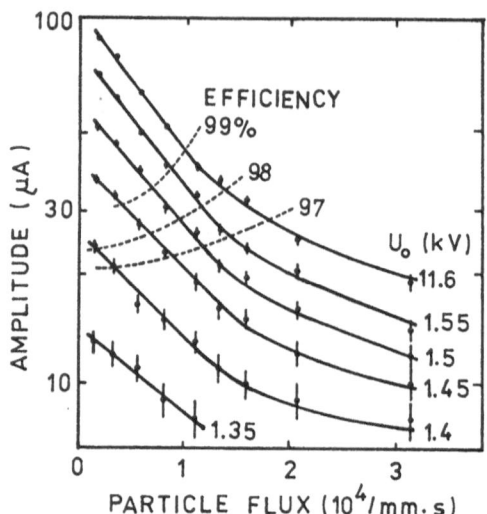

Fig. 3.33. Dependence of the current pulse amplitude on the flux intensity of minimum ionizing particles, per 1 mm of the sense wire of the drift chamber; the dashed curves correspond to the chamber efficiency at a threshold of 5 μA (Breskin et al. 1975)

is presented in Fig. 3.33; the influence of the incident flux on the efficiency of particle registration is also indicated in this figure. The dependence of the gas multiplication factor upon the particle flux intensity Φ is (Sauli 1977)

$$M = M_0 \exp(-K M_0 \Phi) , \tag{3.14}$$

where $K = N_t e \tau_+/(4\pi\varepsilon)$, and τ_+ is the total collection time of ions in the chamber.

3.4.2 Space Charge in the Chamber Volume

The positive ions move slowly towards the drift volume, thus causing distortions of the electric field. In the drift volume, (3.8) transforms into the following one-dimensional equation for the coordinate x:

$$\frac{d}{dx}\left(\varepsilon\frac{d\mathcal{E}}{dx}\right) = 0 , \tag{3.15}$$

the solution of which yields

$$\mathcal{E}_x = C_1(1 + 2C_0 x/C_1^2)^{1/2} ; \tag{3.16}$$

$$U(x) = U_d - (C_1^3/3C_0)(1 + 2C_0 x/C_1^2)^{3/2} , \tag{3.17}$$

where the constants C_0 and C_1 have the form

$$C_0 = \frac{I/L_w}{4\varepsilon r_b \mu_+} ; \quad C_1 = \frac{U_d}{L_d}\left(1 - \frac{C_0 L_d^3}{2U_d^2}\right) \tag{3.18}$$

for $2C_0 L_d/C_1^2 \ll 1$, where U_d is the voltage applied to the drift gap. L_d is the drift-gap length.

The space charge alters the field as if the voltage U_d were reduced by an amount ΔU_d:

$$\Delta U_d = -(I/L_w L_d^3)/8\varepsilon\mu_+ r_b U_d . \tag{3.19}$$

The problem of changes occurring in the characteristics of drift chambers as the incident flux intensity increases is an extremely serious issue, the change in the signal amplitude being largely related to the accumulation of charge in the drift volume, rather than in the immediate vicinity of the wire (Sect. 4.4).

3.5 Changes of Signal in Proportional Detectors and Electronics

3.5.1 Exchange of Charge Between Layers and Channels

In a multilayer drift chamber the signal amplitudes in adjacent channels are correlated. Such correlations especially in the case of long drift lengths are due mainly to the diffusion of electrons from neighboring layers. The charge collected in the j-th layer is

$$Q_j = M_j[(1 - 2a_d)q_j + a_d q_{j+1} + a_d q_{j-1}] , \qquad (3.20)$$

where q_{j-1}, q_j, q_{j+1} are the charges released in the $(j - 1)$-th, the j-th and the $(j + 1)$-th channels, respectively; M_j is the multiplication factor in the j-th channel; and a_d is a correlation parameter depending on the electron drift length in each layer. Thus, for a drift length of 75 cm, $a_d = 0.1$ (Allison et al. 1976).

The correlation due to such charge exchange is termed positive; its influence on the amplitude resolution of the device is insignificant and causes no displacement of the spectrum.

Another type of cross-talk, negative, arises as the signal passes from the chamber to the electronics channels. It is related to the existence of capacitance couplings in the channel circuits and is characterized by the correlation parameter a_c. The resultant charge in the j-th channel is

$$Q'_j = a_c Q_{j-1} + Q_j + a_c Q_{j+1} - b_c \Delta_0 . \qquad (3.21)$$

In accordance with Monte-Carlo calculations for the ISIS chamber $a_c = -0.055\pm 0.01$ (Allison et al. 1976). The minus sign indicates that the cross-talk shifts the spectra of ionization energy loss towards smaller amplitudes. A more significant influence on the resolution may be exercised by the capacitance coupling between the sense wire plane and the high voltage electrodes, which is characterized by the parameter b_c. (For the ISIS chamber $b_c = 0.20 \pm 0.04$.) It causes extension of the trailing edge of the signal and, thus, a shift of the zero level for the subsequent signal, proportional to the amplitude of the preceding signal. Capacitance coupling may become the cause of systematic errors in data processing. It can be removed by (a) reducing the capacitance between the high voltage electrodes and the sense wire plane; and by (b) introduction of a shunting capacitance between the high-voltage electrodes.

As a result, the cross-talk not only causes a broadening of the ionization distribution, but also its shift towards higher values because of the asymmetry in the distribution of energy loss. In practice, taking into account corrections due to cross-talk effects improves the ionization resolution by 5–10 %.

3.5.2 Signal Shape and Ionization Resolution

We mentioned in Sect. 3.2 that certain effects influence ionization resolution during treatment of signals in the electronics (Fig. 3.10). Some of these effects are processes of charge integration, resetting of the zero level, the superposition of signals from close tracks, the non-uniformity of amplification in individual channels, etc.

Integration of Charge. Correct integration of charge is very important. Signals arriving from the chamber not only differ greatly in amplitude, but are also characterized by large variations of shape, because tracks are inclined at different angles and electron drift length varies. Amplitude measurements are usually performed with a fixed threshold, but its adjustment must be carried out with great care. The difficulty is that, in addition to it being necessary to introduce maximum noise suppression, it is also necessary to integrate as fully as possible the charge arriving from a layer. In many experiments charge integration during a time gate is applied. The best approach corresponds to the choice of the lowest possible threshold with subsequent digitization of the signal, so that the necessary cuts, or the choice of integration boundaries, are performed at the stage of data processing (Allison 1982). Utilization of a low threshold imposes serious requirements on the electronics, which must exhibit high quality, low noise and effective suppression of pick up.

Charge integration is closely related to the recognition of close tracks. When a long gate (typically 150–200 ns) is used, good charge integration is achieved, but tracks separated from each other by a distance $x_2 \lesssim 1.5$–2.0 cm merge into a single track.

Shift of the Zero Level. A serious problem is the shift of the zero level due to the superposition of a signal on the trailing edge of the preceding signal. Usually a galvanic decoupling is applied with a capacitance between the sense wire and the amplifier. In this case the "area" of the negative signal equals the "area" of the positive pulse, which has a low amplitude and, therefore, a long duration. Thus, the next signal does not start from zero, but from the current amplitude of the preceding signal. In the case of a high signal frequency this effect results in a noticeable shift of the amplitudes. This non-linear effect is very difficult to eliminate at the stage of signal processing. Allison (1982) proposes two ways of solving this problem. The first consists in constant digitization of the background noise in between the pulses, and taking it into account in calculating the signal amplitude. The second method consists in utilizing the feedback in the electronics for automatic correction of the zero level.

In order to be sure that ionization measurements in a multilayer chamber are correct, the systematic deviations in individual channels must be reduced to a minimum. The influence of the wire positioning precision on the gas amplification has already been discussed in Sect. 3.2 (Table 3.2). In spite of the great effort taken the average amplitudes in individual channels vary by more than $\pm 5\%$. Further reduction is achieved by good calibration of the device and of the electronics, and by consistent introduction of corrections in each channel at the stage of data processing.

3.5.3 Calibration of a Detector

Before an ionization identifier of relativistic charged particles is applied in a physical experiment, it must be thoroughly calibrated in a particle beam. Calibration involves the following steps:

1) *measurement of the relativistic rise of ionization* for various particles (e, π, K, p) as function of p/mc for the gas mixture (Fig. 3.26). These curves serve as the basis for particle identification in physical experiments. A good agreement between the experimental data and the results of calculations was obtained with the EPI (Baruzzi et al. 1983) and the ISIS (Allison et al. 1984) detectors;

2) *determination of the ionization resolution* W_n. The ratio between the computed and experimental quantities reflects the quality of the device; it also is a measure of how well possible effects have been accounted for, and of the performed corrections;

3) *a check on the linearity* of the ratio between the ionization and the signal amplitude. The results of such a test in the EPI identifier are illustrated in Fig. 2.25;

4) *measurement of the gains* for each individual channel. An attempt is always made to equalize them; the remaining small differences are taken into account at the stage of data processing. A method of equalizing the gains on all the sense wires by equalization of the charges on them is described by Saxon (1988). Using this method the surface fields on the sense wires could be equalized to better than $\pm 0.5\%$;

5) *determination of the main characteristics of the device*: electron capture length, drift velocity, influence of the space charge, dependence of the amplitude on the entrance angle of the particle, cross-talk, etc. Methods of determining the most important characteristics of the device and their influence of its ionization resolution were considered in the preceding sections. It must be pointed out that without taking into account all the necessary corrections, a reliable identification of relativistic particles is not to be expected.

3.5.4 Long-Term Stability

Variations of the external pressure and temperature significantly affect the operation of large multilayer chambers (Fig. 3.34). An inverse correlation is observed between pressure and the signal amplitude: when the pressure increases by 1 %, the amplitude falls by (6.5 ± 0.5) %. As the pressure rises, the ratio $\mathcal{E}/P$ in the vicinity of the sense wire falls and the gas amplification decreases. Time variations of the signal amplitude are so significant that without appropriate control of the amplitude, and subsequent introduction of the corresponding corrections, no particle identification is possible.

A damping of amplitude variations is achieved, for example, by automatic regulation of the voltage on the sense wires in accordance with variation of the signal amplitude from a radioactive source. This method was applied successfully at the EPI apparatus, where the oscillations were reduced to ± 0.5 % (Baruzzi et al. 1983). In spite of the significant reduction of the amplitude variations, the ionization resolution exhibited during a long exposure was worse than that obtained in a short time. For example, in the course of calibration of the EPI detector a resolution $W_n = 6$ % was achieved, while the total resolution during a 10-day exposure amounted to 6.6 % (Baruzzi et al. 1983).

Variations of pressure and temperature also cause correlated changes of the drift velocity, but the electron capture coefficient is almost unaffected (Allison et al. 1984).

Great care must be taken to provide constant gas composition and purity. For example, for the ISIS chamber a closed system was constructed for preparing and cleaning the gas mixture. Using this, the composition of the gas mixture was maintained within some parts per thousand, while the volume admixture of oxygen was less than 1 ppm.

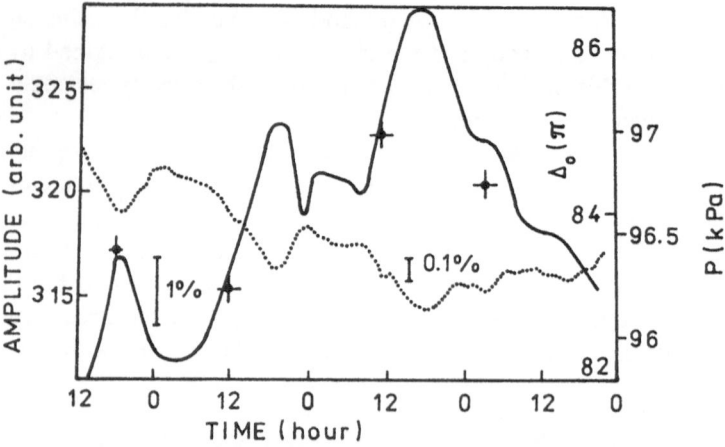

Fig. 3.34. Correlation between the time dependences of the signal amplitude from a radioactive source (solid line) in the EPI chamber and of the atmospheric pressure (dotted line); the points represent measurements of $\Delta_0(\pi)$ (Lehraus et al. 1978)

We shall now analyze some possibilities which would allow continuous control of the identifier and electronics parameters.

a) *Most X-ray sources* cannot be used for ionization measurements, because photons with an energy release close to Δ_0 (2 keV/cm) are subject to strong absorption in the gas. However, utilization of X-ray photons of higher energy (≥ 30 keV) yields an ionization pattern which is due mainly to the Compton effect, and differs significantly from the pattern produced by charged particles (Sadoulet 1981).

b) A potentially interesting method of ionizing a gas with the aid of a *laser beam* (Hilke 1986) cannot be used to calibrate ionization measurements, since long-term stabilization of laser parameters has not yet been achieved (Lehraus 1983). Methods for calibrating chambers as position detectors with a pulsed laser beam are considered in Sect. 4.4.

c) Monitoring the parameters of the electronics with the aid of *test signals* of a certain amplitude is a widely used method. The signals are applied regularly to the input of the amplifier; this allows constant control of each electronics channel and permits the introduction of the necessary corrections. As a result, a long-term stable operation is maintained at a level of 1–2 %.

d) *Photo-electric sources.* The numerous gold tips situated close to the sense wires can serve as sources of photo-electrons. The tips are illuminated by the beam of a pulsed N_2 laser, guided to each tip by an optical fiber. This technique was tested successfully by Fulda-Quenzer et al. (1985). A proposal exists to apply a calibration system in the SLD central detector, based on this utilization of photo-electric sources.

e) *Radioactive sources* established at several points of the chamber are possibilities. ^{55}Fe sources with a photon energy of 5.9 keV are widely applied. Upon their conversion in Ar, about 200 electrons are produced, which is comparable to the number of electrons (about $100 \, \text{cm}^{-1}$) produced by a minimum ionizing particle. Unfortunately, the size of the electron avalanche due to ionization by photon from ^{55}Fe is small (100 μm). Other γ- or β-sources, such as ^{90}Sr, for example, are also used (Lehraus et al. 1978).

f) *Beam or other particles* of known mass and momentum are used. In the course of data taking, tracks of beam particles are recorded; or at the stage of data processing particle tracks of known kinematic parameters are chosen for calibration. Such a method based on the measurement of pion energy loss was successfully applied, for instance, in the ISIS (Allison, Wright 1984) and TPC (Lynch, Hardley 1982) detectors.

Generally, the complexity of ionization measurements requires the simultaneous application of several methods for controlling the operation stability of an identifier.

3.6 Processing of Data in a Multilayer Ionization Detector

The processing of ionization data and the subsequent identification of charged particles is performed by applying a complex set of mathematical programs. Without going into details, we shall present the scheme of the processing procedure and analyze its principal features. The processing of ionization data involves: (1) geometrical reconstruction of tracks in the detector; (2) selection of tracks using information from external coordinate detectors; (3) selection of points on a track for the calculation of energy loss; (4) corrections which take into account systematic effects; (5) statistical analysis of the data; (6) determination of particle masses; (7) determination of the ionization resolution of the device in experimental conditions; (8) estimation of the reliability of particle identification.

The geometrical reconstruction of tracks in an identifier, and determination of their relation to external detectors, is performed by conventional methods of track reconstruction in spectrometers containing multiwire coordinate detectors.

3.6.1 Selection of Points for Ionization Measurements

Of great importance is the selection of points, on a track, suitable for the determination of ionization energy loss. It is necessary to exclude from further processing all the regions where tracks overlap, in which the measured ionization is produced by two particles. In Fig. 3.35 an example is presented of reconstructed tracks in the ISIS chamber. The circles indicate those points on a track which can be used for ionization measurements. The other points along the track are discarded for three reasons: (1) overlapping of tracks (their amount depends upon the double-track resolution – $x_2 \approx 1.5$ cm in ISIS); the track intersection points are indicated by vertical bars; (2) crossing with Compton electrons; (3) noise (these points are indicated by horizontal bars). This filtering procedure means that 99 % of the selected points corresponding to a single particle may be used to determine the ionization.

In the ISIS detector, high-energy particle tracks have, on the average, about 260 points (of the 320 possible ones), and about 230 points can be utilized for ionization measurements (Allison 1982). Tracks with less than 160 points are mainly, due to low-energy particles ($p < 3.5$ GeV/c) In the TPC detector there remained, after filtration, an average of about 150 points per track (of the 183 possible points). A track was subject to processing if the number of points exceeded 120 (Lynch, Hadley 1982).

The number of points on a track remaining after filtration depends on the spatial resolution of the device, on the two-track resolution and on the charged particle multiplicity in an event. Clearly, in devices with a bad spatial resolution, the fraction of discarded points is large. An example is the EPI identifier (with 6×6 cm^2 cells and 128 layers), where tracks with more than 50 "clean" points were used for measurements. In the conditions of this experiment only 30 % of

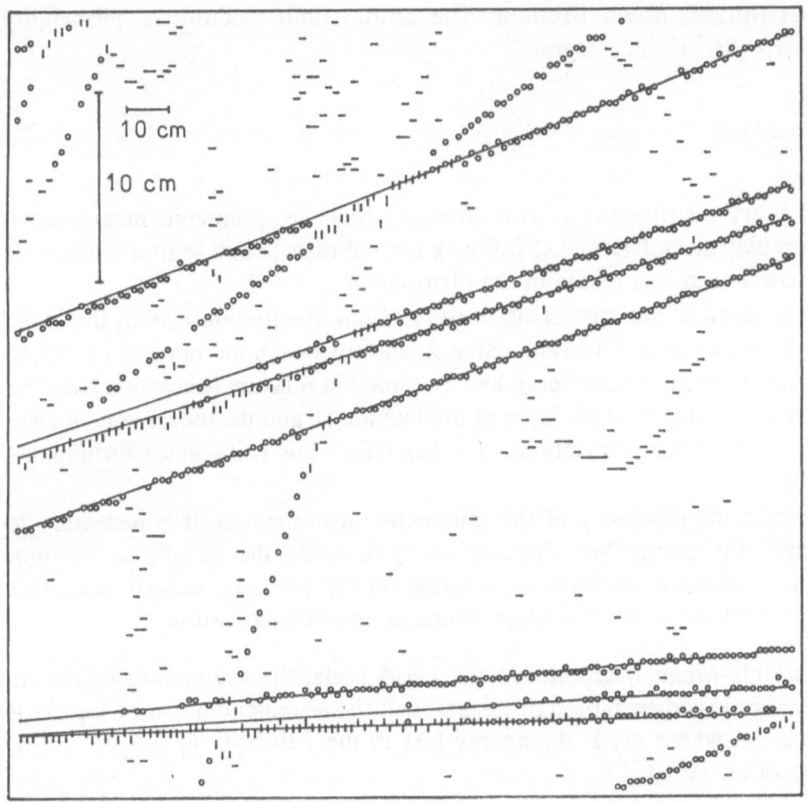

Fig. 3.35. Tracks of secondary particles reconstructed from measurements in the ISIS chamber (Allison et al. 1984). The solid lines correspond to tracks reconstructed by means of data from other coordinate detectors

the tracks complied with this requirement (Baruzzi et al. 1983). We recall that 50 points on a track are not sufficient to identify relativistic charged hadrons (Fig. 3.14).

The next step in the processing procedure is the introduction of corrections which take into account systematic effects in the chamber and in the electronics. These effects have been mentioned previously in this chapter.

3.6.2 Statistical Analysis of Ionization Data from a Multilayer Detector

There exist several effective methods of statistical analysis which allow the probable ionization to be obtained from measurements in a detector containing n independent identical layers (Allison 1981; Merson et al. 1983). To achieve an understanding of the principles and efficiency of these methods, we shall analyze them starting with the most simple cases.

The Arithmetic-Mean Method. The central limit theorem of probability theory asserts that, if the integral

$$\int_{-\infty}^{\infty} \lambda^2 \varphi(\lambda)\, d\lambda \tag{3.22}$$

for an arbitrary distribution $\varphi(\lambda)$ converges, then the arithmetic means for n random samples taken from $\varphi(\lambda)$ follow a normal distribution with a width $n^{1/2}$ times narrower than that of the initial distribution.

This statement is not applicable to the Landau distribution (1.82), the "tail" of which decreases as λ^{-2} towards large λ, and for which the integral (3.22) diverges. In this case the mean energy loss measured in n layers is nearly equivalent to a value measured in a single layer of total width nl, and the measurement error falls very slowly [approximately as $(1 - \ln n/10)^{-1}$] as n increases (Alikhanov et al. 1956).

To enhance the efficiency of the arithmetic-mean method, it is necessary to "symmetrize" the energy loss distribution by lowering the weight of its high-energy part, which carries little information on the probable energy loss. This idea may be realized in several ways which are considered below.

The Geometric-Mean Method. In the $\log \Delta$ scale, the asymmetry of the ionization energy loss distribution decreases, and the arithmetic mean of $\log \Delta_j$ is close to $\log \Delta_0$, where Δ_j is the energy loss in the j-th layer ($j = 1, 2, \ldots, n$) (Alikhanov et al. 1956); i.e.

$$\left(\prod_{j=1}^{n} \Delta_j \right)^{1/n} \approx \Delta_0 . \tag{3.23}$$

The efficiency of this method is also low (Table 3.6).

Table 3.6. Efficiency of statistical methods of ionization data processing in a multilayer detector filled with argon (Aderholz et al. 1974; Bashindzhangyan 1980; Merson et al. 1983)

Method for data processing	Ratio of ionization resolutions in n-layer and single-layer detectors, W_n/W		
Number of layers, n	10	24	62
layer thickness lP (10^5 cm · Pa)	1.5	1.5	4
Arithmetic mean for n measurements	0.74	0.53	
Geometric mean for n measurements	0.55		
The smallest of n measurements	0.59		
Truncated mean	0.46($\alpha = 0.7$)		0.168($\alpha = 0.4$)
Rank statistics method	0.46($j_0 = 4$)	0.26($j_0 = 8$)	0.179($j_0 = 20$)
Maximum likelihood method	0.40	0.24	0.162
Limit ratio W_n/W (equal to $n^{-1/2}$ for a normal distribution)	0.316	0.204	0.127

Truncated Mean Method. If one symmetrizes the fluctuation curve by cutting off its high-energy part, then the integral (3.22) of the resultant distribution converges and the central limit theorem still holds. Therefore the arithmetic-mean of random samples taken from this distribution follows the normal distribution law.

Various approaches have been proposed to allow the use of the truncated mean method. The most effective one involves taking the mean of the n_α lowest Δ_j values from n measurements ($n_\alpha = \alpha n$, $\alpha < 1$) (for example, Alikhanov et al. 1956; Dimcovski et al. 1971; Jeanne et al. 1973; Allison et al. 1979), so that the quantity

$$\Delta_\alpha = \sum_{j=1}^{n_\alpha} \Delta_j / n_\alpha ,$$ (3.24)

which is proportional to Δ_0, serves as a measure of the ionizing power of a particle (Walenta et al. 1979). In this case the relative half-width is better by a factor of $n_\alpha^{1/2}$ compared to the initial distribution:

$$W_n = W n_\alpha^{1/2} ,$$ (3.25)

while the relative standard deviation is $\sigma_n = W_n/2.36$. Within the interval $0.3 \leq \alpha \leq 0.7$ the precision of this method depends weakly upon α (Fig. 3.36). The optimum choice of α depends on the experimental conditions.

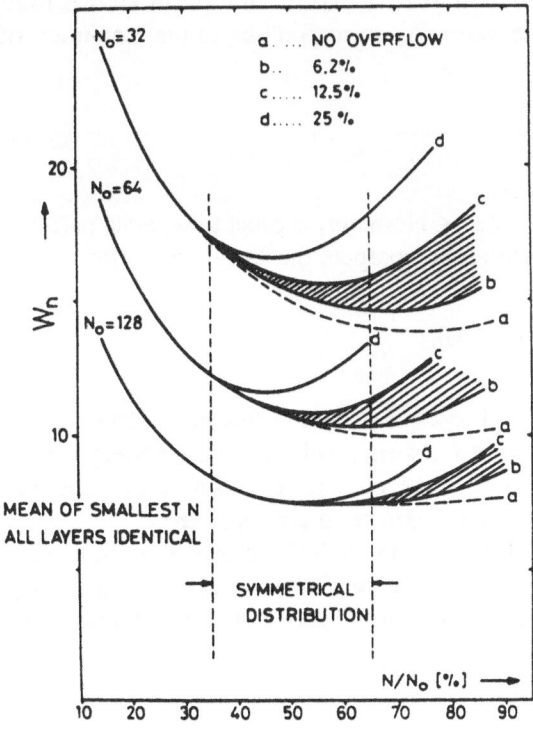

Fig. 3.36. Computed relative half-width W_n of the energy loss distribution by the truncated mean method (Jeanne et al. 1973): the shaded regions correspond to actual experimental conditions; N_0 is the number of layers in the detector

151

The cut-off method is notable for its simplicity and high efficiency. The limiting case of $n_\alpha = 1$, corresponding to the minimum Δ_j, is efficient only when $n < 15$ and was utilized only in early studies for $n = 2$–4.

The Method of Rank (Ordered) Statistics. Let us write down the results of energy loss measurements in an n-layer detector for k particles of the same ionizing power, in the form of a matrix Δ_{ij} ($i = 1, 2, \ldots, k; j = 1, 2, \ldots, n$); in each i-th row the Δ_{ij} are arranged according to rank, i.e. in rising order: $\Delta_{ij} \leq \Delta_{i,j+l}$. Then, at a certain fixed $j = j_0$, which corresponds in each row to the Δ_{ij} that are the closest to Δ_0, the width of the distribution of Δ_{ij} in the j_0-th column will achieve its minimum value (Bashindzhagyan 1980). Thus, for $n = 10$ ($lP = 1.5 \cdot 10^5$ cm $\cdot$ Pa in Ar) the optimum value $j_0 \simeq n/3 \simeq 3$. Here the relative fwhm of the fluctuation curve decreases from 100% down to 46% (Table 3.6). The width of the distribution of rank j_0 varies weakly within the interval $j_0 = n/4 - n/2$, which indicates its insensitivity to the form of the initial distribution. When $n > 20$, it is reasonable to combine the layers into identical groups and to apply within each group the method of rank statistics and then to average the obtained results.

The Maximum Likelihood Method[1]. This is the most general and the most efficient method (i.e. the one with an estimate exhibiting the smallest dispersion) of estimating the parameters of the distribution of a random quantity. In this particular case it allows estimation of the most probable energy loss Δ_0, corresponding to a normalized distribution $f(x, \Delta, \Delta_0)$ of an a priori known form containing Δ_0 as a parameter. The value Δ_0 corresponding to the maximum of the likelihood function

$$\mathcal{L}_0 = \prod_{j=1}^{n} f(x, \Delta_j, \Delta_0) , \tag{3.26}$$

is chosen as the most probable energy loss. However, in order to separate particles of masses m_1 and m_2 it is more simple to compute the likelihood ratio:

$$\mathcal{L}_{12} = \prod_{j=1}^{n} f_1(x, \Delta_j, \Delta_0) / f_2(x, \Delta_j, \Delta_0) , \tag{3.27}$$

where f_1 and f_2 are the respective known distribution functions. The case of $\mathcal{L}_{12} \gg 1$ corresponds to the particle of mass m_1, while $\mathcal{L}_{12} \ll 1$ corresponds to the particle of mass m_2. When $\mathcal{L}_{12} \approx 1$, the uncertainty arises, which depends on how well beam particles of the same momentum are separated.

The advantages of the maximum likelihood method include the utilization of all the available information and a minimum dispersion of the estimate; disadvantages are the sensitivity to the form of the energy loss distribution, the necessity

---

[1] Sometimes called the "universal" method.

152

of absolute calibration of the energy scale and the large amount of computer time required.

The One-Parameter Maximum Likelihood Method. It is free of many of these disadvantages (Allison 1981). It is based on the close similarity between fluctuation curves when $lP > 10^5$ cm · Pa for various particle energies (Fig. 3.37a). If their form $F_0(\Delta/\Delta_0)$ is known, then Δ_0 is found from the maximum of the likelihood function:

$$\mathcal{L}_0 = \prod_{j=1}^{n} F_0(\Delta_j/\Delta_0) . \qquad (3.28)$$

Since F_0 is dependent on the dimensionless ratio Δ_j/Δ_0, such calculations do not require an absolute energy calibration. Also the required computer time is not too large [2 ms per track for $n = 80$ and $lP = 1.5 \cdot 10^5$ cm · Pa of argon (Allison 1981)]. The method is not sensitive to the exact choice of the function F_0, and the track inclination is taken into account simply by introduction of the factor $\cos^{-1} \theta$. In addition to the actual value of Δ_0, its uncertainty is determined. The method

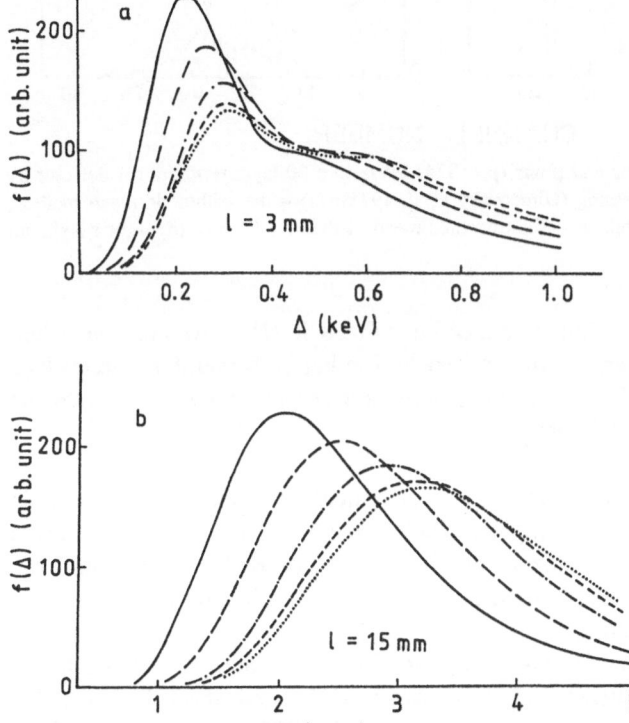

Fig. 3.37. Computed distributions for ionization energy loss spectra in thin layers of argon, (a) – 0.3 cm and (b) – 1.5 cm, $P = 10^5$ Pa (Allison 1981)

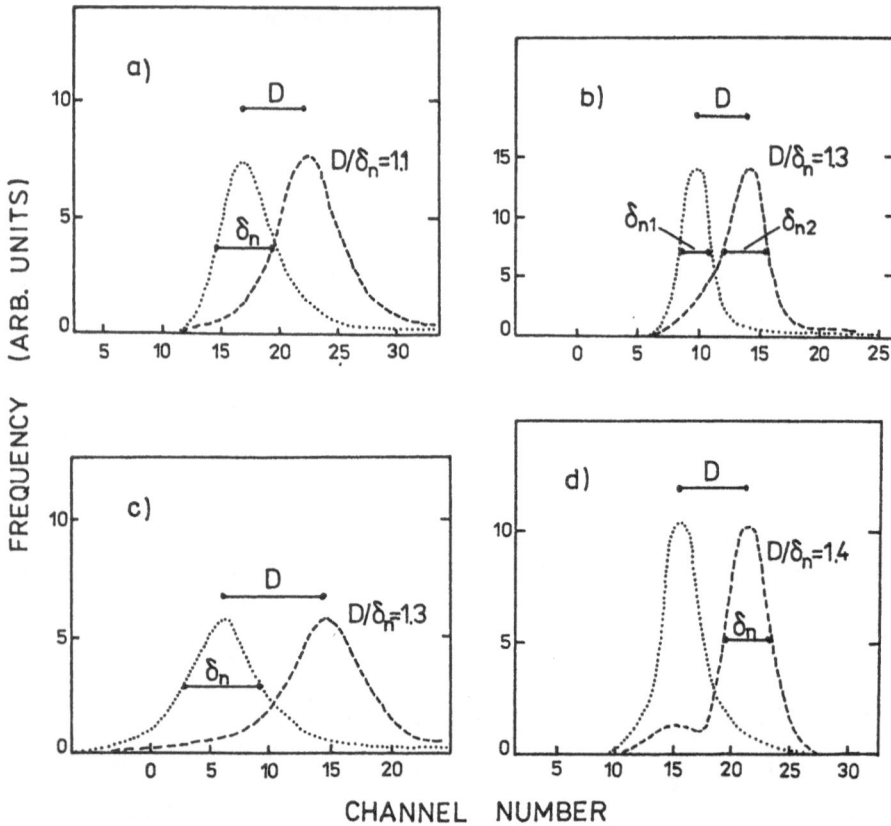

Fig. 3.38. Separation of electrons and pions ($p = 374\,\mathrm{MeV}/c$ in a 30-layer proportional detector by different methods of data processing (Dimcovsky et al. 1971): (a) – the arithmetic mean method; (b) – the geometric mean method; (c) – the truncated mean method ($\alpha = 0.33$); (d) – the maximum likelihood method

has a large dynamic range, but it cannot be utilized if $lP < 10^5\,\mathrm{cm} \cdot \mathrm{Pa}$, when structure effects of the energy transfer distribution begin to manifest themselves (Fig. 3.37b), and the condition that the fluctuation curves remain similar as the particle energy varies is no longer valid.

The efficiencies of various methods of statistical analysis are illustrated in Table 3.6 and in Fig. 3.38. The highest efficiency is exhibited by the maximum likelihood method, followed closely by the method of rank statistics and the truncated mean method. The geometric mean and the arithmetic mean methods are considerably less efficient.

An important condition for proper application of the above methods is the absence of correlations between adjacent layers of the ionization detector; this has been verified experimentally (Dimcovski et al. 1971; Jeanne et al., 1973; Allison et al. 1979). At present the truncated mean method and the maximum likelihood method are most often applied in data processing.

3.6.3 Determining the Mass of a Particle

Determination of the mass of a particle m_i is based on the comparison of the measured value of Δ_0 and the dependence of the ratio $\Delta_0/\Delta_{0m}(p)$ on the particle momentum p (Fig. 3.26), where Δ_{0m} is the probable ionization for the minimum ionizing particle. This dependence can be approximated by the polynomial (Baruzzi et al. 1983)

$$(\Delta_0/\Delta_{0m})_{it} = a_m + b_m \ln(p/m_i c) + c_m [\ln(p/m_i c)]^2 \,, \tag{3.29}$$

where m_i represents the mass of the electron, the pion, the kaon or the proton. The index t indicates that the corresponding value is obtained in a test (calibration) experiment involving particles of the known sort i. Since the distribution of the arithmetic means of Δ_α obtained by the truncated mean method is gaussian, it is possible to compute $\chi^2_{m_i}$ for each possible mass m_i

$$\chi^2_{m_i} = \frac{[(\Delta_\alpha/\Delta_{\alpha m}) - (\Delta_0/\Delta_{0m})_{it}]^2}{\left[\sigma_p \frac{d}{dp}(\Delta_0/\Delta_{0m})_{it}\right]^2 + [\sigma_{ni,\exp}(\Delta_\alpha/\Delta_{\alpha m})]^2} \tag{3.30}$$

and to calculate the probability $W_i(\chi^2_{m_i})$ that the measured particle is of mass m_i. In (3.30) $(\Delta_0/\Delta_{0m})_{it}$ is the relative most probable ionization loss of a particle of mass m_i in the test (t) measurement; σ_p is the r.m.s. deviation in the track momentum measurement; $\sigma_{ni,\exp}$ is the relative ionization resolution for particles of known momentum and mass. The denominator of (3.30) contains the sum of the squared uncertainties, which arise in momentum measurements and in the measurement of Δ_α, in the actual experimental conditions.

There also exist other methods for determining the mass of the particle which is being identified. Allison et al. (1984) recommend that the quantities $\ln(\Delta_0/\Delta_{0m})$, which are distributed normally, should be substituted into (3.30) instead of Δ_0/Δ_{0m}. The EPI group (Baruzzi et al. 1983) made use of the weighting method for particle identification.

Identification of secondary particles in an experiment is rendered more complicated by very different numbers N_i of particles of each sort (where $i = e, \pi, K, p$). Therefore it is necessary to know the ratio of the average number $\overline{N}_i$ of particles of a certain kind to the average number $\overline{N}_{ch}$ of all secondary charged particles, as a function of their momentum: $f_i(p) = \overline{N}_i(p)/\overline{N}_{ch}(p)$. Clearly, the behavior of function $f_i(p)$ varies from experiment to experiment. The probability that the particle being identified has mass m_i is the following:

$$W_{f_i}(p) = f_i(p)W_i \,. \tag{3.31}$$

3.6.4 Reliability of Particle Identification

The reliability of identification depends on the ionization resolution of the device, the rate at which particles of mass m_i are produced in events, and on other factors. It is convenient to calculate the *identification efficiency* (i.e. the relative number

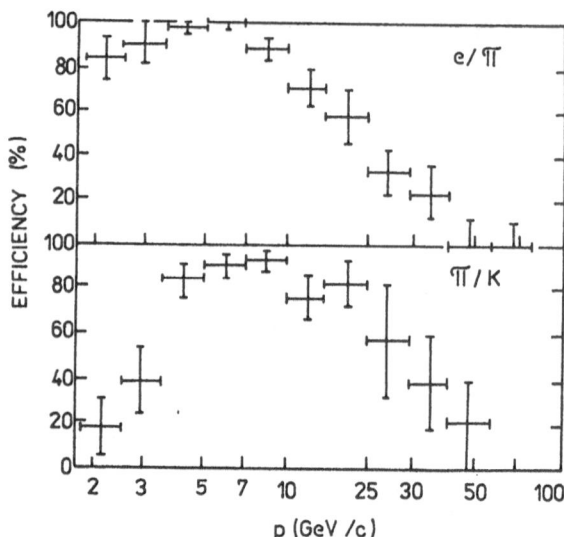

Fig. 3.39. Efficiency of e/π and π/K separation as a function of the momentum at an identification confidence level of 99 % (Allison et al. 1984)

of correctly identified particles) for a given confidence level. For example, in the ISIS chamber, for a confidence level of 0.99 (i. e. the competing hypothesis had a probability ≤ 1 %) 77 % of the electrons and 82 % of the pions were correctly recognized; only 38 % of the protons, however, were separated from the kaons (Allison et al. 1984). Identification efficiency depends strongly on the particle momentum (Fig. 3.39).

3.7 Calculated and Experimental Ionization Resolutions

Achieving the calculated ionization resolution in a multilayer identifier represents a difficult task. One must introduce corrections for all the effects leading to measurement uncertainties in large multilayer chambers, as noted in the preceding sections.

3.7.1 Resolution in Test Conditions

The results of tests depend strongly upon the conditions in which they are performed. The experimental situation is reflected, in part, in Table 3.7, in which the calculated separation coefficient S_n and the separation coefficient S_{nt}, obtained from test measurements, are presented. The ratio S_n/S_{nt} varies from 1 to 1.56 due to differing conditions in test measurements: e. g., differences in momentum of the beam particles, in the beam intensity etc.

The ionization resolution obtained from test measurements may be quite close to the computed value in the case of very special conditions. This happened, for instance, in the ISIS–1 (Table 3.7), JADE (Wagner 1982) and ARGUS (Hasemann 1982) chambers (Table 3.8), in studies, for example, of cosmic muons

Table 3.7. Comparison of computed and experimental particle separation coefficients S_n for various identifiers (partly Gabathuler 1981)

Identifier	Separated particles	p [GeV/c]	S_n Computed (S_n)	S_n Test (S_{nt})	S_n/S_{nt}
ISIS–1	e/π	0.5	8.1	8.0	1.01
CRISIS	e/p	100	10.7	7.9	1.35
CRISIS	π/K	100	5.7	4.9	1.16
EPI	π/p	50	6.1	5.1	1.2
TPC (test)	e/π	0.8	18.0	11.8	1.36
JADE	e/π	0.45	7.8	5.0	1.56
HRS (test)	e/π	4	4.1	3.2	1.32
CLEO	e/π	0.45	11.1	8.25	1.36
SLAC (test)[1]	e/π	4	3.7	2.9	1.27
CERN (test)[2]	e/π	15	3.2	3.0	1.07
	π/p	15	6.6	6.0	1.1

[1] Va'vra et al. (1982)
[2] Lehraus et al. (1982a)

Table 3.8. Computed W_n, test W_{nt}, and experimental $W_{n,\exp}$ resolutions for several identifiers

Identifier	W_n [%]	W_{nt} [%]	$W_{n,\exp}$ [%]
EPI[a]	6	6.6	7.6–11
TPC	5.9	8.9	9.4
ISIS	6.5	7.4	8.3
ISIS–1	14	18.6	
CRISIS	7.8	7.6 (p); 7.9 (π)	10
JADE	10.6	10.8	
		13.4 B.e.[b]	22
CLEO	11.8	13.6	15.8
UA–1	12	14.2	
HRS (test)	17.4	19.5	
AFS	20	27	
ARGUS	10	10.4 (propane)	
		16.4 (Ar+8 % CH$_4$)	
MARK II	16.3	17.0 B.e.	
SLD	14.6	20.6	
DELPHI	13.0	14.4	
OPAL	8.2	8.0	
TOPAZ	10.8	10.8	

[a] References can be found in the preceding text.
[b] B.e. stands for Bhabha electrons.

traversing the entire sensitive volume of the detector; this permitted comparison of ionization measurements in the two halves of tracks passing close to the chamber axis, which was demonstrated in the TPC (Lynch, Hadley 1982). In the ISIS (Allison et al. 1984), EPI (Baruzzi et al. 1983), CRISIS (Toothacker et al. 1988), OPAL (Breuker et al. 1987) and TOPAZ (Shirahashi et al. 1988) chambers, a resolution close to the computed value was achieved in "natural" conditions with beam particles, but at a very low beam intensity. For instance, in the EPI chamber the calculated resolution $W_n = 6\%$ was achieved in such a test, but when the test measurements were prolonged up to three weeks, the resolution deteriorated to 6.6%. In Table 3.8 the calculated ionization resolutions W_n are presented, together with the resolution values W_{nt} obtained in test measurements in the beam.

The quality of particle separation in a multilayer identifier in test conditions is illustrated in Fig. 3.40.

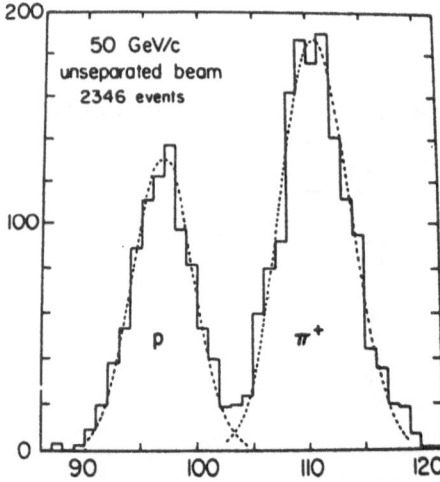

Fig. 3.40. Protons and pions at 50 GeV [corrected data, measured in counter No. 2 of EPI (Lehraus et al. 1978)]

3.7.2 Resolution in a Physical Experiment

When secondary particles are being identified, the ionization resolution in most cases differs strongly from the computed value (Table 3.8). The main reason for this is the large reduction of the number of points on a track upon filtration. The resolution (%), after filtration of secondary particle tracks, becomes (Baruzzi et al. 1983):

$$W_{n,\,\exp} = W_{nt}(n_{\exp}/n_t)^{-0.45} + W_c, \tag{3.32}$$

where $n_{\exp}$ and n_t are the numbers of points along a track used for ionization measurements in experimental and test conditions, respectively; and W_c takes into account non-statistical effects related to the background. Estimations for the EPI detector (Baruzzi et al. 1983) give $W_c = 2.6(n_t - n_{\exp})/n_{\exp}\%$.

The number of points remaining on the track depends strongly on the experimental conditions, particularly on the multiplicity of the secondary charged particles. Hence there are large variations in $W_{n,\exp}$ (Table 3.8). For example, the large discrepancy between $W_{n,\exp}$ and W_{nt} in the JADE chamber is because the resolution $W_{n,\exp} = 22\%$ was obtained in an experimental study of hadron jets, when the charged particle multiplicity was high and, consequently, the ratio $n_{\exp}/n_t$ was much lower then one.

In conclusion we shall present some recommendations, worked out in practice, concerning the utilization of multilayer proportional detectors for the identification of relativistic charged particles.

1) There is an inevitable decrease of the number of points on a secondary particle track in experimental conditions, which leads to a deterioration of the ionization resolution by 30–50 %, on the average. This must be foreseen and taken into account. The most simple way to do this is to enhance the number of layers of the detector by 30–50 %.

2) An important role is played by the double-track resolution x_2; it must be improved as much as possible. The best value of x_2 is achieved in drift chambers operated at elevated pressure with short dead time electronics.

3) A small difference between the computed and experimental resolutions is achieved in chambers in which all points are measured by a single set of wires (the ISIS and CRISIS chambers); in these chambers there exist less systematic effects. Unfortunately, it is almost impossible to fulfill this requirement, for instance, in chambers operating in collider beams.

4) It is desirable to work at a low gas amplification because of the influence of space charge on the signal amplitude. Regrettably, this requirement is contrary to achieving a good spatial resolution.

5) It is reasonable to optimize the composition of the gas mixture and its pressure in accordance with the sort of particles being identified and with their momentum range.

More useful recommendations may be found, for example, in Allison (1981), Allison et al. (1984), Lehraus (1983), Lehraus et al. (1982a), Breuker et al. (1987) and Toothacker et al. (1988).

3.7.3 Application of Proportional Detectors in Physical Experiments

Ionization measurements in multilayer proportional detectors with the aim of particle identification have been performed repeatedly either in fixed-target experiments, or in colliding beams. Table 3.9 presents examples of the application of multilayer ionization identifiers in experiments. In spite of the large differences between the actual conditions of various experiments, a common feature of the application of multilayer chambers is a good three-dimensional reconstruction of events and, in many cases, reliable identification of secondary particles.

Table 3.9. Examples of application of multilayer proportional detectors for particle identification in physical experiments

Chamber	Detector	Experiment	Separated particles
EPI	External particle identifier behind the BEBC bubble chamber (Baruzzi et al. 1983)	1. $K^- p \to h + X$ (110 GeV) 2. $K^+ p \to \pi^+ + X$ (70 GeV) 3. $K^+ p \to K^+ + X$ (70 GeV)	$\pi/K/p$ (20–100 GeV/c)
ISIS	Identifier in the European hybrid spectrometer (CERN) (Allison 1982; Allison et al. 1984)	1. Search for charmed particles in hadron beams (CERN, experiment NA-16) (observation of decays such as $\overline{D}^0 \to K^+\pi^+\pi^-\pi^-$)	$\pi/K/p$
JADE	Central detector of JADE at the e^+e^-–collider PETRA at DESY (Wagner 1982)	1. $e^+e^- \to$ hadrons 2. Search for free quarks	$\pi/K/p$ in non-relativistic region; $\pi/(K+p)$ in relativistic region Fractional charge / total charge
CLEO	Central detector of CLEO at the CESR e^+e^-–collider (Kubota et al. 1983)	3. $e^+e^- \to e^+e^-\eta'$ 1. $e^+e^- \to$ hadrons 2. Pion yield in the continuum $\Upsilon(3S)$ and $\Upsilon(1S)$ particles	e/π K/p in non-relativistic region e/π
ARGUS	Central drift chamber of ARGUS used for $(-dE/dx)$ measurements and with TOF for particle identification (ARGUS Collaboration 1989)	$e^+e^- \to$ hadrons. Process used, for example, for identification of $D^0 \to \overline{K}{}^0_s\phi$ decay products	$e/\pi/K/p$ in non-relativistic region
TPC	Central detector of PEP-4 at the e^+e^- PEP at SLAC collider (Lynch, Hadley 1982; Aihara et al. 1983)	1. $e^+e^- \to$ hadrons ($\sqrt{s} =$ 29 GeV)	$e/\pi/K/p$
CRISIS	Identifier in FNAL experiment E 597 (Toothacker et al. 1988)	1. $p + A \to h^-$ 2. $\pi^\pm + A \to h^-$ 3. $p\bar{p}$ annihilation	$\pi/K/p$

3.8 Cluster Counting in a Chamber with Longitudinal Electron Drift

Secondary electrons produced by a single primary electron on a particle track are well localized in a cluster. Particle identification based on the measurement of primary ionization, e. g. on the number of clusters, is more efficient than total ionization measurement characterized by large fluctuations. The mass resolution computed by Allison and Cobb (1980), corresponding to measurement of the pri-

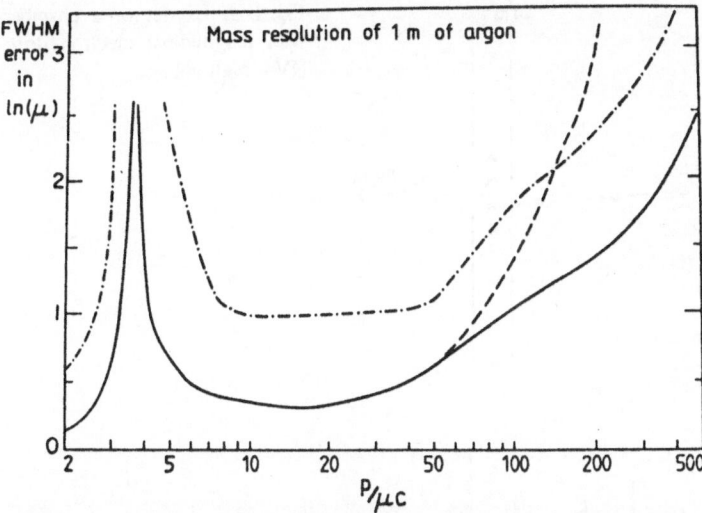

Fig. 3.41. Dependence of computed separation of particles of masses m_1 and m_2 upon p/mc in 1 m of argon at $P = 10^5$ Pa (Allison, Cobb 1980): - - - - - primary ionization; ········· total ionization; ——— primary ionization involving measurement of the cluster amplitude

mary and total ionizations in "ideal" conditions, i.e. without taking into account experimental errors and systematic effects, is presented in Fig. 3.41. When $\gamma = 4$–100, measurement of the primary ionization provides better particle separation than measurement of the total ionization. The most reliable results are obtained when both methods are applied simultaneously, i.e. when both the number and charge of clusters are measured.

Until recently the primary ionization was measured only in condensation and streamer chambers, and also using low-efficiency gas-discharge detectors (Chaps. 5, 6). Note that because of difficulties in recording the primary ionization in a streamer chamber, and the use in these chambers of gases with low ionization densities (He, Ne), the length of a streamer chamber required for identification of relativistic particles must be at least 4–5 m.

3.8.1 Cluster Counting Method

A new method for determining the primary ionization by counting clusters in a chamber with longitudinal electron drift was proposed by Walenta (1979). Such a chamber consists of two main parts, separated by a grid (Fig. 3.42). In the first part, the drift volume, a weak electric field (about 100 V/cm) is applied, while in the second part, the proportional chamber, the field is higher. Particles pass through the chamber in a direction perpendicular to the sense wire plane. The electrons drift slowly along the particle trajectory in the weak field. They then pass through the grid into the volume of higher field where, in the vicinity of the sense wires, gas amplification takes place.

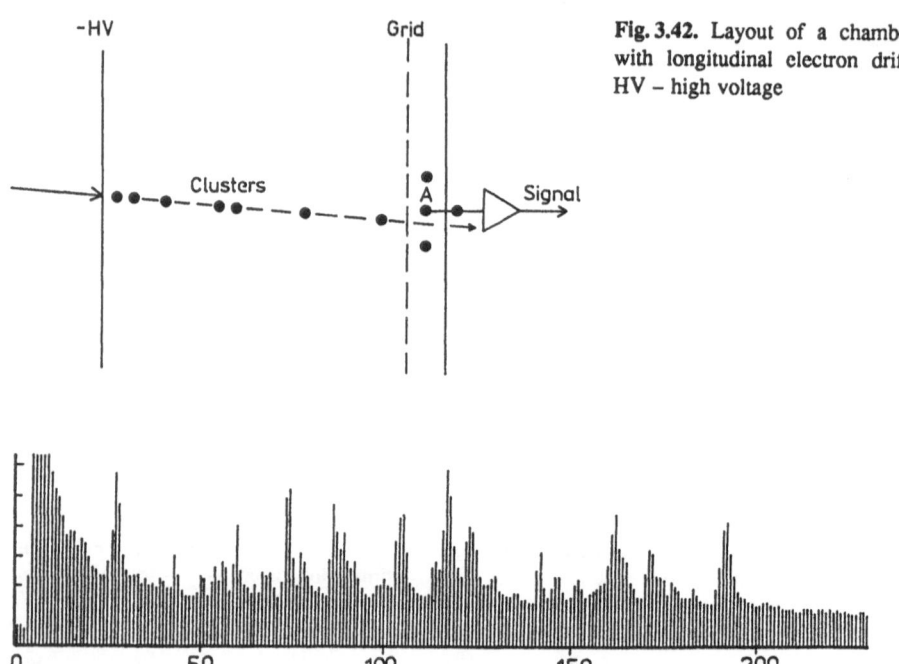

Fig. 3.42. Layout of a chamber with longitudinal electron drift: HV – high voltage

Fig. 3.43. Cluster spectrum in a chamber with longitudinal drift of electrons (Lehraus 1983)

In conventional drift, or proportional, chambers all the electrons from the particle track arrive at a sense wire at almost the same time; upon gas amplification the entire charge is integrated, resulting in the appearance of a large single signal. On the contrary, in a chamber for cluster counting, they drift slowly towards the sense wires and upon gas amplification a series of signals appears in the sense wire, each of which corresponds to a single cluster (Fig. 3.43). Thus, when the electron drift velocity is about $1\,\text{cm}/\mu\text{s}$ and $dN_1/dx \simeq 25\text{--}30\,\text{cm}^{-1}$, the average time interval between pulses from clusters is about 30–40 ns. Upon amplification the signals can be counted using a fast counter. Thus, this method permits the measurement of the number of clusters along the track, i.e. it determines the primary ionization.

3.8.2 Measurement Conditions

For cluster counting the gas mixture must provide for : 1) a small drift velocity in a field of sufficiently high field strength; 2) a small longitudinal diffusion coefficient; 3) minimum electron capture; 4) a large relativistic rise of primary ionization; 5) an optimum number of clusters per unit track length.

To count clusters in the chamber it is necessary to utilize *"slow gases"* (Sect. 3.3). These include carbon dioxide, methylal, isobutane and others. The drift velocity in these gases is several millimeters per μs for a field strength of

300–500 V/cm (Fig. 2.5), and the characteristic energy ε_c is close to the thermal limit kT. For chambers with longitudinal electron drift, Piuz (1983) recommends an 80% Ar+20% CO_2 mixture, in which in a drift field of about 200 V/cm, the drift velocity is about 8 mm/μs, and the spatial resolution (r.m.s.) $\sigma_x \leq 200\,\mu$m. Mixtures of CO_2+5% isobutane and CO_2+10% Ar are also recommended (Becker et al. 1983). Kruglov and Nikitin (1988) counted clusters in isobutane with a several per cent admixture of acetone.

It must be noted that the requirements of a gas mixture used for cluster counting are quite contradictory. The mixture must not only allow a slow electron drift and a small diffusion, but most importantly a *large relativistic rise of primary ionization* R_{N1} which cold gases usually do not exhibit. Therefore Walenta (1979), Rehak and Walenta (1980) and Budagov et al. (1984) have utilized a mixture consisting of a gas with a large R_{N1} (argon) and of a cool gas (methylal vapor and isobutane). From the point of view of a large R_{N1}, the most useful gas is neon $[R_{N1}(\mathrm{Ne}) = 1.58]$ which was used for cluster counting by Polyakov and Rykalin (1987). In neon, however, electrons undergo large diffusion, and the cluster density is low. Clearly, there still remains much work to be done before a "perfect" gas mixture for cluster counting is found.

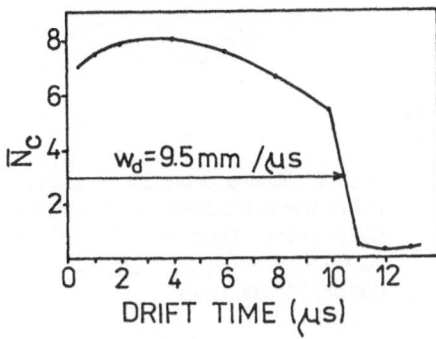

Fig. 3.44. Mean number of clusters recorded in 1 cm of a muon track versus the electron drift time for a drift length up to 10 cm (Budagov et al. 1984b)

The purity of the gas mixture plays an important role in cluster counting, since the loss of a single electron signifies the loss of an entire cluster. The life time of an electron has to be significantly larger than its drift time. For this reason admixtures of O_2 and H_2O must not exceed ppm concentrations. The situation in a chamber with longitudinal electron drift is illustrated in Fig. 3.44 (Budagov et al. 1984b). The measured mean cluster number $\overline{N}_c$ increases as the drift length increases up to $L_d \approx 4$ cm. The reason for this is electron diffusion, which results in the enhancement of the cluster sizes and to the "decay" of some of them into subgroups, even to single electrons. When the drift length is 4 to 10 cm, electron capture by admixtures exhibiting electron affinity becomes the dominant process, which leads to a decrease in the mean number of clusters.

The *gas amplification* in the chamber must be significant to provide for effective registration of single electrons composing the main part of clusters (Table 3.10). Cluster counting efficiency is plotted against the gas multiplication factor

Table 3.10. The fraction of clusters with a given number of electrons in argon for $\gamma = 4$ (Lapique, Piuz 1980)

Number of electrons in a cluster	1	2	3	4	5	6	7	8-20	> 20
Percentage of clusters	80.2	7.7	2.0	1.3	0.8	0.6	0.5	5.3	1.6

M in Fig. 3.45. It can be seen that M may reach 10^7; a further rise leads to no significant increase in $\overline{N}_c$, because of the accumulation of space charge in the vicinity of the wire.

Another necessary condition for efficient cluster counting is the *effective transport of electrons through the grid* separating the drift and proportional volumes of the chamber. The transport efficiency sometimes amounts to 80–90 % (Polyakov, Rykalin 1987).

The main advantage of a particle identifier based on cluster counting consists in reduction of the size of the device and the simple electronics required.

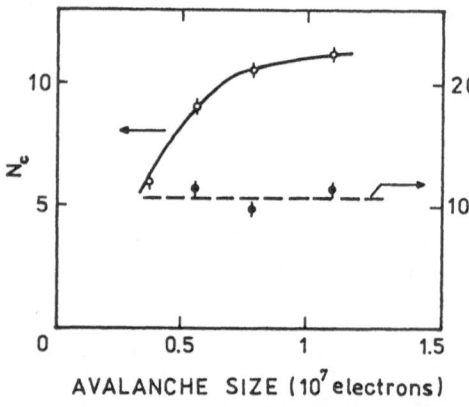

AVALANCHE SIZE (10^7electrons)

Fig. 3.45. Number of clusters N_c and relativistic rise as a function of the avalanche size (for pions). The plotted relativistic rise $R_c = [\overline{N}_c(\pi) - \overline{N}_c(p)]/\overline{N}_c(p)$ for 4 GeV (Rehak, Walenta 1980)

3.8.3 Electronics for Chambers with Longitudinal Electron Drift

The electronics for cluster counting (Fig. 3.46), and for ionization measurement in very thin gas samples in a chamber with longitudinal electron drift, have much in common with the electronics applied for high-precision drift chambers (Sect. 4.2): high input sensitivity, low noise, large amplification and high speed. An electronics channel for a single sense wire consists of a preamplifier, a discriminator and a fast counter.

When a particle passes through the chamber with longitudinal electron drift, a signal of high amplitude (10–100 pC) is produced at the input of the amplifier (Fig. 3.43). This signal corresponds to ionization in the vicinity of the sense wire. There then arrives at the input a series of pulses from clusters with a mean charge of the order of 0.1 pC and 10–20 ns long. The average distance between

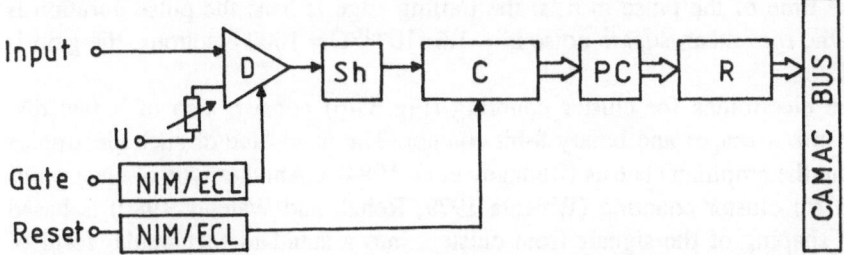

Fig. 3.46. Functional scheme of an electronics channel for cluster counting: U – reference voltage, D – discriminator, Sh – shaper, C – counter, PC – photo coupler, R – register (Budagov 1984b)

them is about 30 ns. Thus, an amplifier for cluster counting has to exhibit large amplification, high speed, low noise and a wide dynamic range, and it must not be overloaded by large initial signals.

An amplifier fulfilling all these requirements (Boie et al. 1981) consists of a preamplifier with an equivalent root-mean-square noise lower than 2000 electrons, a rise time of 2 ns, and a time constant of 10 ns, followed by a shaper in which the long trailing edges are cut off, and by a gaussian filter providing output signals of a gaussian shape. Somewhat modified amplifiers of this type were utilized in drift chambers for cluster counting (Rehak, Walenta 1980), in chambers with longitudinal electron drift applied for ionization measurement in very thin samples of gas (Arai et al. 1983; Imanishi et al. 1983), and in high-current drift chambers.

Approximately the same results can be achieved with the aid of a more simple amplifier, for example, a VA733 shown in Fig. 3.47 (Budagov et al. 1984b). Its feedback time constant is $C_6 R_8 \approx 5$ ns; the equivalent input capacitance is 20 pF;

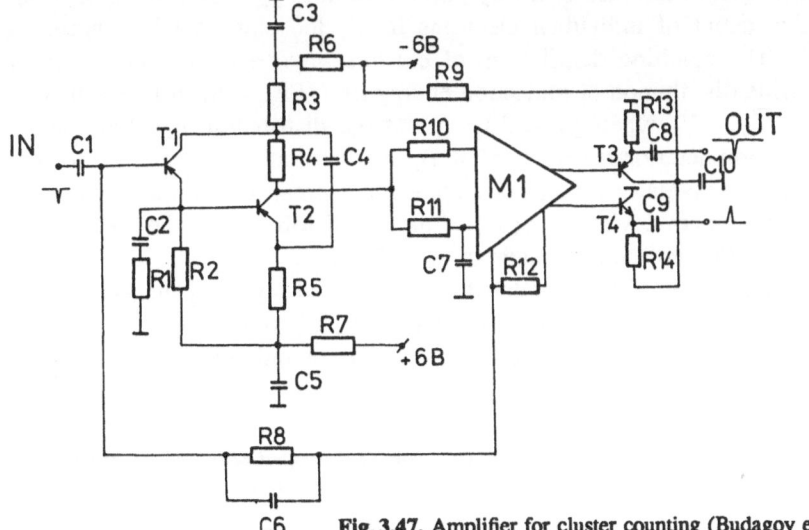

Fig. 3.47. Amplifier for cluster counting (Budagov et al. 1984b)

165

the rise time of the pulse in 6 ns; the trailing edge is 5 ns; the pulse duration is 15 ns; the root-mean-square noise is $\sim 1.6 \cdot 10^{-16} C = 1000$ electrons; the gain is 1 V/pC.

The electronics for cluster counting (Fig. 3.46) consist, also of a fast discriminator, a shaper and binary 8-bit counter. The dead time of such electronics (without the amplifier) is 6 ns (Budagov et al. 1984b). Another type of electronics system for cluster counting (Walenta 1979; Rehak and Walenta 1980) is based on the shaping of the signals from clusters into a standard rectangular form of regular amplitude and length; the subsequent integration and measurement of the total charge correspond to the number of generated signals.

3.8.4 Factors Limiting the Cluster Counting Method

The conditions of cluster counting in a chamber with slow electron drift have been analyzed in detail by Lapique and Piuz (1980). The dead time of the electronics τ_m is determined by the speed of the preamplifier and the counter, and for a fixed electron drift velocity w_d it determines the spatial resolution of the chamber, $\varrho_x = \tau_m \cdot w_d$. Thus, for $\tau_m = 10$ ns and $w_d = 1$ cm/μs, one has $\varrho_x = 0.1$ mm. In Ar the mean number of clusters for a relativistic particle amounts to $\overline{N}_1 = 25$–35 cm^{-1}, depending on its momentum, and the corresponding mean distance between the clusters is $d_c \approx 0.28$–0.4 mm. Owing to statistical fluctuations the distance between clusters may turn out to be smaller than the resolution ϱ_x, and when $\varrho_x = 0.1$ mm the fraction of counted clusters, i. e. the counting efficiency, is $\eta_c = N_c/\overline{N}_1 = 0.75$, where N_c is the number of counts recorded by the counter. In the case of slow electron drift, diffusion begins to play an significant role, since it may lead to two clusters merging into one, or to multi-electron clusters disintegrating into several clusters, even into individual electrons. We recall that in Ar, 80 % of the clusters actually consist of single electrons (Table 3.10). Calculations reveal that in the case of a very good resolution $\varrho_x < 50\,\mu$m ($\tau_m < 5$ ns), the effective count of individual electrons inside the clusters takes place, so $\overline{N}_c > \overline{N}_1$. The resulting distribution of clusters broadens and approaches in shape the wide distribution of ionization energy loss (Fig. 3.48). It is worthwhile noting that $\overline{N}_c = \overline{N}_1$ when $\varrho_x \approx 50\,\mu$m, and the distribution is a Poissonian. As the resolution deteriorates ($\varrho_x > 50\,\mu$m), the distribution becomes somewhat narrower, but the average number of counts $\overline{N}_c$ decreases. The minimum relative root-mean-square deviation $\sigma_c(\varrho_x)$ of the N_c distribution in the presence of diffusion ($\sigma_D = 200\,\mu$m), corresponds to $\varrho_x \approx 100\,\mu$m.

Lapique and Piuz (1980) calculated the number of layers of the identifier n required for separation of a proton and a kaon with $S_n(p/K) = 2$ (Fig. 3.49).

The separation coefficient in the case of cluster counting is

$$S_n(p/K) = [\overline{N}_c(p) - \overline{N}_c(K)]/\overline{N}_c(K)\sigma_n \qquad (3.33)$$

where $\sigma_n = \sigma_c/\sqrt{n}$. It must be noted that the assumed identification reliability for $S_n(p/K) = 2$ is quite low. For comparison, Fig. 3.49 presents the results of identifications of p and K from $(-dE/dx)$ measurements.

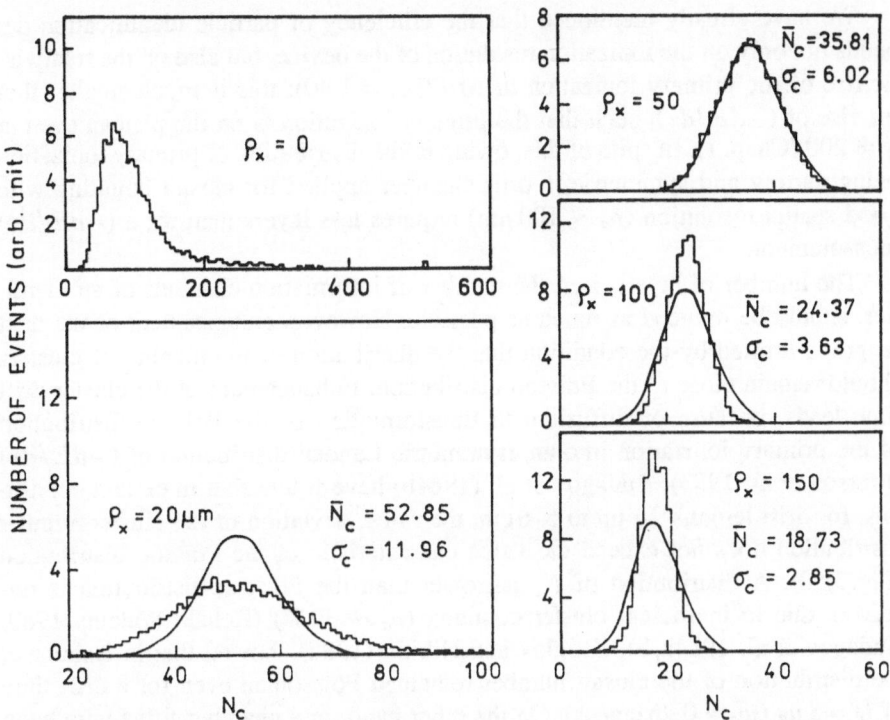

Fig. 3.48. Computed distributions of counted clusters in a 1 cm layer of argon for various spatial resolutions ϱ_x and $\sigma_D = 200\,\mu m$ (Lapique, Piuz 1980). The solid line corresponds to a poisson distribution with a mean number of clusters equal to $\overline{N}_c$

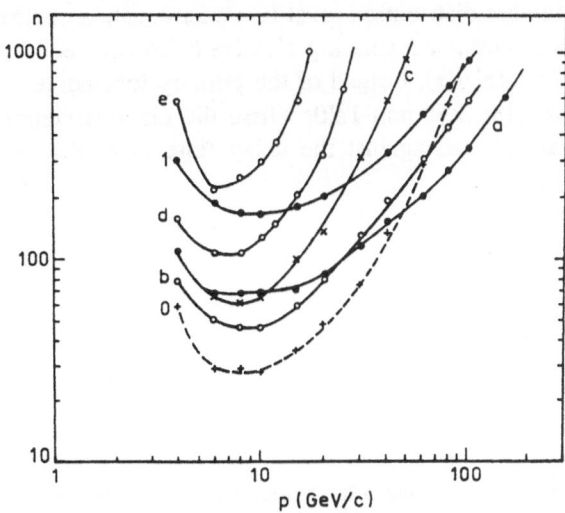

Fig. 3.49. Number n of layers of thickness $l = 1$ cm required for p/K separation of $S_n(p/K) = 2$, by cluster counting in argon at $P = 10^5$ Pa (Lapique, Piuz 1980) and for various values of ϱ_x: curves a, b, c, d, e correspond to $\varrho_x = 20, 50, 100, 150, 200\,\mu m$, respectively, and to $\sigma_D = 200\,\mu m$; curve 0 to $\varrho_x = 0$, $\sigma_D = 0$; curve 1 to a $(-dE/dx)$ measurement

We have already mentioned that the efficiency of particle identification depends not only on the ionization resolution of the device, but also on the relativistic rise of the primary ionization in Ar (R_{N1} = 1.40); this is much smaller that the rise of ($-dE/dx$), such that the primary ionization is on the plateau even at $\gamma \approx 200$ (Chap. 1). In spite of this, owing to the distribution of primary ionization being narrow and symmetric, a drift chamber applied for cluster counting with good spatial resolution ($\varrho_x \leq 100\,\mu$m) requires less layers than for a ($-dE/dx$) measurement.

The number of layers, i.e., the number of information channels of an identifier, should be reduced as much as possible. However, enhancement of the drift length is limited by the condition that the distribution of the number of clusters should remain close to the Poisson distribution. Enhancement of the cluster drift time leads, because of diffusion, to transformation of the Poisson distribution of the primary ionization into an asymmetric Landau distribution of ($-dE/dx$) (Merson et al. 1983). Budagov et al. (1984b) have found that in certain conditions, for drift lengths L_d up to 5–6 cm, the r.m.s. deviation of the cluster number distribution does not exceed the value characteristic of the Poisson distribution (Fig. 3.50). A distribution of N_c narrower than the Poisson distribution is observed, due to inefficient cluster counting ($\eta_c \sim 30\%$) (Rehak, Walenta 1980; Budagov et al. 1984a, b). Kruglov and Nikitin (1988) showed that in isobutane, the distribution of the cluster number remained Poissonian even for a drift time of t_d = 5 μs (w_d = 0.25 cm/μs). On the other hand, in a chamber filled with neon (Polyakov, Rykalin 1987) a Poisson distribution occurs when L_d = 1 cm and $w_d \sim 1$ cm/μs, but becomes noticeably wider at L_d = 5 cm. Thus it follows that the drift length appropriate for cluster counting depends strongly on the composition of the gas mixture and on the electron drift velocity. It must be stressed that the width of the cluster number distribution must be under constant control, since diffusion may cause the electrons constituting clusters to be counted, thus resulting in a measurement of ($-dE/dx$), instead of the primary ionization.

It is interesting to compare Figs. 3.50 and 1.20, where the r.m.s. deviations of the number of streamers are plotted against the delay time τ_d of the high

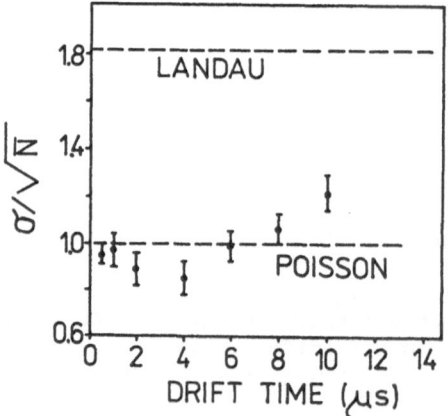

Fig. 3.50. Dependence of the standard deviation $\sigma_c/\sqrt{N_c}$ of the cluster number distribution on the electron drift time t_d: ● from (Budagov et al. 1984b), ○ from (Rehak, Walenta 1980)

voltage pulse applied across the streamer chamber. In drift and streamer chambers, transformation of the Poisson distribution into the Landau distribution is due to the disintegration of clusters into individual electrons by thermalization and diffusion. However, in a drift chamber this process proceeds more slowly because of a small longitudinal diffusion; in a streamer chamber the thermalization length and the diffusion coefficient are greater, since these processes take place in absence of the electric field.

The efficiency of particle identification by cluster counting is also restricted by the experimental relativistic rise of the number of clusters differing from the computed value. The measured value of the difference between the numbers of clusters produced by a pion and by a proton, respectively, of momentum $4\,\mathrm{GeV}/c$, $[\overline{N}_c(\pi) - \overline{N}_c(p)]/\overline{N}_c(p) \approx 10\,\%$ (Rehak, Walenta 1980), differs greatly from the computed value ($\approx 25\,\%$). In an argon-isobutane mixture this difference is smaller ($\sim 6\,\%$) (Rehak, Walenta 1980; Budagov et al. 1984; Kruglov, Nikitin 1988). This issue is analyzed by Kruglov and Nikitin (1988). The possible reasons for the weak relativistic rise of the measured amount of clusters are the following:

– a low efficiency ($\sim 30\,\%$) of cluster counting;
– the influence of diffusion;
– miscounting related to the dead time of the electronics;
– accumulation of space charge in the vicinity of the sense wires;
– the presence of additional signals probably related to photo-ionization.

Erroneous reduction of the relativistic rise in cluster counting is a crucial issue which must be studied thoroughly.

The above difficulties may be overcome by an application of alternative methods of cluster recording. Two methods have been proposed: cluster counting in an electro-luminescent chamber (Polyakov, Nikitin 1987), or in a multistep chamber (Breskin, Chechik 1986).

The principle of operation of the *electro-luminescent chamber* (ELC) is dealt with in Sect. 6.3. In the ELC a light output sufficient for effective registration of individual electrons can be achieved for a gas gain $M \simeq 10^4$–10^5 and a pulse duration $< 10\,\mathrm{ns}$ (Polyakov, Rykalin 1987). ELCs operate with a significantly lower gas gain than multiwire proportional chambers (MWPC), which strongly reduce the influence of space charge on the cluster counting. Moreover, ELCs exhibit a larger rate capability than MWPCs. The resulting efficiency of cluster counting in an ELC is high. Actually, in a chamber filled with neon $[(dN_1/dx)_{min} = 10.9\,\mathrm{cm}^{-1}]$, an efficiency of cluster counting of $\sim 80\,\%$ was achieved (Polyakov, Rykalin 1987). Monte-Carlo calculations based on experimental results reveal that in a 50-layer ELC 1 m long, effective hadron identification can be performed in the momentum region 3 to $30\,\mathrm{GeV}/c$ (Polyakov, Rykalin 1987). In our opinion, however, it would be difficult to construct a large identifier based on an ELC with a small amount of matter along the particle trajectories, since the light signal from each wire is extracted with the aid of a light guide involving a wavelength shifter.

Operation of the *low pressure multistep chamber* (MC) is based on the division of the gas amplification process into two steps: preamplification in a gap between parallel grids, and subsequent amplification in a parallel plate avalanche chamber (PPAC) or in a proportional chamber (Breskin et al. 1979). In a low-pressure MC a gas amplification of $M \sim 10^8$ is achieved, which allows almost full registration efficiency of single electrons (Breskin, Chechik 1986). The advantages of low-pressure MCs for cluster counting are the following:

- a high gas amplification and, consequently, a high efficiency of cluster counting;
- an enhancement of the relativistic rise of the number of clusters at low pressure;
- at low pressures the Fermi plateau starts at higher γ;
- a minimum probability of the production of δ-electrons.

The disadvantages of low-pressure MCs for cluster counting are:

- a small number of clusters (for example, $\overline{N}_c \approx 4\,\mathrm{cm}^{-1}$ in isobutane at a pressure of $5 \cdot 10^3$ Pa), resulting in a large length of the identifier;

- complicated structure in the detector, involving a large number of metal grids; this leads to an undesirable enhancement of the amount of matter inside the sensitive volume of the chamber;

- low-pressure MCs must be separated from other detectors operating at atmospheric pressure by thick walls, which in many cases is highly undesirable.

Low-pressure MCs and ELCs exhibit a number of advantages from the point of view of cluster counting. Until now, only initial steps have been taken in their development. Thorough studies of their properties have still to be carried out, to demonstrate their suitability for constructing large identifiers of relativistic charged particles.

The first step is to construct a device from which a large identifier (larger than $1\,\mathrm{m}^3$ in size) is obtained by simple enlargement of the prototype. The wire diameter of a large identifier must be at least $20\,\mu$m for a wire length ≥ 1 m. A detector of this type was investigated by Budagov et al. (1984a,b). The main features of the detector were the following: the sense wire diameter was $20\,\mu$m; the anode-cathode distance was $4\,\mathrm{mm}$; the drift volume was $10\,\mathrm{cm}$ wide; an argon-isobutane mixture was used; the drift velocity was $\sim 1\,\mathrm{cm}/\mu$s. Investigations revealed the following:

a) in such a detector the achievable efficiency of cluster counting is as good as that in the Walenta chamber (Rehak, Walenta 1980);
b) when cool gases and sufficiently high fields are used, the thickness of the drift volume can be enhanced up to 5 cm;
c) it is possible to use for cluster counting inexpensive electronics, which are suitable for mass production and provide a low dead time (10–15 ns).

The method of cluster counting, as compared with $(-dE/dx)$ measurements in multilayer drift chambers, exhibits the following advantages: a much smaller length of the identifier and a lower price for the electronics channels; simplicity, reliability, a smaller flow of information; and enhanced rate capability. The disadvantages of the method include a large number of electronics channels, worse spatial resolution, and the impossibility of identifying particles of large momenta $(p/mc \geq 200)$.

3.9 Ionization Measurement in Very Thin Samples of Gas

The reliability of relativistic particle identification increases with the number of layers for a given length of the identifier (Fig. 3.13). Achievement of the maximum separation coefficient S_n for a minimum identifier length is extremely important for detectors in colliding beams, since their dimensions are limited. However, the thickness of an identifier sample (the distance between the sense wires) has a technical (and economic) limit. Because of this, a sample thickness of several millimeters can be used at the "cost" of a large number of electronics channels (for instance 47400 channels in the ALEPH detector). The way out of this situation is to apply chambers with longitudinal electron drift.

3.9.1 Ionization Measurement in a Chamber with Longitudinal Electron Drift

In a chamber with longitudinal drift the electrons travel in the electric field along the particle trajectory (Fig. 3.42). The chamber is of the same construction as a chamber for cluster counting. In contrast to the case of straightforward cluster counting, here we are interested in the number of electrons in each cluster as well as the number of clusters; i.e. we are interested in their amplitudes. The information on the number of clusters and on the charge they carry provides the entire pattern of ionization in the gas, which allows maximum particle identification reliability to be achieved. In addition, this method substantially raises the upper momentum limit for the identified particles (the solid line in Fig. 3.41).

Signals corresponding to clusters arrive from the chamber with longitudinal electron drift at the input of a flash analog-to-digital converter (ADC) operating at a frequency of 50–100 MHz. At the output of the ADC a pattern is obtained which corresponds to the positioning of the clusters along the particle track (Fig. 3.43). Each point in Fig. 3.43 represents an interval of 25 ns, which corresponds, in this case, to a distance of 0.3 mm in the drift direction.

The ionization in very thin gas samples was first measured by Ludlam et al. (1981) with a 100 MHz ADC, which corresponded to a sample thickness of 0.25 mm.

The main task in successful ionization measurement in very thin samples of gas, is to achieve a high efficiency in the recording of clusters by the ADC. This efficiency still remains low (Fig. 3.43): of 30 clusters per 1 cm track length in Ar only about five are recorded.

There exist several methods for enhancing the cluster recording efficiency, of which the main one is reduction of the gas thickness for sampling. The analysis performed in the preceding section showed that the optimal sample thickness is 0.1 mm. It may be achieved using a 100 MHz ADC and an electron drift velocity of about 1 cm/μs.

An electronics channel for ionization measurement in very thin gas samples usually consists of a preamplifier, an amplifier-shaper and a flash ADC (Fig. 4.32). These electronics are described in greater detail in Chap. 4. An important part of the electronics system is the amplifier, which has to exhibit low noise and to generate very short gaussian-shaped signals. At present the best amplifier involving a gaussian shaper (Boie et al. 1981) exhibits a dead time of 12–15 ns. For effective cluster recording, a flash ADC with 10 ns time sampling is necessary. Ludlam et al. (1981), Arai et al. (1983), Imanishi et al. (1983) and Lehraus et al. (1983a), showed that a FADC with only 4 bits is quite acceptable. The results of measurements, presented in Fig. 3.51, show that the separation coefficients for relativistic charged particles S_n (3.1), for 4-bit and 6-bit FADCs are almost the same.

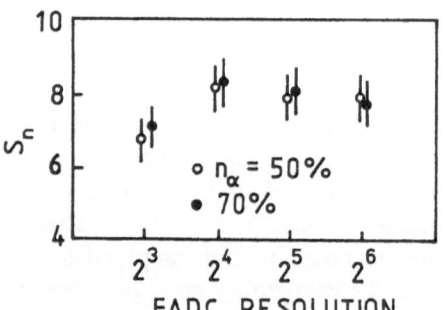

Fig. 3.51. Dependence of the separation coefficient S_n on the FADC resolution (Imanishi et al. 1983)

Measurements of cluster amplitudes with flash ADCs are very sensitive to the base level shift. Each signal has a long trailing "tail" (the time constant is $\sim 10\,\mu$s), upon which the following signals are superimposed. Shifts of the base line occur with a consequent systematic rise of the amplitudes. To avoid this harmful effect, special electronics are applied which restore the amplifier base level. Even when such measures are applied, a shift of amplitude by 10–20 % is observed (Boie et al. 1981; Ludlam et al. 1981, 1982). The latter effect can be removed by using a scheme for dynamical base line restoration (Ludlam et al. 1981). Another method involves restoration of the base line by software during data processing (Ludlam et al. 1981, 1982; Imanishi et al. 1983).

Diffusion significantly influences ionization measurements in very thin gas samples. We have mentioned that diffusion causes clusters to merge together, or to disintegrate into individual electrons. Therefore, in measuring the cluster charge, it is necessary to use cold gases (CO_2, isobutane, NH_3, dimethylether), in which diffusion is minimal and the electron drift velocity is low. These gases must be extremely pure to minimize electron capture during their drift time. To achieve a high efficiency of cluster recording, the gas amplification must also be optimized to avoid the accumulation of a spatial charge around the sense wire.

In transition to very thin samples of gas it is necessary to take into account changes in the distribution of ionization energy loss. Calculations by Allison (1981) have revealed a significant deviation of the spectrum from the usual Landau distribution for layer thicknesses $l < 1$ cm (Fig. 3.37). Such a change in the form of the energy loss distribution is related to the structure of the energy transfer spectrum in individual collisions (Fig. 1.12).

3.9.2 Relativistic Rise in Thin Gas Samples

Arai et al. (1983), Ludlam et al. (1981) and Imanishi et al. (1983) have shown that in very thin gas samples, the relativistic rise of ionization is enhanced owing to the more effective ejection of δ-electrons from the outer shells in the atom. Therefore it follows that R_E is strongly dependent on the cut-off coefficient α (Fig. 3.52). It seems that the increase in the relativistic rise and the enhancement of the cut-off efficiency are related to the change in the distribution of ionization energy losses in thin samples of the gas (Fig. 3.37).

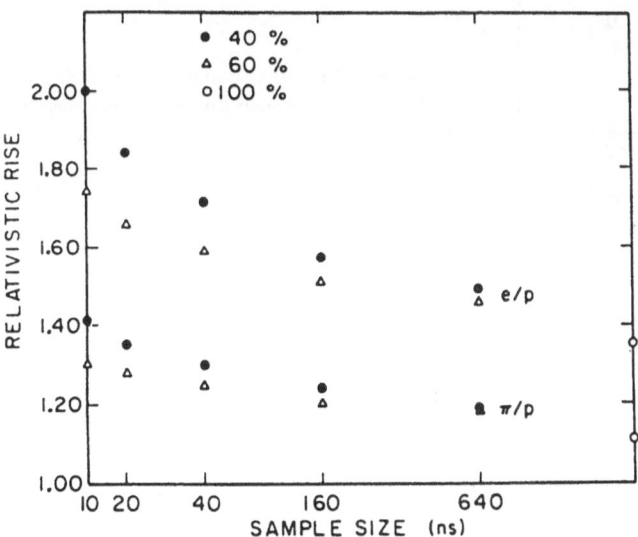

Fig. 3.52. Relativistic rise versus sample size and truncation parameter α, for $p = 3.5$ GeV/c (Ludlam et al. 1981)

3.9.3 Reliability of Identification by Ionization Measurements in Thin Gas Samples

Enhancement of the relativistic rise, with the reduction of the gas sample thickness, is compensated by the increase in width of the distribution of ionization energy loss (Fig. 3.53). Nevertheless, the more thorough ionization measurements enhance the reliability of particle identification in very thin gas samples; this is reflected in the rise of the separation coefficient S_n as l decreases (Fig. 3.54). The separation of electrons and protons becomes more effective as the number of measurement layers increases for a fixed particle track length (Fig. 3.55), and even when the identifier reaches 1 m, very good separation is achieved.

The enhancement of the separation coefficient in thin-layer devices could be even greater, if there would not exist the influence of the strong correlations shown in Fig. 3.56 (Arai et al. 1983; Imanishi et al. 1983). The open circles are obtained from n measurements of a single 1 m long track (with correlations present); the full circles correspond to the case when correlations are excluded. These latter points were obtained by Imanishi et al. (1983) by a random choice of samples along tracks of identical particles, from which an artificial track 1 m long

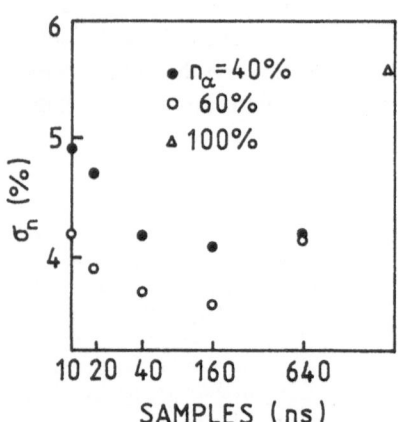

Fig. 3.53. Ionization resolution σ_n on a 1 m track versus sample size for various truncation parameters n_α (Ludlam et al. 1981)

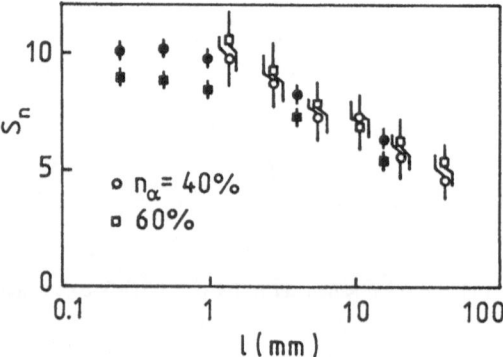

Fig. 3.54. Dependence of the separation coefficient for a 1 m track on the sample length: filled circles and squares (Ludlam et al. 1981); open symbols (Imanishi et al. 1983)

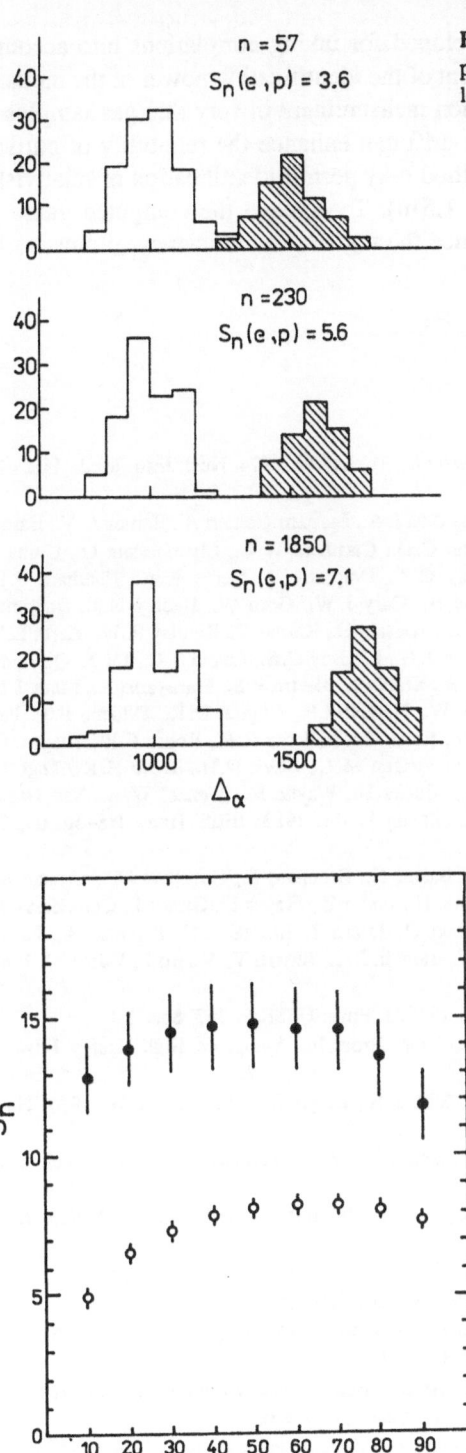

Fig. 3.55. The truncated means for electrons and protons for various sample sizes on a track 1 m long (Ludlam et al. 1982)

Fig. 3.56. The separation coefficient in a chamber with longitudinal electron drift on a 1 m track vs. number of samples left after truncation n_α. The open circles correspond to real tracks, and the full circles correspond to artificial tracks, composed of randomly picked up samples from various tracks (Imanishi et al. 1983)

was made up. Methods must be developed for taking correlations into account, since this promises a large enhancement of the identification power of the method.

To conclude, we note that ionization measurement in very thin gas samples in a chamber with longitudinal electron drift can enhance the reliability of particle identification. Application of this method may permit identification of relativistic hadrons using a small length (about 1.5 m). To achieve the computed value of S_n, it is first of all necessary to enhance the registration efficiency of clusters by fast ADCs.

References

Aderholz M., Lazeyras P., Lehraus I., Matthewson R., Tejessy W. 1974: Nucl. Instr. Meth. **118**, 419; Ibid. 1975: **123**, 237. Sects. 3.2, 6

Aihara H., Alston-Garnjost M., Badtke D.H., Bakken J.A., Barbaro-Galtieri A., Barnes A.V., Barnett B.A., Blumenfeld B., Bross A., Buchanan C.D., Carithers W.C., Chamberlain O., Chiba J., Chien C.-Y., Clark A.R.. Dahl O.I., Day C.T., Delpierre P., Derby K.A., Eberhard P.H., Fancher D.L., Fujii H., Fujii T., Gabioud B., Gary J.W., Gorn W., Hadley N.J., Hauptman J.M., Heck B., Hilke H., Huth J.E., Hylen J., Iwasaki H., Kamae T., Kenney R.W., Kerth L.T., Koda R., Kofler R.R., Kwong K.K., Layter J.G., Lindsey C.S., Loken S.C., Lu X.-Q., Lynch G.R., Madansky L., Madaras R.J., Majka R., Mallet J., Martin P.S., Maruyama K., Marx J.N., Matthews J.A.J., Melnikoff S.O., Moses W., Nemethy P., Nygren D.R., Oddone P.J., Park D., Pevsner A., Pripstein M., Robrish P.R., Ronan M.T., Ross R.R., Rouse F.R., Shapiro G., Shapiro M.D., Shen B.C., Slater W.E., Stevenson M.L., Stork D.H., Ticho H.K., Toge N., Urban M., Van Dalen G.J., Van Tyen R., Videau H., Wayne M., Wenzel W.A., Van Daslen Wetters R.F., Yamauchi M., Zeller M.E., Zhang W.-M. 1983: IEEE Trans. NS-30, 63; 76. Sects. 3.1, 7

Alard J.P., Arnold J., Augerat J., Babinet R., Bastid N., Brochard F., Costilhes J.P., Cronau M., De Marco N., Drouet M., Dupoeux P., Fanet H., Fodor Z., Frayss L., Girard J., Gorodetzky P., Gosset J., Laspalles C., Lamaire M.C., L'hote D., Lucas B., Montarou G., Papineau A., Parizet M.J., Poitou J., Racca C., Schimmerling W., Tamain J.C., Terrien Y., Valero J., Valette O. 1987: Nucl. Instr. Meth. A **261**, 379. Sect. 3.1

Alikhanian A.I., Alikhanov A.I., Nikitin S.Ya. 1945: J. Phys. USSR **9**, 367. Sect. 3.1

Alikhanov A.I., Lubimov V.A., Eliseev G.P., 1956: Proc. Int. Symp. on High Energy Phys. **2**, CERN, Geneva, 87. Sects. 3.1, 6

Allison W.W.M., Brooks C.B., Bunch J.N., Cobb J.H., Lloyd J.L., Pleming R.W. 1974: Nucl. Instr. Meth. **119**, 499. Sects. 3.1, 3

Allison W.W.M., Brooks C.B., Bunch J.N., Pleming R.W., Yamamoto R.K. 1976: Nucl. Instr. Meth. **133**, 325. Sects. 3.2, 5

Allison W.W.M., Brooks C.B., Lyons L., Romaya A.M., Shield P.D., McPherson 1979: Nucl. Instr. Meth. **163**, 331. Sects. 3.1, 6

Allison W.W.M., Cobb J.H. 1980: Ann. Rev. Nucl. Part. Sci. **30**, 253. Sects. 3.2, 3, 8

Allison W.W.M. 1981: Physica Scripta **23**, 348. Sects. 3.1, 2, 6, 7, 9

Allison W.W.M. 1982: Proc. Int. Conf. on Instrumentation for Colliding Beam Physics, Stanford Linear Accelerator Report, 61. Sects. 3.1, 2, 3, 5, 6, 7

Allison W.W.M., Brooks C.B., Shield P.D., Aguilar-Benitez M., Willmott C., Dumarchez J., Schouten M. 1984: Nucl. Instr. Meth. **224**, 396. Sects. 3.1, 3, 5, 6, 7

Allison W.W.M., Wright P.R.S. 1984: in *Formulae and Methods in Experimental Data Evaluation*, Vol. 2, ed. by Bock R.K. (Europ. Phys. Soc. Geneva). Sects. 3.1, 3, 5

Anashin V. V., Anashkin E. V., Aulchenko V. M., Barkov L. M., Baru S. E., Blinov A. E., Blinov
V. E., Blinov G. A., Bondar A. E., Bukin A. D., Vorobev A. I., Volkov A. E., Groshev V. R.,
Zhilich V. N., Zholents A. A., Klimenko S. G., Kozlov V. N., Kolachev G. M., Kuzmin A. S.,
Lebedev P. K., Lelchuk M. Y., Minakov G. D., Minakov M. D., Mishnev S. I., Nagaslaev V. P.,
Nomerotski A. B., Onuchin A. P., Panin V. S., Pestov Y. N., Petrov V. V., Pril Y. V., Protopopov
I. J., Purlats T. A., Root N. I., Rodyakin V. A., Romanov L. V., Rylin A. V., Savinov G. A.,
Sidorov V. A., Skovpen Y. I., Skrinski A. N., Smakhtin V. P., Sukhanov S. V., Tayurski V. A.,
Telnov V. I., Temnykh A. B., Tikhonov Y. A., Tumaykin G. M., Undrus A. E., Chilingarov A. G.,
Shamov A. G., Shvarts B. A., Shusharo A. I., Eidelman S. I., Eidelman Y. I., Frabetti P. L., Cassata
A. M., DeMartinis K., Palombo F., Sala A., Liverali V., Maloberti F., Manfredi P. F., Marioli
D., Massetti P., Pe V., Speciali V., Vercelli L., 1987: Proc. Int. Symp. on Position Detectors in
High Energy Physics, Dubna. Sect. 3.1
Arai R., Bensinger J., Boerner H., Fakushima Y., Hayashi K., Ishihara N., Inaba S., Kohriki T.,
Nakamura S., Ogawa K., Takasaki F., Unno Y., Watase Y., 1983: Nucl. Instr. Meth. 214, 209.
Sects. 3.2, 8, 9
ARGUS Collaboration 1989: Nucl. Instr. Meth A 275, 1. Sects. 3.1, 7
Ash W. W., Band H. R., Bloom E. D., Bosman M., Camporesi T., Chadwick G. B., Delfino M. C.,
De Sangro R., Ford W. T., Gettner M. W., Goderre G. P., Godfrey G. L., Groom D. E., Hurst
R. B., Johnson J. R., Lau K. H., Lavine T. L., Leedy R. E., Lippi I., Maruyama T., Messner R. L.,
Moromisato J. H., Moss L. J., Muller F., Nelson H. N., Peruzzi I., Piccolo M., Prepost R., Ryrlik
J., Qi N., Read, Jr. A. L., Ritson D. M., Rosenberg L. J., Shambroom W. O., Sleeman J. C., Smith
J. G., Venuti J. P., Verdini P. G., Von Goeler E., Wald H. B., Weinstein R., Wiser D. E., Zdarko
R. W. 1987: Nucl. Instr. Meth. A 261, 399. Sect. 3.1
Atwood W. B., Carr J., Chadwick G., Csorna S., Hansl-Kozanecka T., Hodges C., Nauenberg U.,
Nielsen B. S., Panvini R. S., Prescott C. Y., Reeves T. W., Rochester L. S., Simpson K., Steiner
A., Young C. C., 1986: Nucl. Instr. Meth. A 252, 295. Sect. 3.1
Babaev A. I., Barski D. A., Boris S. D., Brakhman E. V., Varlamov L. I., Galaktionov Yu. B., Gorodkov
Yu. V., Danilov M. V., Eliseev G. P., Zeldovich O. J., Ilin M. A., Kamyshkov Yu. A., Kravcov
A. I., Laptin L. P., Lubimov V. A., Nagovicin V. V., Nozik V. Z., Popov V. P., Semenov Yu. A.,
Sopov V. S., Tichomirov I. N., Cvetkova T. N., Chudakov V. N., Shevchenko V. G., Shumilov
E. V. 1978: Prep. Institute of Theoretical and Experimental Physics, ITEP–103, Moscow (in
Russian). Sect. 3.1
Bai Jing-zhi, Ma Ji-mao, Mao Hui-shun, Xie Pei-pei, Yan Jie, Liu Rong-guang, Mao Ze-pu, Rong
Gang, Shen Ben-wei, Song Xiao-fei, Wang Yao-hut, Wang Yun-yong, Zhou Jie 1987: Proc. Int.
Symp. on Position Detectors in High Energy Physics, Dubna. Sect. 3.1
Barbaro-Galtieri 1982: Proc. Int. Conf. on Instrumentation for Colliding Beam Physics, Stanford
Linear Accelerator Report. Sect. 3.1
Baruzzi V., Carosio R., Crijns F., Gerdyukov L., Goldschmidt-Clermont Y., Grant A., Johnson D.,
Kröner F., Lehraus I., Matthewson R., Milstene C., Nikolaenko V., Oren Y., Petrovikh Y.,
Ross R. T., Sixel P., Spyropoulou-Stassinaki M., Stergiou A., Tejessy W., Theocharopoulos P.,
Vassiliadis, Wright P. R. S., Zoll J. 1983: Nucl. Instr. Meth. 207, 339. Sects. 3.1, 5–7
Bashindzhagyan G. L., Sarycheva L. I., Sinev N. B. 1977: Proc. 15th Int. Cosmic Ray Conf. 9, 185.
Sect. 3.1
Bashindzhagyan G. L. 1980: Thesis SRINP Moscow State Univ., Moscow. Sect. 3.6
Becker Ch. Weihs W., Zech G. 1983: Nucl. Instr. Meth. 213, 243. Sect. 3.4
Becker U., Cappel M., Chen M., White M., Ye C. H., Yee K., Fehlman J., Seiler P. G. 1983: Nucl.
Instr. Meth. 214, 525. Sects. 3.3, 8
Blum W. 1984: Nucl. Instr. Meth. 225, 557. Sect. 3.1
Blum W. (ed.) 1989: The ALEPH Handbook, CERN ALEPH Report 89–77, Note 89–03. Sect. 3.1
Boie R. A., Hrisoho A. T., Rehak P. 1981: IEEE Trans. Nucl. Sci. NS–28, 603. Sects. 3.8, 9
Breskin A., Charpak G., Sauli F., Atkinson M., Schultz G. 1975: Nucl. Instr. Meth. 124, 189. Sect. 3.4
Breskin A., Charpak G., Demierre C., Majewski S., Policarpo A., Sauli F., Santiard J. C. 1977: Nucl.
Instr. Meth. 143, 29. Sect. 3.4

177

Breskin A., Charpak G., Majewski S., Melchart G., Petersen G., Sauli F. 1979: Nucl. Instr. Meth. **161**, 19. Sect. 3.8

Breskin A., Chechik R. 1986: Nucl. Instr. Meth. A **252**, 488. Sect. 3.8

Breuker H., Fischer H.M., Hauschild M., Hartmann H., Wunsch B., Boerner H., Burckhart H.J., Dittmar M., Hammarstrom R., Heuer R.D., Michelini A., Plane D.E., Runolfsson O., Schaile D., Weisz S., Zankel K., Ludwig J., Mohr W., Rohner F., Runge K., Schaile O., Schwarz J., Stier H.E., Weltin A., Bock P., Heintze J., Igo-Kemenes P., Lennert P., Wagner A. 1987: Nucl. Instr. Meth. A **260**, 329. Sects. 3.1, 7

Brooks C.B., Bullock F.W., Cashmore R.J., Devenish R.C., Foster B., Fraser T.J., Gibson M.D., Gilmore R.S., Gingrich D., Harnew N., Hart J.C., Heath G.P., Hiddleston J., Holmes A.R., Jamdagni A.K., Jones T.W., Llewellyn T.J., Long K.R., Lush G.J., Malos J., Martin N.C., McArthur I., McCubbin N.A., Mc.Quillan D., Miller D.B., Mobayyen M.M., Morgado C., Nash J., Nixon G., Parham A.G., Payne B.T., Roberts J.H.C., Salmon G., Saxon D.H., Sephton A.J., Shaw D., Shaw T.B., Shield P.B., Shulman J., Silvester I., Smith S., Strachan D.E., Tapper R.J., Tkaczyk S.M., Toudup L.W., Wallis E.W., Wastie R., Wells J., White D.J., Wilson F.F., Yeo K.L. 1989: Nucl. Instr. Meth. A **283**, 215. Sect. 3.1

Bryman D.A., Leitch M., Navon I., Numao T., Schlatter P., Dixit M.S., Hargrove C.K., Mes H., MacDonald J.A., Skegg R., Spuller J., Burnham R.A., Hasinoff M., Poutissou J.M., Azuelos G., Depommier P., Martin J.P., Poutissou R., Blecher M., Gotow K., Carter A.L. 1985: Nucl. Instr. Meth. A **234**, 42. Sect. 3.1

Budagov Yu.A., Hlinka V., Semenov A.A., Sergeev S.V., Sitar B., Feschenko A.A., Spalek J. 1984a: Prib. Tekh. Exp. **1**, 62 (English transl.: Instr. Exper. R **27**, 59). Sect. 3.8

Budagov Yu.A., Hlinka V., Nagaitsev L.P., Omelyanenko A.A., Omelyanenko M.N., Semenov A.A., Sitar B., Spalek J. 1984b: Prep. Joint Institute for Nuclear Research JINR 13–84–337, Dubna (in Russian). Sect. 3.8

Cassel D.G., Desalvo R., Dobbins J., Gilchriese M.G.D., Gray S., Hartill D., Mueller J., Peterson D., Pisharody M., Riley D., Kinoshita K. 1986: Nucl. Instr. Meth. A **252**, 325. Sects. 3.1, 2

Chapin T.J., Cool R.L., Goulianos K., Jenkins K.A., Silverman J.P., Snow G.R., Sticker H., White S.N. 1984: Nucl. Instr. Meth. **225**, 550. Sect. 3.1

Cockerill D., Fabjan C.W., Frandsen P., Hallgren A., Heck B., Hilke H.J., Hogue R., Jeffreys P., Jensen H.B., Killian T., Kreisler M., Lindsay J., Ludlam T., Lissauer D., Molzon W., Nielsen B.S., Oren Y., Queru P., Rosselet L., Rosso E., Rudge A., Scire M., Wang D.W., Wang Ch.J., Willis W.J., Botner O., Boggild H., Dahl-Jensen E., Dahl-Jensen I., Dam Ph., Damgaard G., Hansen K.H., Hooper J., Møller R., Nielsen S.Ø., Schistad B., Akesson T., Almehed S., von Dardel G., Henning S., Jarlskog G., Lorstad B., Melin A., Mjornmark U., Nilsson A., Albrow M.G., McCubbin N.A., Evans W.M. 1981: Physica Scripta **23**, 649. Sect. 3.1

Calvetti M., Cennini P., Centro S., Chesi E., Cittolin S., Cnops A.A., Dumps L., Di Bitonto D., Geer S., Haynes B., Jank W., Jorat G., Karimaki V., Kowalski H., Kinnunen R., Lacava F., Maurin G., Norton A., Piano Mortari G., Pimia M., Placci A., Queru P., Rijssenbeek M., Rubbia C., Sadoulet B., Sumorok K.S., Tao C., Vuillemin V., Vialle J.P., Verweij H., Zurfluh E. 1982: Proc. Int. Conf. on Instrumentation for Colliding Beam Physics, Stanford Linear Accelerator Report. Sect. 3.1

Delpierre P. 1984: Nucl. Instr. Meth. **225**, 566. Sect. 3.1

Dimcovski Z., Favier J., Charpak G., Amato G. 1971: Nucl. Instr. Meth. **94**, 151. Sect. 3.6

Ehrlich R.D. 1982: Proc. Int. Conf. on Instrumentation for Colliding Beam Physics, Stanford Linear Accelerator Report. Sect. 3.1

Erskine G.A. 1972: Nucl. Instr. Meth. **105**, 565. Sect. 3.2

Etkin A., Eiseman S.E., Foley K.J., Hakenburg R.W., Longrace R.S., Love W.A., Morris T.W., Platner E.D., Saulys A.C., Lindenbaum S.J., Chan C.S., Kramer M.A., Hallman T.J., Madansky L., Bonner B.E., Buchanan J.A., Chiou C.N., Clement J.M., Corcoran M.D., Krishna N., Kruk J.W., Miettinen H.E., Mutchler G.S., Nessi-Tedaldi F., Nessi M., Phillips G.C., Robert J.B., 1989: Nucl. Instr. Meth. A **283**, 352. Sect. 3.1

Fulda-Quenzer F., Haissinksi J., Jean-Marie B., Pagot J. 1985: Nucl. Instr. Meth. A **235**, 517. Sect. 3.5

Gabathuler E. 1981: Physica Scipta **23**, 590. Sect. 3.7

Garabatos C. representing the NA36 Collaboration 1989: Nucl. Instr. Meth. A **283**, 357. Sect. 3.1

Goloskie D., Kistiakowsky V., Oh S., Pless I. A., Stoughton T., Suchorebrow V., Wadsworth B., Murphy O., Steiner R., Taft H. D. 1985: Nucl. Instr. Meth. A **238**, 61. Sect. 3.1

H1 Collaboration 1986: Technical Report for the H1 Detector, DESY (Berlin). Sect. 3.1

Hanson G. G. 1986: Nucl. Instr. Meth. A **252**, 343. Sect. 3.1

Hasemann H. 1982: Proc. Int. Conf. on Instrumentation for Colliding Beam Physics, Stanford Linear Accelerator Report. Sects. 3.2, 4, 7

Hendricks R. W. 1969: Rev. Sci. Instrum. **40**, 1216. Sect. 3.4

Heuer R. D., Wagner A. 1988: Nucl. Instr. Meth. A **265**, 11. Sect. 3.1

Hilke H. 1986: Nucl. Instr. Meth. A **252**, 169. Sect. 3.5

Hilke H. 1987: Proc. Int. Symp on Position Detectors in High Energy Physics, Dubna, p. 35. Sect. 3.1

Huk M., Igo-Kemenes P., Wagner A. 1988: Nucl. Instr. Meth. A **267**, 107. Sect. 3.3

Imanishi A., Ishii T., Ohshima T., Okuno H., Shiino K., Naito F., Matsuda T. 1983: Nucl. Instr. Meth. **207**, 357. Sects. 3.8, 9

Jeanne D., Lazeyras P., Lehraus I., Matthewson R., Tejessy W., Aderholz M. 1973: Nucl. Instr. Meth. **111**, 287. Sects. 3.1, 6

Kamae T., Aihara H., Enomoto R., Fujii H., Fujii T., Itoh R., Kusuki N., Miki T., Shirahashi A., Takahashi T., Tanaka M., Du Bois A., Jared R. C., Kirsten F., Salz P., Ikeda H., Iwasaki H., Iwata S., Matsuda T., Nakamura K., Yamamoto A., Yamauchi M. 1986: Nucl. Instr. Meth. A **252**, 423. Sect. 3.1

Kruglov V. V., Nikitin M. V. 1988: Prep Joint Institute for Nuclear Research JINR 13–88–882; 13–88–883, Dubna. Sect. 3.8

Kubota Y., Morrow F., Stone S., Wilcke R. 1983: Nucl. Instr. Meth. **217**, 249.

Lapique F., Piuz F. 1980: Nucl. Instr. Meth. **175**, 297. Sect. 3.8

Lehraus I., Matthewson R., Tejessy W., Aderholz M. 1978: Nucl. Instr. Meth. **153**, 347. Sects. 3.1, 5, 7

Lehraus I., Matthewson R., Tejessy W. 1982a: Nucl. Instr. Meth. **196**, 361. Sects. 3.1, 2, 7

Lehraus I., Matthewson R., Tejessy W. 1982b: Nucl. Instr. Meth. **200**, 188. Sects. 3.1, 3

Lehraus I., Matthewson R., Tejessy W. 1982c: Proc. Int. Conf. on Instrumentation for Colliding Beam Physics, Stanford Linear Acclerator Report. Sect. 3.1

Lehraus I., Matthewson R., Tejessy W. 1983a: IEEE Trans. Nucl. Sci. NS–30, 44. Sects. 3.1, 9

Lehraus I., Matthewson R., Tejessy W. 1983b: IEEE Trans. Nucl. Sci. NS–30, 50. Sect. 3.1

Lehraus I. 1983: Nucl. Instr. Meth. **217**, 43. Sects. 3.2, 3, 5, 7, 8

Ludlam T., Platner E. D., Polychronakos V. A., Lindenbaum S. J., Kramer M. A., Teramoto Y. 1981: IEEE Trans. Nucl. Sci. NS–28, 439. Sect. 3.9

Ludlam T., Platner E. D., Polychronakos V. A., Lindenbaum S. J., Rosso E., Queru P., Kramer M. A., Teramoto V., Akesson T., Jarlskog G., Roennborn L. 1982: Proc. Int. Conf. on Instrumentation for Colliding Beam Physics, Stanford Linear Accelerator Report. Sect. 3.9

Lynch G. R., Hadley N. J. 1982: Proc. Int. Conf. on Instrumentation for Colliding Beam Physics, Stanford Linear Accelerator Report. Sects. 3.1, 5–7

Merson G. I., Sitar B., Budagov Yu. A. 1983: Fiz Elem. Chastits At. Yadra **14**, 648 (English transl.: Sov. J. Part. Nucl. **14**, 270). Sects. 3.1, 6, 8

Mori Ch., Uno M., Watanabe T. 1982: Nucl. Instr. Meth. **196**, 49. Sect. 3.4

Nygren D. 1981: Physica Scripta **23**, 584. Sect. 3.1

OPAL Detector Technical Proposal 1983: CERN/LEPC/83–4. Sect. 3.1

Peisert A., Sauli F. 1984: Prep. CERN 84–08. Sect. 3.3

Piuz F. 1983: Nucl. Instr. Meth. **205**, 425. Sect. 3.8

Polyakov V. A., Rykalin V. I. 1987: Proc. Int. Symp. on Position Detectors in High Energy Physics, Dubna, p. 223. Sect. 3.8

Rehak P., Walenta A. H. 1980: IEEE Trans. Nucl. Sci. NS–27, 54. Sect. 3.8

Roehrig J., Einsweiler K., Hutchinson D., Mc.Nerney E., Odian A., Skarpaas K., Unno Y., Villa F., Grancagnolo F., Rowe W., Seiden A., Baltrusaifis R., Hauser J., Richman J., Russel J. J., Blaylock G., Thaler J., Wisniewski W. 1984: Nucl. Instr. Meth. **226**, 319. Sect. 3.1

Sadoulet B. 1981: Physica Scripta **23**, 434. Sect. 3.5

Sauli F. 1977: Prep. CERN 77–09. Sects. 3.2–4

Sauli F. 1984: in *The Time Projection Chambers* (ed. Mac Donald J. A.) (Amer. Inst. Phys., New York). Sects. 3.1, 3, 4

Sauli F. 1988a: Nucl. Instr. Meth A **273**, 805. Sect. 3.1

Sauli F. 1988b: Z. Phys. C **38**, 339. Sect. 3.1

Saxon D. H. 1988: Nucl. Instr. Meth. A **265**, 20. Sects. 3.1, 5

Shirahashi A., Aihara H., Itoh R., Kamae T., Kusuki N., Tanaka M., Fujii H., Fujii K., Ikeda H., Iwasaki H., Kobayashi M., Matsuda T., Miyamoto A., Nakamura K., Ochiai F., Tsukamoto T., Ynauchi M. 1988: IEEE Trans. Nucl. Sci. NS–35, 414. Sects. 3, 7

Sipilä H., Vanha-Honko V. 1978: Nucl. Instr. Meth. **153**, 461. Sects. 3.4

Sitar B. 1985: Proc. 8th Conf. Czechoslovak. Phys., Bratislava, p. 227. Sect. 3.1

Sitar B. 1987: Fiz. Elem. Chastits At. Yadra **18**, 1080 (English transl.: Sov. J. Part. Nucl. **18**, 460). Sect. 3.1

SLD Design Report 1984: Stanford Linear Accelerator Report 273. Sect. 3.1

TPC 1976: Proposal for PEP Facility Based on the Time Projection Chamber, Stanford Linear Accelerator Report, PUB–5012. Sect. 3.1

Toothacker W. S., Trumpinski S., Whitmore J., Lewis R. A., Gress J., Ammar R., Coppage D., Davis R., Kankel S., Kwak N., Elcombe P. A., Hill J. C., Neale W. W., Kowald W., Robertson W. J., Walker W. D., Lucas P., Voyvodic L., Bishop J. M., Biswas N. N., Cason N. M., Kenney V. P., Mattingly M. C. K., Ruchti R. C., Shepard W. D., Ting S. J. Y. 1988: Nucl. Instr. Meth. A **271**, 97. Sects. 3.1, 7

Ueno K. 1988: Prep. University of Rochester UR–1050, New York. Sect. 3.1

UCD 1988: Proposal for Experiments on The Universal Calorimetric Detector, Serpukhov. Sect. 3.1

Va'vra J., Roberts L., Freytag D., Clancey P. 1982: Nucl. Instr. Meth. **203**, 109. Sects. 3.1, 7

Wagner A. 1981: Physica Scripta **23**, 446. Sect. 3.1

Wagner A. 1982: Proc. Int. Conf. on Instrumentation for Colliding Beam Physics, Stanford Linear Accelerator Report. Sects. 3.1, 7

Wagner R. L. 1988: Nucl. Instr. Meth. A **265**, 1. Sects. 3.1

Walenta A. H. 1979: IEEE Trans. Nucl. Sci. NS–26, 73. Sect. 3.8

Walenta A. H., Fischer J., Okuno H., Wang C. L. 1979: Nucl. Instr. Meth. **161**, 45. Sects. 3.2, 3, 6

Walenta A. H. 1981: Physica Scripta **23**, 354. Sects. 3.1, 2

ZEUS Collaboration 1986: The ZEUS Detector; Technical Proposal. Sect. 3.1

4. Spatial Resolution and Electronics of Multilayer Proportional Detectors

Multilayer drift chambers are primarily applied for visualization of the trajectories of charged particles. As shown in Fig. 3.9, track images in MDCs are reminiscent of bubble chamber pictures and are of high quality even in the case of large charged particle multiplicity. The good spatial resolution of MDCs permits extremely precise determination of secondary particle entrance angles and, in the presence of a magnetic field, measurement of their momenta. We shall consider the main factors determining the coordinate precision provided by drift chambers.

The most common constructions of proportional and drift chambers are shown in Fig. 4.1. The layout of a proportional chamber is presented in Fig. 4.1a. The scheme without field-shaping electrodes (Fig. 4.1b) is usually applied in narrow-gap drift chambers, used for precise determination of trajectory coordinates in particle beams of high intensity. The drift chamber with field-shaping wires, at a linearly dropping potential (Fig. 4.1c), is characterized by an electric field of high homogeneity. Such geometry is adopted in most plane drift chambers serving as coordinate detectors. Electrode configurations like those in Fig. 4.1d, are utilized in some cylindrical drift chambers. The arrangement of electrodes

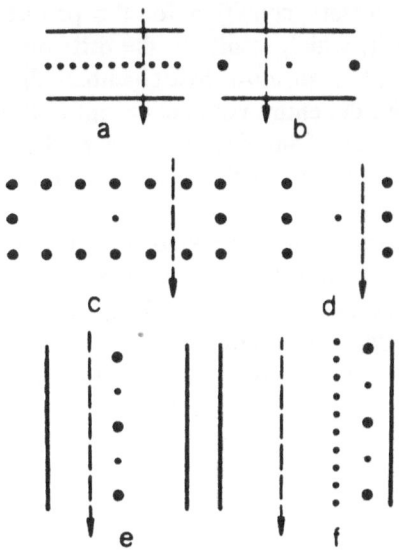

Fig. 4.1. Arrangement of wires in various types of chambers: (a) proportional chamber; (b) planar drift chamber; (c) drift chamber with field-shaping electrodes, (d) element of drift chamber with symmetric cells; (e) element of the Jet or of the SSC chamber; (f) TPC. ● – field-shaping electrodes, · – sense wires, solid line – cathode, dashed line – particle trajectory (Jaros 1980)

shown in Fig. 4.1e is typical of multilayer drift chambers (for example, cylindrical chambers of the Jet and SSC types). To provide a higher field homogeneity, the drift volume in a TPC is separated from the region of signal wires by an additional wire plane (Fig. 4.1f).

4.1 Spatial Resolution of Drift Chambers

A track coordinate x, in the direction perpendicular to the anode wire plane, is determined in a drift chamber by measuring the time interval between the time of passage of the particle through the chamber t_0, and the moment t_1 that the electron swarm arrives at the anode:

$$ x = \int_{t_0}^{t_1} w_d dt . \tag{4.1} $$

For a constant drift velocity w_d is

$$ x = w_d t_d \tag{4.2} $$

where the electron drift time is $t_d = t_1 - t_0$.

A constant drift velocity within a certain interval of electric field strength can be provided by an appropriate gas mixture (Sect. 3.3). A number of gas mixtures provide a constant drift velocity within a wide range of field strengths (Fig. 3.17). Drift velocities for a large number of mixtures can be found in Peisert and Sauli (1984), and Fehlmann and Viertel (1983). Another condition which allows a drift chamber to operate with high precision is the homogeneity of the electric field throughout the drift volume [$\mathcal{E}(x) \approx$ const].

A *constant electron drift velocity* is a necessary condition for the precise measurement of coordinates in a drift chamber, which results in the drift time depending linearly upon the track coordinate (4.2). In most drift chambers this condition can be fulfilled accurately throughout the entire volume, except in the regions close to the sense wires and at the edges of the chamber, where there exists a strong electric field gradient. Condition (4.2) usually holds for particle tracks perpendicular to the direction of electron drift.

During long-term operation of a drift chamber the drift velocity must be kept constant; in order to achieve this the gas composition, its pressure and its temperature are stabilized. The relative change in drift velocity $\Delta w_d / w_d$ with temperature is plotted in Fig. 4.2. For a number of organic gases in a weak electric field the ratio $\Delta w_d / w_d = 3.4 \cdot 10^{-3} /°C$ is independent of the field strength (Schultz, Gresser 1978).

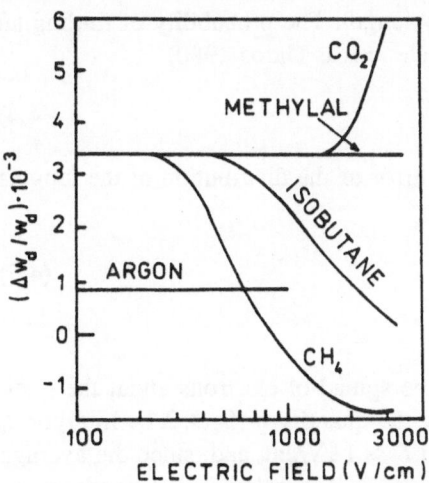

Fig. 4.2. Relative change in drift velocity, $\Delta w_d / w_d$, for a rise in temperature of 1°C under normal conditions, versus $\mathcal{E}$ (Schultz, Gresser 1978)

4.1.1 Factors Contributing to Spatial Resolution

The root-mean-square error σ_x, of coordinate measurement on a particle trajectory in a drift chamber, is determined by the expression

$$\sigma_x^2 = \sigma_s^2 + \sigma_{D'}^2 + \sigma_a^2 , \tag{4.3}$$

which takes into account the uncertainties due to the actual cluster distribution along the particle trajectory (σ_s), diffusion of the electron swarm in the electric field ($\sigma_{D'}$), and instrumental errors (σ_a), which include the resolution of the device, fluctuations of the gas amplification, etc.

Let us now estimate these errors. Consider a particle having passed at a certain distance x from the wire, and having produced N_t electrons per 1 cm of its track length in the gas (Fig. 4.3). The first electron will reach the wire in a

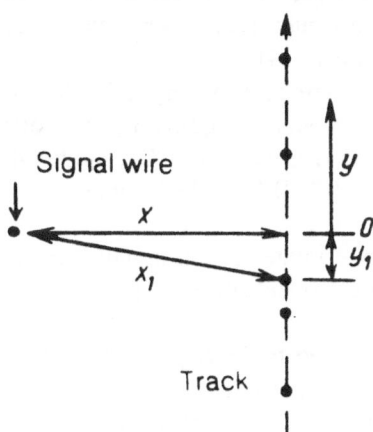

Fig. 4.3. Distribution of electrons along a particle track in a drift chamber

time $t_1 = x_1/w_d$, where x_1 is its trajectory length. The probability of finding an electron at a distance y_1 from the coordinate axis is (Jaros 1980)

$$P(y_1) = 2N_t \exp(-2N_t y_1) , \tag{4.4}$$

from which follows the root-mean-square error of the distribution of the lengths x_1:

$$\sigma_s = \sqrt{\frac{5}{16}} \cdot \frac{1}{N_t^2 \cdot x} \tag{4.5}$$

for $x \gg 1/N_t$.

A contribution to σ_s is also given by the spread of electrons about the track upon their thermalization, characterized by the quantity l_t (Sect. 2.1). In argon l_t is of the order of several microns in a field $\mathcal{E} \approx 1\,\mathrm{kV/cm}$, and, since the average distance between clusters is $\sim 300\,\mu\mathrm{m}$, $l_t \ll \sigma_s$. The calculated dependence of σ_s upon the drift length x is presented in Fig. 4.4. We recall that in argon $\overline{N}_t \approx 100\,\mathrm{cm}^{-1}$.

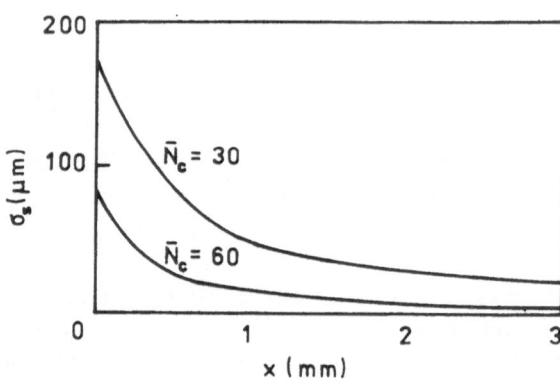

Fig. 4.4. Dependence of σ_s on the electron drift length x (Jaros 1980)

When the drift length is large, the main contribution to the spatial resolution is due to the *diffusion term* $\sigma_{D'}$. The values of $\sigma_{D'}$ measured in drift chambers differ significantly from those σ_D determined from (2.14).

The dependences of σ_D for a number of gases are given in Fig. 4.5. The σ_D values differ noticeably because of differences in the energy acquired by electrons in the electric field (Sect. 2.2); thus, in cool gases (for example CO_2, methylal, isobutane, etc.), σ_D is close to the thermal limit [(2.30)], while in other gases (for instance, Ar, CH_4) it is significantly larger.

Owing to diffusion, the spatial resolution deteriorates proportionally to the square root of the drift length $\sqrt{x}$ (Fig. 4.6), but it is enhanced as the pressure of the gas increases.

The difference between σ_D and $\sigma_{D'}$ is also related to the degree of *isochronism*, i.e. to the fact that electrons from different sections of a track do not arrive at a sense wire simultaneously. Indeed, the electric field in the vicinity of a wire

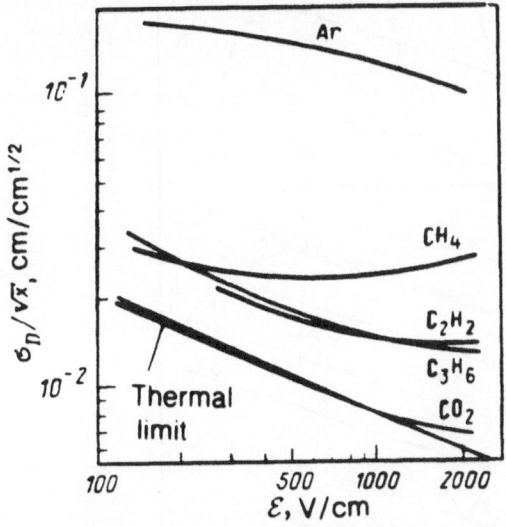

Fig. 4.5. Dependence of $\sigma_D/\sqrt{x}$ on electric field strength (Palladino, Sadoulet 1975)

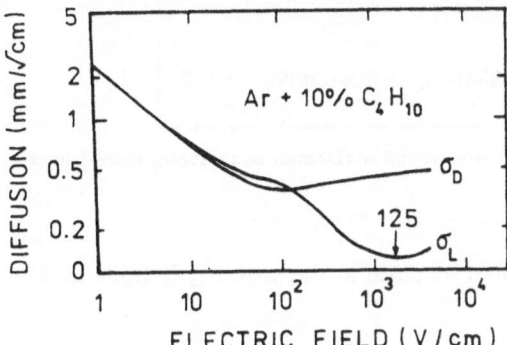

Fig. 4.6. Calculated transverse and longitudinal diffusion for a 1 cm drift and a pressure of 10^5 Pa (Fehlmann et al. 1983b)

exhibits cylindrical symmetry, so that the path of an electron coming from the edge of a cell to the sense wire is longer than the path of an electron arriving from the center of the cell. The solid lines in Fig. 4.7 indicate the trajectories of electrons in the JADE chamber, and the dashed lines represent isochrones. It can be seen that electrons from different sections of the particle trajectory arrive at the sense wire at different times. Deviation from isochronism leads to substantial deterioration of the spatial resolution of a drift chamber.

The resolution of the device σ_a depends mainly on the accuracy with which the chamber is constructed and on the quality of the electronics system. For high-precision drift chambers $\sigma_a = 20$–$50~\mu m$.

4.1.2 Timing Methods

Spatial resolution also depends on the method of recording an electron avalanche at a sense wire. If the electronics system is triggered by the leading edge of the signal and a minimum detection threshold is used, then the spatial resolution

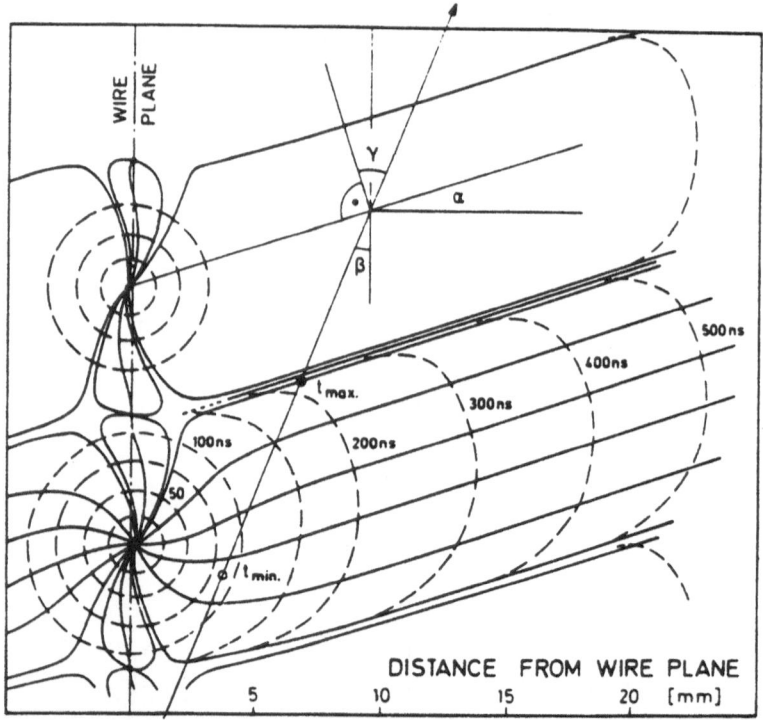

500 ns

400 ns

300 ns

200 ns

100 ns

50

$t_{max.}$

$\phi / t_{min.}$

γ

α

β

DISTANCE FROM WIRE PLANE

5   10   15   20   [mm]

Fig. 4.7. Trajectories of electrons in the Jade drift chamber (Drumm et al. 1980); dashed lines –
isochrones

corresponding to the arrival of the first electron at the wire σ_{1e} is (Schultz and
Gresser 1978):

$$\sigma_{1e} = \frac{1.28 \cdot \sigma_L(1)}{\sqrt{2 \cdot \ln N_t} \cdot \sqrt{P}} \qquad (4.6)$$

where $\sigma_L(1)$ represents the resolution due to single-electron diffusion (2.28) and
N_t is the number of electrons. This equation takes into account pressure P,
registration efficiency, electron capture and other effects occurring in the chamber
volume.

Another method is based on the use of the centroid of the collected charge
as the time marker. In this case the root-mean-square error in measuring the
coordinate x is

$$\sigma_G = \frac{\sigma_L(1)}{\sqrt{N_t} \cdot \sqrt{P}} . \qquad (4.7)$$

The dependences $\sigma_{1e}(x)$ and $\sigma_G(x)$ for $\sigma_L(1) = \sqrt{2\varepsilon_c x / e\mathcal{E}}$ ($\varepsilon_c = kT/e$; e is
the electron charge) are given in Fig. 4.8. This figure illustrates that in a real
detector the dependence $\sigma_{D'}(x)$ lies somewhere between the extreme cases σ_{1e}
and σ_G, because the electronics system is usually triggered by the arrival of
several electrons at a sense wire . The dispersion σ_j due to arrival of the j-th

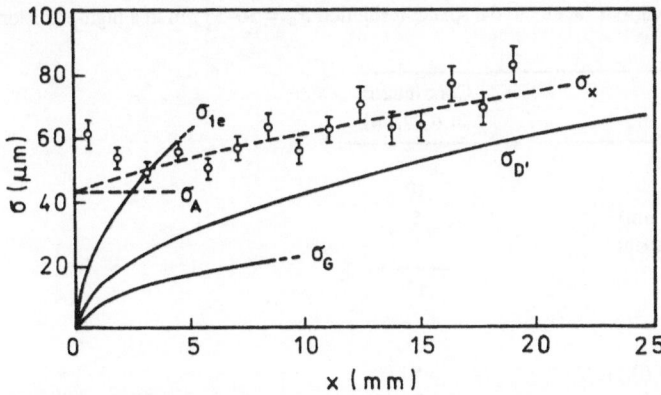

Fig. 4.8. Dependences of σ_{1e}, σ_{G}, $\sigma_{D'}$ on drift length. The points indicate experimental data for pions of momentum $p = 150\,\text{GeV}/c$ (Schultz, Gresser 1978)

electron at the wire is given for $j \ll N_t$ by the following approximate relation (Schultz, Gresser 1978):

$$\sigma_j = \frac{\sigma_L(1)}{\sqrt{2 \cdot \ln N_t} \cdot \sqrt{P}} \sqrt{\sum_{j}^{N_t} \frac{1}{n^2}} . \tag{4.8}$$

The resulting value of $\sigma_{D'}$ turns out to be significantly lower than σ_D.

4.1.3 Resolution in Real Drift Chambers

The total effect of all the factors considered above determines the root-mean-square error σ_x of coordinate measurements in a drift chamber (4.3). The dependence of σ_x on the drift length is given in Fig. 4.9, and it reveals that at a distance of several millimeters from the wire, the precision with which the coordinate is determined is about $50\,\mu\text{m}$.

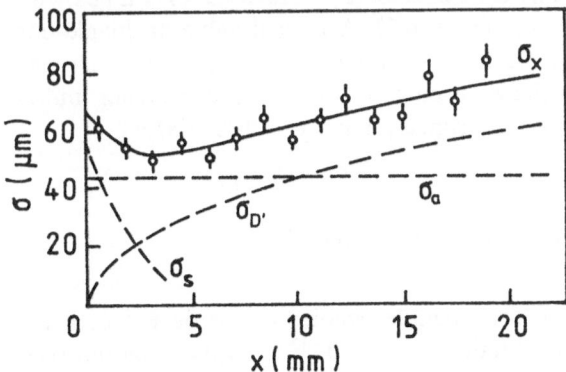

Fig. 4.9. Spatial resolution σ_x of a drift chamber and its components $\sigma_{D'}$, σ_a, σ_s as functions of electron drift length (Filatova et al. 1977)

187

Table 4.1. Contribution of various factors to the spatial resolution $\sigma_x = 50$–$55\,\mu$m in a high-precision drift chamber (Jaros 1980)

	Contribution to σ_x [μm]
1. *Mechanical tolerances*	
Positioning of a wire	10
Gravitational sag (40 μm)	5
Electrostatic displacement	5
	12
2. *Timing uncertainties*	
Start-time (0.2 ns)	10
Particle time of flight (0.1 ns)	5
Signal propagation (0.1 ns)	5
Calibration of TDC (0.1 ns)	5
Resolution of TDC (0.2 ns)	10
	17
3. *Parameters of chamber*	
High voltage variations (10 V)	10
Temperature changes (1°C)	10
Variations of pressure (10^4 Pa)	10
Cell-to-cell gain variations	5
	18
4. *Drift time accuracy*	
Intrinsic resolution	40
Uncertainty in the space-time relation	20
	45
Resolution of a drift chamber	50–55

The conditions to be fulfilled for achieving a resolution $\sigma_x \approx 50$–$55\,\mu$m are presented in Table 4.1, where the contributions of various factors to the total resolution in a drift chamber are also given.

The spatial resolution in large drift chambers is usually 2 to 3 times worse than the limit quantity $\sigma_x \sim 50\,\mu$m (Sect. 4.3). A typical value of double-track resolution is $x_2 \sim 1$ cm. If the electronics records one signal from each sense wire, then x_2 is given by the wire spacing. In chambers employing multihit electronics, x_2 depends mainly on the signal duration (~ 100–200 ns).

4.2 High Precision Gas Coordinate Detectors

A number of physical problems, for example, studies of the decays of beauty and charmed hadrons exhibiting lifetimes of $\sim 10^{-13}$ s, require high-precision coordinate detectors [e. g. Damerell (1987)]. For the solution of such problems a resolution of $\sigma_x < 50\,\mu$m and a double-track resolution of $x_2 \sim 500\,\mu$m are necessary. In many laboratories great efforts are made to develop such detectors (Va'vra 1986a; Manfredi, Ragusa 1986; Damerell 1987; Hayes 1988; Ritson 1988; Villa 1988). The main types of detectors are silicon detectors, scintillating fiber detectors and gas coordinate detectors. The principal advantages of gas coordinate detectors compared to semiconductor detectors (Manfredi, Ragusa 1986; Damerell 1987; Ritson 1988) are low cost, the possibility of constructing large-size detectors, and their high radiation resistance.

High-resolution gas coordinate detectors include the following:

- high resolution drift chambers;
- time expansion chambers;
- multidrift chambers;
- electroluminescence drift chambers;
- induction chambers;
- high resolution cathode readout MWPC (multiwire proportional chamber);
- high resolution avalanche chambers.

4.2.1 High Resolution Drift Chambers

To achieve high precision the following measures are applied in drift chambers;

- slow electron drift;
- cool gases;
- high pressure;
- sample restricting and focusing geometry;
- measurement of the drift time of clusters in a time expansion chamber;
- multiple measurements of track coordinates in jet or multidrift chambers.

In high-precision drift chambers several of the above measures are usually applied. For details we refer the reader to reviews by Charpak and Sauli (1984) and by Va'vra (1986a).

Slow electron drift of the order of $\sim 5\ \mu m/ns$ (a fast drift would be $\sim 50\ \mu m/ns$) is often used in precision drift chambers. Application of slow drift results in the reduction of the contributions of electronics time resolution and time fluctuations in the process of gas amplification to the spatial resolution of the drift chamber. The double-track resolution also improves greatly in the case of a low drift velocity.

A disadvantage of the slow drift is deterioration of the count rate capability of the drift chamber and the increase of the contribution given by electron diffusion to the spatial resolution.

Slow electron drift permits applications of available 100 MHz waveform digitizers, which can provide a spatial resolution of the order of 20–50 μm (Va'vra 1986a; Bock et al. 1986; Hayes 1988; Anderhub et al. 1988).

Cool Gases. The main contribution to the spatial resolution of drift chambers is given by the diffusion of electrons in the gas (Sect. 4.1). In a high-resolution drift chamber it is reasonable to apply gases exhibiting small diffusion coefficients. The cool gases mentioned in Sect. 2.2 satisfy these conditions: CO_2, NH_3, dimethylether CH_3OCH_3, methylal $[(CH_3O)_2CH_2]$, isobutane (iso-C_4H_{10}) and others.

Electron transport in cool gases is characterized by certain peculiarities which are discussed in detail by Christophorou (1970), Dolgoshein et al. (1984) and Sosnovtsev (1987):

– a large momentum transfer cross section, $\sigma_m \sim 10^{-14}$–10^{-13} cm^2 (Huxley, Crompton 1974);

– a large cross section for excitation by electrons of rotational transitions involving a change of angular momentum of the gas molecules. This cross section is only about twice as small as σ_m, i.e., almost every electron collision involves transfer of energy to a molecule (Crawford et al. 1967);

– the characteristic energy transferred in an electron collision with a molecule is $\Delta\varepsilon \approx \sqrt{r_m kT}$, where r_m is the rotational constant of the molecule. The energy transfer coefficient $\Delta\varepsilon/\varepsilon_e$ in cool gases is of the order of 10^{-2}–10^{-1}, while in inert gases it is $\sim 10^{-5}$–10^{-4}.

As a result, electrons in cool gases remain thermal even in high electric field, and their characteristic energy is $\varepsilon_c = kT/e$. The electron drift velocity is not high (several μm/ns) even when $\mathcal{E} \sim 1$ kV/cm, and σ_D is close to the thermal limit (2.30). The dependence $\sigma_D(\mathcal{E})$ in some cool gases is presented in Fig. 4.10. Utilization of slow electron drift in a chamber filled with dimethylether

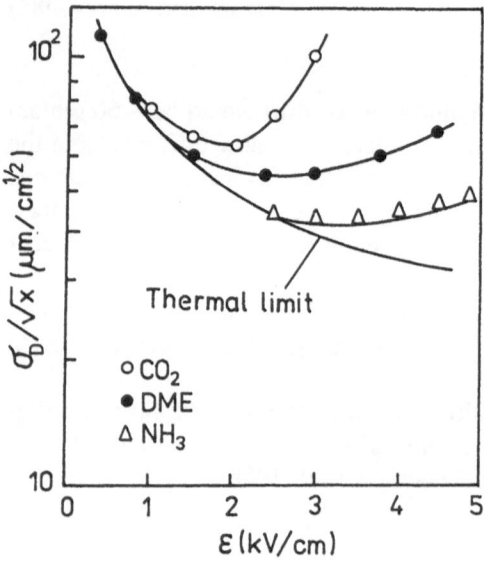

Fig. 4.10. The diffusion (r.m.s.) for 1 cm drift and 10^5 Pa pressure as a function of the electric field in cool gases (Bobkov et al. 1984)

at atmospheric pressure allows a resolution of 16 μm (r.m.s.) to be achieved (Villa 1983). It must be noted that this result is not typical (Basile et al. 1985; Huth, Nygren 1985; Bari et al. 1986; Bobkov et al. 1984).

Let us compare the qualities of operation of cool and hot gases. In accordance with calculations by Va'vra (1986a) it is possible to achieve, in cool gases, a $\sigma_L \sim 50$–$80\,\mu$m (2.28), for $P = 10^5$ Pa, a drift length 1 cm and a first electron timing. For hot gases in the same conditions $\sigma_L \sim 120$–$200\,\mu$m. The advantages of using slow drift in cool gases, however, are reduced significantly by system parameters, which affect the resolution (Va'vra 1986a). The following systematic effects can be observed: distortions in the mechanical construction, wire sag under electrostatic forces, the electric field gradient in the vicinity of the sense wires and near the edges of the chamber, variations in the electron drift velocity due to changes in temperature, pressure and composition of the gas mixture, variations in time of parameters of the electronics, etc. It turns out that when the chamber operates on the "plateau" of the drift velocity in hot gases ($w_d \sim 50\,\mu$m/ns), the influence of systematic effects is much smaller than in the case of cool gases, where the drift velocity rises linearly with the field strength, and $\Delta w_d / w_d \approx \Delta \mathcal{E} / \mathcal{E}$. As a result, the experimental resolutions in cool and hot gases, at high pressures (2–$4 \cdot 10^5$ Pa) do not differ significantly. Thus, for example, in a hot 75 % C_3H_8 / 25 % C_2H_4 gas mixture, at a pressure of $4 \cdot 10^5$ Pa and for a 4 mm drift length, the mean resolution achieved in a Jet drift chamber was 23 μm (Belau et al. 1982).

High Pressure. The spatial resolution is enhanced with the increase of pressure, if an increase of electric field strength $\mathcal{E}$ is provided such that the reduced field $\mathcal{E}/P = $ const. Then from (2.14, 29, 30)

$$\sigma_D = \sqrt{\frac{2 \cdot D(\mathcal{E})x}{\mu_e \mathcal{E}}} = \sqrt{\frac{2kT}{e\mathcal{E}/P}} \cdot \sqrt{\frac{x}{P}} \cdot \tag{4.9}$$

As the pressure increases, not only does diffusion reduce as $1/\sqrt{P}$, but the δ-electron range also decreases, which also contributes to the enhancement of the spatial resolution. The resolution for various gas pressures is shown in Fig. 4.11. The dependence of the spatial resolution on the pressure is presented in Fig. 4.12.

In a drift chamber filled with a cool gas at a pressure of 4–$5 \cdot 10^5$ Pa it is possible to achieve a spatial resolution $\sigma_{1e} \sim 10$–$20\,\mu$m/$\sqrt{\text{cm}}$ (Va'vra 1986a) and a resolution of adjacent tracks of $x_2 \sim 200\,\mu$m (Bobkov et al. 1984). In hot gases in the same conditions $\sigma_{1e} \sim 20$–$50\,\mu$m/$\sqrt{\text{cm}}$ can be achieved.

At very high pressure the relation $\sigma_L \sim 1/\sqrt{n_a}$, where n_a is the atomic gas density, is violated due to density effects (Sosnovtsev 1987; Khrapak, Yakubov 1981; Atrazhev, Yakubov 1977) and a further increase of n_a gives almost no enhancement of the resolution. In Table 4.2 the calculated diffusion coefficients and spatial resolution at high pressures are presented.

Table 4.2. Longitudinal and transverse diffusion coefficients and computed resolutions σ_L (2.28) and σ_D (2.14) for high pressure and a drift length of 1 cm (Sosnovtsev 1987; Wagner et al. 1967; Warren, Parker 1962; Crompton et al. 1967)

Gas	P [10^5 Pa]	$\mathcal{E}$ [kV/cm]	D_L/μ_e [V]	D_T/μ_e [V]	σ_L [μm]	σ_D [μm]
Ar	100	100	1	7	45	120
He	100	50	0.4	0.9	40	63
N_2	100	150	0.5	0.8	26	33
H_2	100	150	0.2	0.4	16	23
CO	50	127	0.19	0.5	17	27
CO_2	20	10	0.03	0.03	24	24

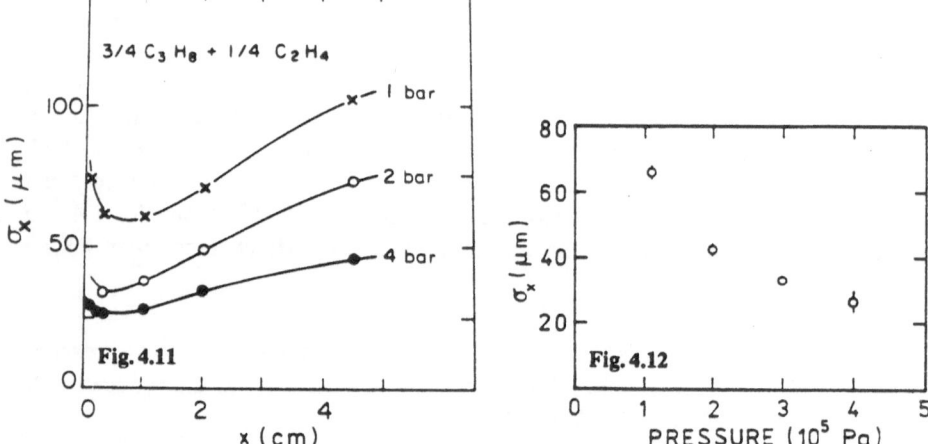

Fig. 4.11. Spatial resolution (r.m.s.) versus drift length for different gas pressures in the JADE chamber (Drumm et al. 1980)

Fig. 4.12. Average spatial resolution over a 2 cm drift distance versus pressure (Alexander et al. 1986)

Sample Restricting and Focusing Geometry. The isochronism of a drift chamber (Fig. 4.7) can be strongly enhanced by using only a small section of the track close to the sense wire. The choice of the required track segment can be carried out either mechanically (Fig. 4.13) or with the aid of a focusing electric field (Fig. 4.14). In a chamber known as a drift precision imager (Bobkov et al. 1984) sample restricting geometry was used, by application of a 0.8 mm slit which allowed $\sigma_s \leq \sigma_L$, for $x = 2\,\text{mm}$ to be achieved. At a pressure of $4 \cdot 10^5$ Pa and $x = 10\,\text{mm}$ a resolution of $\sim 30\,\mu\text{m}$ and a double-track resolution of $\sim 200\,\mu\text{m}$ were achieved. Sample restriction not only enhances the spatial resolution, but also increases requirements on the electronics system. In Fig. 4.14 various electrode configurations in the gas amplification region of vertex drift chambers are

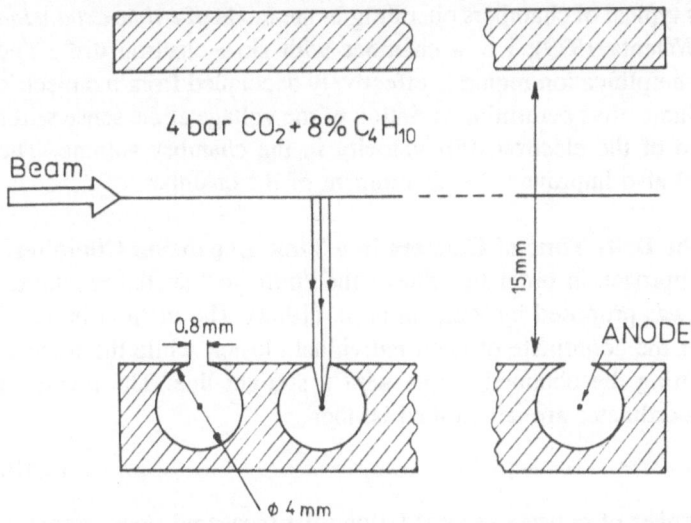

$4 \, bar \, CO_2 + 8\% \, C_4H_{10}$

Beam

15 mm

0.8 mm

ANODE

$\phi \, 4 \, mm$

Fig. 4.13. Sample restricting geometry in a drift precision imager (Bobkov et al. 1984)

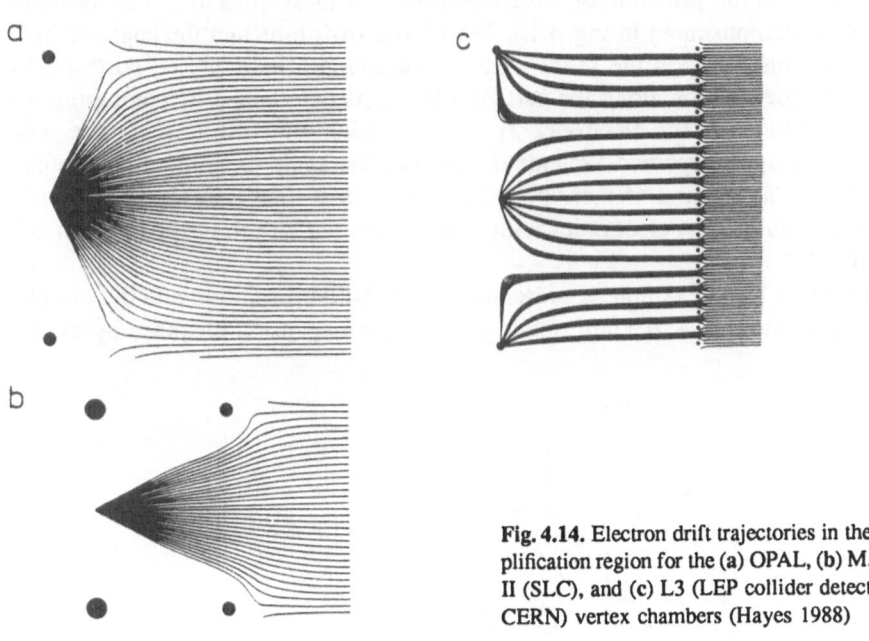

a

c

b

Fig. 4.14. Electron drift trajectories in the amplification region for the (a) OPAL, (b) MARK II (SLC), and (c) L3 (LEP collider detector at CERN) vertex chambers (Hayes 1988)

presented. Chamber (a) has no grid separating the drift and amplification regions. Such a chamber is characterized by strong non-isochronism.

Focusing geometry (b) enhances the isochronism and the operation stability of the sense wires. This allows the sense wire spacing to be decreased, which leads to improved double-track resolution and capability of the chamber to operate in intense beams.

Geometry (c) is typical of chambers operating in the mode of a *time expansion chamber* (TEC) (Walenta 1979), i.e. a chamber with slow electron drift. The electric field in the amplification region is effectively decoupled from the electric field in the drift volume, thus permitting variation of the voltage at the sense wires without any change of the electron drift velocity in the chamber volume. The presence of the grid also improves the isochronism of the chamber.

Measurement of the Drift Time of Clusters in a Time Expansion Chamber.
The most radical approach in order to achieve the "ultimate" spatial resolution in a drift chamber was proposed by Walenta et al. (1982). The method is based on determination of the coordinate of each individual cluster, while the track is reconstructed by fitting the obtained points with a straight line. The precision with which track coordinates are determined is, then,

$$\sigma_x = \sigma_{1e}/\sqrt{N_c} \tag{4.10}$$

where N_c is the number of clusters recorded along the measured track segment. Thus, in a standard gas mixture, where $N_c \approx 30\,\mathrm{cm}^{-1}$, a 3–5-fold improvement of the spatial resolution can be achieved in ideal conditions.

The operation principle of a drift chamber for measurement of cluster coordinates is demonstrated in Fig. 4.15. The cluster drift time and the angle of incidence of clusters upon the sense wire are measured. The angle α_c corresponds to the cluster position along the particle track. It is determined by comparing the charges induced in the electrodes P_l and P_r, which are close to the sense wire (A). Applying this method Walenta determined the angle α_c with an uncertainty of $\pm 2.3°$. The method of measuring the cluster drift time is extremely complicated. It requires fast electronics with a dead time $\sim 15\,\mathrm{ns}$, described in Sect. 3.9 and 4.5.

In Fig. 4.16 an example is presented of the spatial resolution and the double-track resolution $x_2 \sim 0.3\,\mathrm{mm}$ obtained in a time expansion chamber by the L3

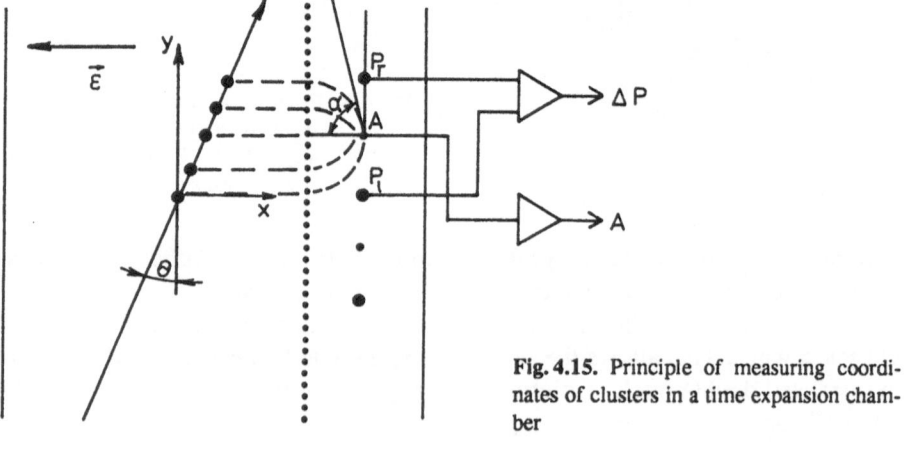

Fig. 4.15. Principle of measuring coordinates of clusters in a time expansion chamber

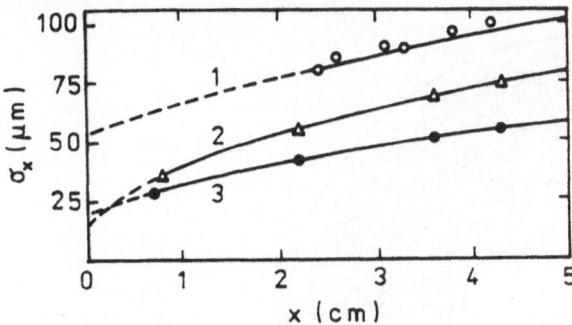

Fig. 4.16. Dependence of spatial resolution σ_x on drift length in a time expansion chamber; the curves correspond to $\sigma_x = (\sigma_{D'}^2 x + \sigma_a^2)^{1/2}$; *curve 1:* $\sigma_{D'} = 40$, $\sigma_a = 54$, *curve 2:* $\sigma_{D'} = 36$, $\sigma_a = 14$, *curve 3:* $\sigma_{D'} = 25$, $\sigma_a = 19$. Units of $\sigma_{D'}$ are $\mu m/cm^{1/2}$; of σ_a μm (Aachen-Siegen-Zürich Vertex group 1984)

collaboration (Anderhub et al. 1988). The drift times of individual clusters were not measured in this chamber.

The idea of measuring the cluster drift time was developed by Huth and Nygren (1985) who proposed the concept of a *radial drift chamber* (RDC). An RDC is similar, in principle, to the scheme presented in Fig. 4.15, but large angles θ of the particle entrance to the chamber are assumed. The pick-up electrodes P_l and P_r are continuously scanned by flash ADC, providing information on the positions of clusters along the particle track and providing a resolution corresponding to (4.10). Huth and Nygren (1985) assume that in an RDC filled with dimethylether (DME) at a pressure of $1.5 \cdot 10^5$ Pa, it is possible for a resolution in individual cluster positioning of $\sigma_c \sim 40\,\mu m$, to obtain a tracking accuracy of $8\,\mu m$.

Drift Chambers with Symmetric Cells. The volume of such a chamber is partitioned into numerous symmetric cells, usually of a square or hexagonal profile (e. g. Fig. 3.4 for CDCs). The advantage of such geometry is simplicity of construction, a low number of field wires, and weak dependence of the resolution upon the track inclination angle. The main disadvantages are the drift velocity not being constant, owing to the strong inhomogeneity of the field in a cell, and a significant non-isochronism of the arrival of electrons at the anode wire. Vertex detectors with symmetric cells were applied in the MARK II (PEP) (Jaros 1988) detector, where a resolution of 85 μm for Bhabha electrons was achieved and in the ARGUS (ARGUS Collaboration 1989) detector with $\sigma_x' \sim 40$–$80\,\mu m$. In the new ARGUS Micro-vertex drift chamber with $5.2 \times 5.3\,mm^2$ cells, a resolution of 20 μm was obtained (Michel et al. 1989) in a 90 % $CO_2 + 10$ % C_3H_8 mixture at a pressure of $4 \cdot 10^5$ Pa (or in DME at $2.5 \cdot 10^5$ Pa) and for a 1.5–2 mm drift length.

Planar drift chambers with symmetric cells (Fig. 4.1d) are also applied in fixed target experiments; for example, in a drift chamber with $6 \times 6\,mm^2$ cells at atmospheric pressure, a 50 μm resolution was achieved (Chirikov-Zorin et al. 1987).

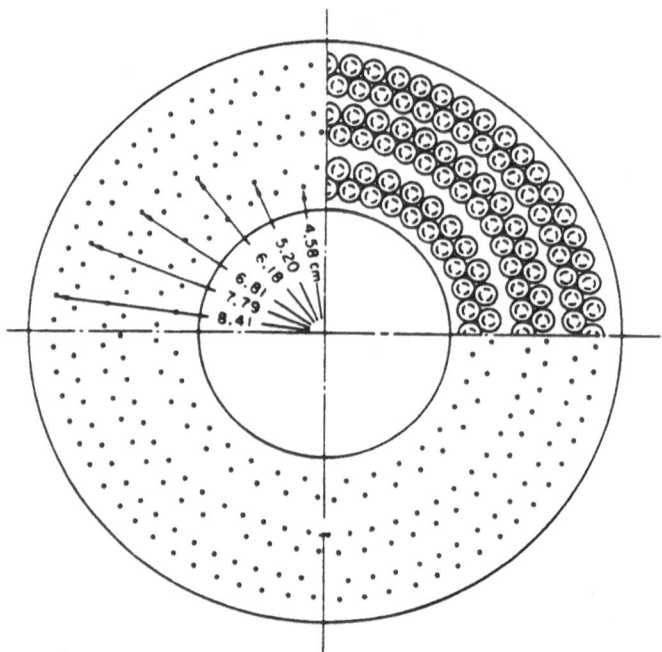

Fig. 4.17. Scheme of the MAC vertex detector assembly of drift tubes (Nelson 1988)

Multidrift Tubes. A very important feature of vertex detectors is their operation reliability and the possibility of rapid removal of defects during physical experiments. Such properties are exhibited by modular detector systems. In the MAC detector inexpensive, readily constructible, drift tubes, so-called "straws", were applied with success (Nelson 1988). The cathodes of the tubes are made of thin mylar, thus removing problems of shaping the drift field. Drift tubes are used for the assemblage of vertex detectors (Fig. 4.17). In the MAC vertex drift chamber a good resolution of 20–25 μm was achieved at a pressure of $4 \cdot 10^5$ Pa. This is presented in Fig. 4.18.

A group at CERN has developed the idea of drift tubes and has combined several tens of hexagonal tubes into a single module (Bouclier et al. 1988). In this way a convenient and compact multidrift tube was obtained, which can be used for assembling detectors of various configurations.

The advantages of multidrift tubes consist not only in their modularity, but also in their small radial dimension, which allows their application in a high-rate environment, and facilitates the disentangling of complex multitrack events. They are less subject to radiation damage.

The disadvantages of drift tubes are: inhomogeneity of the electric field; it is not reasonable to use multihit electronics, multiple scattering and conversion of photons in the tube walls. Long tubes are subject to deformation.

Vertex TPC. A time projection chamber was used as a vertex detector in the CDF detector (Wagner 1988). Its construction is similar to the TPC presented in

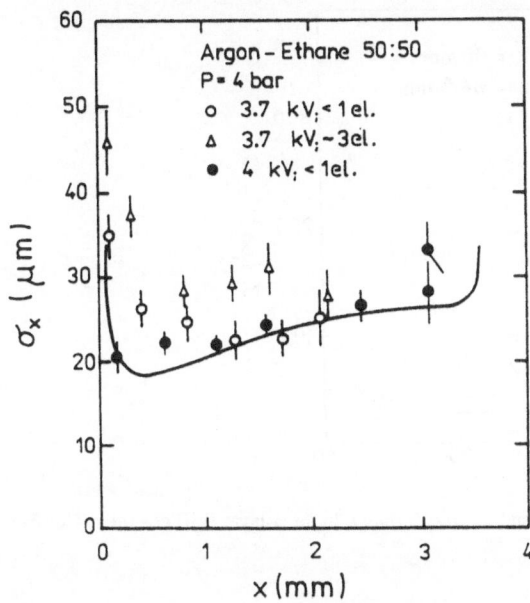

Fig. 4.18. Spatial resolution versus drift distance for a variety of thresholds in the MAC vertex drift detector (Nelson 1988). The line represents a Monte-Carlo calculation

Fig. 3.5. In test measurements a moderate resolution of 200–500 μm was obtained in the drift direction and 300 μm was achieved along the sense wires.

An advantage of the TPC used as a vertex detector is three-dimensional track reconstruction with a good spatial resolution along the beam axes of the collider; in most vertex chambers this coordinate is measured badly (several cm). A disadvantage of the TPC is the insufficiently high spatial resolution (even along one of the coordinates) and the long time required for collecting the electrons. TPCs will hardly overcome the competition of other types of vertex detectors.

Electroluminescence Drift Chamber. In these chambers the arrival time of electrons at the anode is recorded, utilizing the photons produced in electroluminescence of the gas in the course of gas amplification (Charpak et al. 1975; Baskakov et al. 1979; Simon, Braun 1983; Volkov et al. 1985). The operational principles of electroluminescence drift chambers (ELDCs) are dealt with in Sect. 6.3. The main advantage of the ELDC is its speed and operation at low gas amplification, which allows it to be utilized in a high-rate environment. An excellent resolution of 16 μm was achieved in an ELDC filled with xenon at a pressure of $2 \cdot 10^6$ Pa (Baskakov et al. 1979). The high speed of these chambers permits recording individual clusters on a particle track (Volkov et al. 1985; Polyakov, Rykalin 1987). The disadvantages of ELDCs include their complicated construction, with light guides containing a wavelength shifter for transformation of the spectrum. This makes it difficult to produce large electroluminescence chambers, and to our knowledge no such large system is yet under operation.

Multiple Measurement of Track Coordinates. Localization of the trajectories of high-energy particles characterized by low multiple scattering can be improved

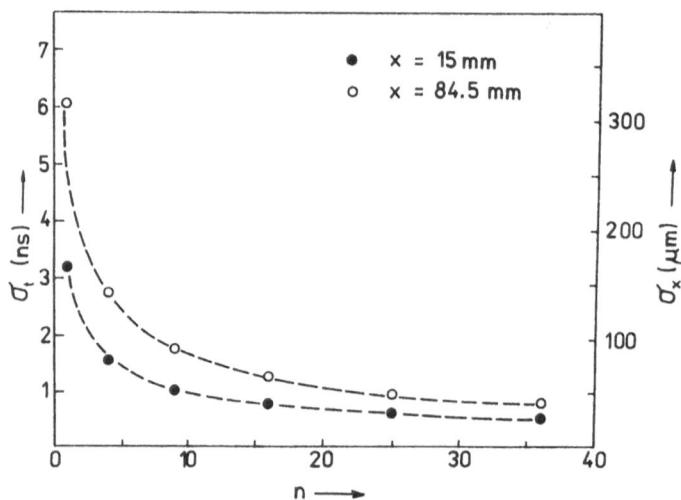

Fig. 4.19. Spatial and time resolutions σ_x and σ_t, respectively, in an n-layer drift chamber (Duinker et al. 1982)

significantly by multiple measurement of their coordinates in a multilayer drift chamber. This approach was tested experimentally (Duinker et al. 1982), and it turned out that the resolution $\sigma_x(n)$ is enhanced with the square root of the number n of measured points: $\sigma_x(n) = \sigma_x/\sqrt{n}$. The $\sigma_x(n)$ dependence shown in Fig. 4.19 is valid when the systematic errors are reduced to the minimum. When the number of measurements is large, the resolution no longer improves with increasing n and does not follow the $1/\sqrt{n}$ law.

The principle of multiple coordinate measurements is applied mainly in high-precision *Jet drift chambers*. The layout of a Jet chamber is presented in Figs. 3.6 and 3.7. Jet drift chambers, with a number of layers from 20 to 40, are applied as vertex detectors in the MARK II (SLC) (Alexander et al. 1986; Jaros 1988), (OPAL Detector 1983; Carter et al. 1989) and UA–1 (Muller 1986) spectrometers. The L3 detector has no central drift chamber, but it involves a high-precision Jet drift chamber with 70 layers (Anderhub et al. 1988). The spatial resolution of measurements of individual points on tracks in these chambers is 20–50 μm (Fig. 4.20).

Vertex Drift Chamber with "Ultimate" Parameters. In high-precision drift chambers several of the measures mentioned above are applied simultaneously in order to achieve a high spatial resolution: cool gases at high pressures (3–$4 \cdot 10^5$ Pa), slow electron drift, multiple measurement of coordinates on tracks, sample restricting and focusing geometries. The scheme of a version of a vertex drift chamber with "ultimate" parameters proposed by Va'vra (1986a) is presented in Fig. 4.21. This chamber would be characterized by good isochronism and resolution, presented in Table 4.3.

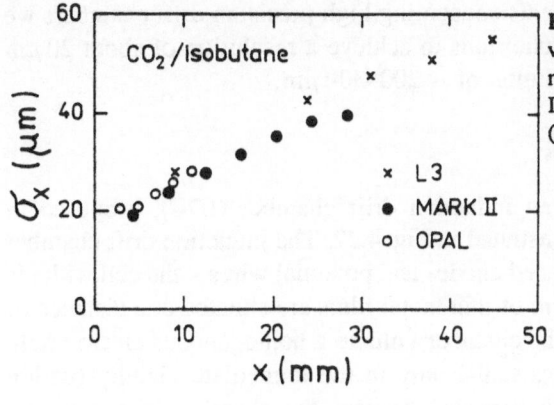

Fig. 4.20. Spatial resolution (r.m.s.) versus drift distance in the CO_2-isobutane mixture of the OPAL, L3 and MARK II (SLC) vertex Jet drift chambers (Hayes 1988)

CO_2/Isobutane

× L3
● MARK II
○ OPAL

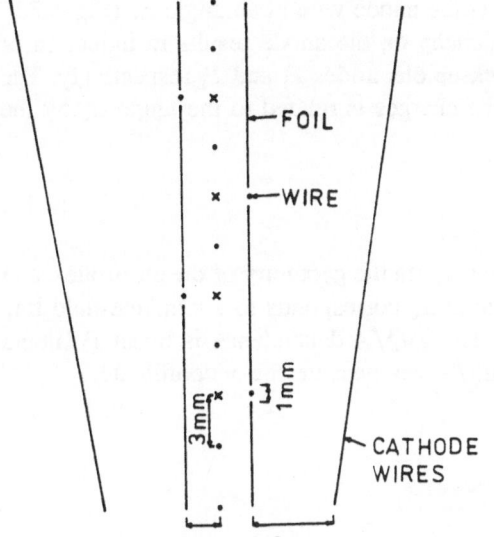

FOIL

WIRE

CATHODE WIRES

3 mm   1 mm

2 mm   < 2 cm

Fig. 4.21. A version of an "ultimate" vertex drift chamber design using the TEC concept with sample restricting and focusing geometry (Va'vra 1986a)

Table 4.3. Calculated drift chamber (Fig. 4.21) resolution for a 92 % CO_2 + 8 % C_4H_{10} gas mixture at a pressure of $4 \cdot 10^5$ Pa, 12 mm drift distance and $w_d = 4.0–4.2\,\mu$m/ns (Va'vra 1986a)

Timing method	Resolution [μm]	
	$\theta = 0°$	$\theta = 10°$
First electron timing with infinitely fast electronics	20 ± 3	34 ± 3
Centroid timing with infinitely fast electronics (first 100 ns)	7 ± 1	14 ± 2
Leading edge timing with realistic pulses (threshold 2–3 % of the average amplitude)	20 ± 3	48 ± 4
Centroid timing with realistic pulses and 100 MHz digitizer	20 ± 1	27 ± 2

To conclude with the arguments concerning high precision drift chambers we note that is is possible in real conditions to achieve a resolution of about 20 μm (r.m.s.) and a double-track resolution of $\sim$200–300 μm.

4.2.2 Induction Drift Chamber

The principle of operation of an induction drift chamber (IDC), proposed in Roderburg et al. (1986) is demonstrated in Fig. 4.22. The induction drift chamber consists of a plane of closely spaced anodes and potential wires – the cell width is 600 μm. The cathodes in the form of thin metal films are situated at a distance of 6 mm from each other. Within the chamber volume a homogeneous electric field is created, and the field becomes radial only in the immediate vicinity (within $\sim$0.5 mm) of the sense wire. A particle traverses the chamber in a direction perpendicular to the anode wire plane. Electrons drift away from the particle track along the field lines and arrive at the anode wire at an angle α_c (Fig. 4.22). The development of an electron avalanche on the anode results in induction of charges Q_1 and Q_r in the adjacent pick-up electrodes P_1 and P_r respectively. The difference $\Delta Q = Q_1 - Q_r$ between the charges is related to the angle α_c by the relation

$$\alpha_c = \arcsin\left(\frac{1}{K_1} \cdot \frac{\Delta Q}{A}\right) , \tag{4.11}$$

where K_1 is a factor which depends mainly on the geometry of the electrodes, and A is the charge on the anode. The angle α_c corresponds to a concrete field line and, thus, to a certain coordinate x. The $\Delta Q/A$ dependence is linear (Walenta et al. 1988), so that by measuring $\Delta Q/A$ we measure the x coordinate.

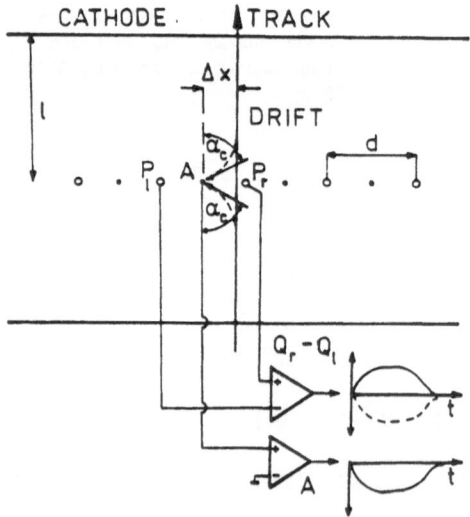

Fig. 4.22. Schematic cross section and principle of operation of the induction drift chamber

In the prototype IDC a resolution $\sigma_x \sim 25~\mu\mathrm{m}$ has been achieved in a $300~\mu\mathrm{m}$ wide drift gap. At a distance of $x \sim 50~\mu\mathrm{m}$ from the sense wire a $15~\mu\mathrm{m}$ resolution was obtained (Walenta et al. 1988).

In an IDC there exists no left-right ambiguity characteristic of other drift chambers. A great advantage of the induction drift chamber is its capability to operate at high beam rates, due to the small sense wire pitch. The rate capability of the IDC is up to 10^6 particles/mm$^2 \cdot$ s.

A disadvantage of the IDC is that the angular dependence of its resolution is quite strong

$$\sigma_x(\theta) = \sqrt{\sigma_0^2 + (A_b \theta)^2} \tag{4.12}$$

where $\sigma_0 = 29~\mu\mathrm{m}$ and $A_b = 6.25~\mu\mathrm{m}$/degree (Walenta et al. 1988).

An IDC is under construction as the vertex chamber of the ZEUS detector (Walenta et al. 1988). The authors intend to achieve a resolution of $\sim 20~\mu\mathrm{m}$.

4.2.3 High Precision MWPC with Cathode Readout

The scheme of an MWPC with cathode readout is given in Fig. 4.23. Particles travel in a direction perpendicular to the anode wire plane. An avalanche produced at an anode wire induces in the cathodes a charge of quasi-lorentzian shape. The width of the cathode strips is chosen to satisfy the condition that most of the induced charge be collected by three strips. The signal from the strips goes to charge-sensitive amplifiers followed by stretchers. The y coordinate of the avalanche along the sense wire is

$$y = \sum_i Q_i y_i \Big/ \sum_i Q_i \tag{4.13}$$

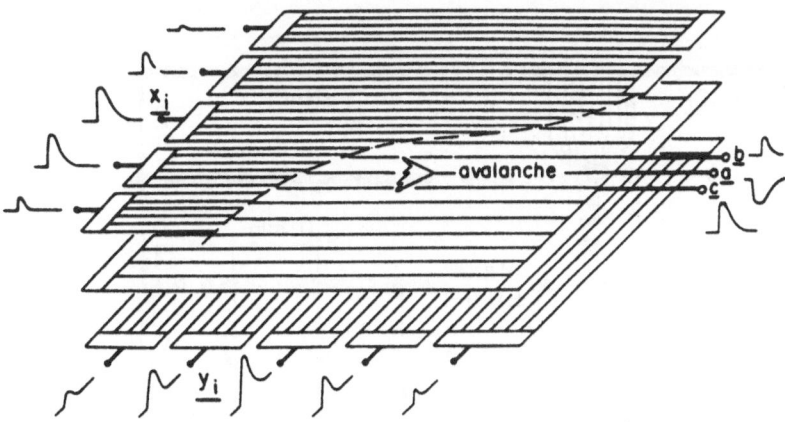

Fig. 4.23. Principle of the center-of-gravity localization method by readout of the induced charges on cathode strips (Charpak, Sauli 1984)

where Q_i is the charge in the y_i-th strip. The method of cathode readout is discussed in detail, for example, by Charpak and Sauli (1984).

It turns out that in an MWPC involving sophisticated readout electronics (Fischer et al. 1986) it is possible to achieve the excellent position resolution of $6\,\mu$m (r.m.s.) in recording 8 keV X-rays. The parameters of the MWPC are the following: anode-cathode gap – 0.4 mm, wire pitch – 1.4 mm, wire diameter – 8 μm, pitch of the cathode strips – 0.575 mm, gas – 90 % Xe + 10 % CO_2, pressure – 10^6 Pa.

Fischer et al. (1986) analyzed the contribution of various factors to the resolution of 14 μm (fwhm): electronic noise $\sim 4\,\mu$m, diffusion $\sim 7\,\mu$m, the spread of the photon beam $\sim 7\,\mu$m, the electron range $\sim 6\,\mu$m. The contribution of avalanche fluctuations can be neglected. From the above it follows that, if electron ranges are not taken into account all these factors contribute less than 10 μm (fwhm) to the resolution. Our opinion is that in a high precision MWPC filled with Xe + CO_2 at a pressure of 5–10·10^5 Pa one may achieve a resolution $\sim 10\,\mu$m (r.m.s.) in recording weakly ionizing particles. Thus, MWPCs with cathode readout may become the most precise gas coordinate detector. To our knowledge, the best resolution for minimum ionizing particles obtained in MWPCs is $\sim 25\,\mu$m (r.m.s.) (Bondar et al. 1983).

4.2.4 Parallel Plate Avalanche Chamber

Avalanche detectors exceed other gas coordinate detectors in an unusually high count rate capability of $\sim 10^7$ particles/mm$^2 \cdot$ s (Gaukler et al. 1977), and an excellent time resolution of ~ 0.14 ns (Breskin, Zwang 1977) providing a good coordinate resolution of $\sim 25\,\mu$m (Potter 1984). These characteristics were obtained for heavy ions.

A parallel plate avalanche chamber (PPAC) consists of two parallel plane electrodes (Fig. 4.24) close to each other. High voltage is applied to the electrodes, to create a uniform electric field sufficient for ionization via electron collisions.

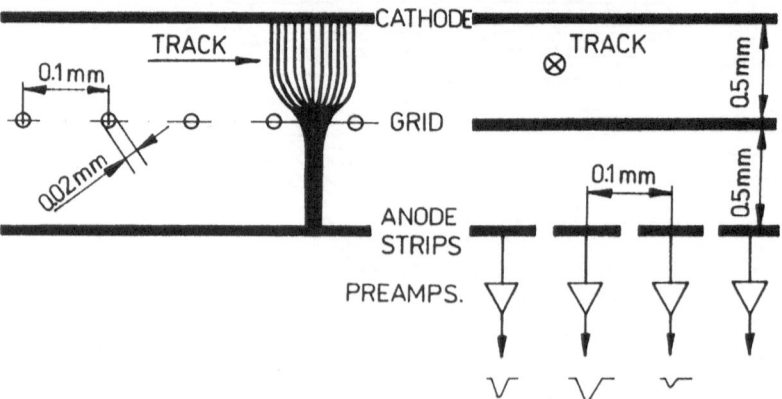

Fig. 4.24. Layout of a vertex detector based on PPAC

In a PPAC a well localized avalanche forms on the particle track. The number of secondary electrons in an avalanche is given by (2.68). The efficiency of particle registration depends on the gas gain (2.68) and on the number of electrons on the track N_t; therefore PPACs are very efficient in recording strongly ionizing particles (Breskin, Zwang, 1977; Gaukler et al. 1977; Potter 1984).

For weakly ionizing particles the number of primary electrons in a detector 1–2 mm thick at atmospheric pressure is small ($N_t \sim 10$), so in PPACs difficulties arise owing to large fluctuations of the avalanche size, which depends exponentially on the distance of the primary electron from the anode. To achieve efficient registration of weakly ionizing particles in PPACs, large gas gains should be used. If a strongly ionizing particle happens to pass through the PPAC, a powerful avalanche arises which results in a spark discharge. The frequent appearance of breakdowns in PPACs represents the main disadvantage of such detectors.

These difficulties may be overcome in two ways: by enhancement of the pressure (Potter 1984; Karpenko et al. 1987; Khazins 1989) or by supplementing the PPAC with a drift gap (Charpak, Sauli 1984; Bellazzini et al. 1986; Hendrix, Lentfer 1986; Peisert, Sauli 1986; Khazins 1989). In the drift gap (the upper part of the chamber in Fig. 4.24) a weak electric field ($\sim 1\,\text{kV/cm}$) is created in which electrons drift towards the grid. Below the grid the electric field is very high ($> 10\,\text{kV/cm} \cdot 10^5\,\text{Pa}$), and it is in this field that gas amplification takes place. Frequently detectors are used which involve two or more parallel-plate amplifying elements, the so-called multistep avalanche chambers (Charpak, Sauli 1978, 1984).

Bellazzini et al. (1986) obtained a resolution of $60\,\mu\text{m}$ (r.m.s.) for weakly ionizing particles in a PPAC filled with 90 % Ar + 10 % CH_4 at a pressure of $10^5\,\text{Pa}$. In 88 % Ne + 10 % He + 2 % CH_2 at $0.5 \cdot 10^5\,\text{Pa}$ a resolution of $\sim 25\,\mu\text{m}$ (r.m.s.) was obtained using optical readout (Potter 1984).

Khazins (1989) has proposed a high-precision PPAC for measuring the coordinates of relativistic particles (Fig. 4.24). The anode-cathode distance in the PPAC is 1 mm. The grid of 0.1 mm pitch is made of $20\,\mu\text{m}$ wires and allows separation of the drift and amplification regions and focusing of the electron trajectories. The anode is divided into strips with a 0.1 mm spacing that are oriented in a direction perpendicular to the grid wires. The PPAC is filled with argon or xenon and small admixtures (several percent) of molecular gases up to a pressure of $5 \cdot 10^5\,\text{Pa}$. About 80 electrons are produced in the drift gap, which greatly reduces the fluctuations of the signal amplitudes. The average signal amplitude is $\sim 20\,\text{fC}$, and its duration is 20–30 ns. The coordinates of a particle track along the sense wires are determined from measurement of the charge Q_i, collected in the strips using (4.13), with a computed resolution of $\sim 20\,\mu\text{m}$ (r.m.s.) in argon and of $\sim 11\,\mu\text{m}$ in xenon (Karpenko et al. 1987). The double-track resolution x_2 should be $\sim 200\,\mu\text{m}$. The collection rate of the positive charge provides a maximum count rate of $\sim 10^7$ particle/mm$^2 \cdot$ s. Spark breakdowns for a flux of $\sim 10^8$ particles/s over the entire detector area reduce the efficiency by about 10 %. The resistance of the detector to radiation damage is estimated to be up to $\sim 5 \cdot 10^7\,\text{rad}$.

Thus, PPACs have good prospects as high precision coordinate detectors, mainly due to their application in very high count rate environments and also their high resistance to radiation damage.

4.2.5 Gas Coordinate Detectors Versus Silicon and Scintillating Fiber (SCIFI) Detectors

A comparison of the parameters of certain vertex detectors is given in Table 4.4. It must be noted that not all the figures given in Table 4.4 have been obtained in identical conditions. The spatial resolution for drift chambers were obtained in test measurements, while for some detectors only estimates are given. The "ultimate" resolution of drift chambers is estimated to be 15–20 μm. The most precise gas detectors seem to be electroluminescence drift chamber, the cathode readout MWPC and the parallel-plate avalanche chamber, in which it may be possible, after some work, to achieve a resolution of $\sim 10\,\mu$m. These detectors

Table 4.4. Spatial and double-track resolutions and radiation hardness for some vertex detectors

Vertex detector	Spatial resolution [μm]	Double track resolution [μm]	Lifetime [10^6 rad]
Drift chambers:			
L3/MARK J TEC[a] (Anderhub et al. 1988)	30	600	0.2
MARK II (SLC) Jet DC (Jaros 1988)	30	900	0.2
DPI (Bobkov et al. 1984)	30	300	0.2
ARGUS SC DC (Michel et al. 1989)	20–100	5000[b]	0.3
Electroluminescence DC (Baskakov et al. 1979)	16	$\sim$200	$\sim$10.0
Other gas detectors:			
MAC DT (Nelson 1988)	25	14000[b]	0.3
Multidrift tubes (Bouclier et al. 1988)	30	600	0.5
ZEUS IDC (Walenta et al. 1988)	25	600[b]	1.0
MWPC (Fischer et al. 1986)	$\sim$10	$\sim$1000	1.0
PPAC (Khazins 1989)	15–20	$\sim$200	$\sim$50.0
Silicon detectors:			
MSD (Damerell 1987; Ritson 1988)	3	40	$\geq$0.1
CCD (Damerell 1987; Ritson 1988)	5 $\times$ 5	40 $\times$ 40	0.1
SIDC (Rehak et al. 1985)	20	$\sim$200	0.1
Scintillation detectors:			
SCIFI – glass (Kirkby 1988; Manfredi, Ragusa 1986)	$\sim$10[c]	$\sim$20	10.0
SCIFI – plastic (Kirkby 1988)	$\sim$500[c]	$\sim$1000	1.0

[a] TEC – time expansion chamber; DC – drift chamber; DPI – drift precision imager; SC DC – drift chamber with symmetric cells; DT – drift tubes; IDC – induction drift chamber; MWPC – proportional chamber with cathode readout ; PPAC – parallel plate avalanche chamber; MSD – multistrip silicon detector; CCD – charge coupled device; SIDC – silicon drift chamber; SCIFI – scintillating fiber detector
[b] cell diameter
[c] fiber diameter

are capable of operation in a high count rate environment, and they can be used in detectors at future colliders and in fixed target experiments.

A comparison of gas and silicon detectors reveals that although the first exhibit a lower spatial resolution, their resistance to radiation damage is much higher. This is important for detectors to be used at the new generation of colliders where, for example, at the SSC and the LHC, a radiation dose of 10^6 rad/year (for $r = 10$ cm) is expected, while at the upgraded LHC with a 5 ns bunch spacing it will even be $\sim 10^7$ rad/year. It must be noted that the figures given for radiative hardness in Table 4.4 are only tentative. In the literature extremely diverse values are encountered, since estimation of the operation capability of a detector having been exposed to radiation damage is subjective.

Another advantage of gas detectors is that detectors of large areas and volumes may be assembled at a low price per unit area.

4.3 Spatial Resolution of MDCs

A multilayer drift chamber (MDC) (Figs. 3.1 to 3.8) represents a large gas volume in which electrons drift from particle tracks toward a system of numerous sense wires, where their arrival time is recorded. Multilayer drift chambers are also termed pictorial chambers; some of them are time projection chambers (TPC). They provide true three-dimensional reconstruction of events with high track multiplicity and permit the identification of secondary charged particles based on multiple measurements of the ionization energy loss (Chap. 3). The images of tracks in a MDC are of the very high quality illustrated by Fig. 3.9. In this section we would like to discuss the spatial resolution of different MDCs.

The idea of constructing a large MDC was put forward at the beginning of the 1970s (Allison et al. 1974; TPC Proposal 1976). The first chambers [(ISIS (Allison et al. 1984), TPC (TPC Proposal 1976), JADE (Wagner 1982), UA–1 (Calvetti et al. 1982)] went into operation at the beginning of the 1980s.

There exist two main kinds of MDC (Sect. 3.1): planar MDCs are usually employed in fixed target experiments, and cylindrical chambers are used at colliders (Wagner 1981; Sauli 1984; Sitar 1985, 1987; Sauli 1988a; Saxon 1988).

4.3.1 Planar MDCs

Planar multilayer drift chambers are sometimes used as external identifiers of charged relativistic particles in magnetic spectrometers with fixed targets. Some examples are the ISIS (Allison et al. 1984), CRISIS (Toothacker et al. 1988) and ASTRON (Babaev et al. 1978) chambers mentioned in Chap. 3. MDCs are also used as vertex detectors, for example, by the European muon collaboration (EMC) (Braun 1983) at CERN, the ICS (Budagov et al. 1989) chamber in the HYPERON spectrometer at Dubna, or the TPC chamber (Benso et al. 1983) of the OMEGA spectrometer at CERN. MDCs are also built for measuring the ranges of heavy ions (Artukh et al. 1991). The main parameters of some plane

MDCs are presented in Table 4.5. All the chambers represented in Table 4.5 are time projection chambers.

TPCs can be recommended as devices with good prospects for relativistic heavy ion experiments (Sauli 1988b). Such chambers are under construction at Brookhaven National Lab (Etkin et al. 1989) and CERN (Garabatos 1989). Their

Table 4.5. The main parameters and the space resolution of some planar MDCs

Detector	Chamber length [cm]	Number of layers	Maximum drift length [cm]	Wire length [cm]	Number of electronics channels	Magnetic field [Tesla]	σ_x [mm]	σ_y [mm]	x_2 [mm]	Method of determining coordinates along a wire
ISIS	510	320	200	200	320		3		12	
CRISIS	300	192	25.4	103	384		2		10–12	
ASTRON	450	250	50	100	330		0.5	7	8–10	Additional wires
EMC	155	325	30	60–145	6260	1.5	1	0.25	20	Cathode pads
ICS	160	80	35	85	240	1.5	0.2–0.5	1–1.5	4.5	Delay lines
Omega-TPC	20	16	16	40	224	1.2	0.25			Cathode pads
BNL E 810	47 × 3	36	60	1	256 × 36	0.5	0.5	0.6	5	Short wires
NA 36 TPC	100	40	50	1	196 × 40	2.7	1.2	0.73	10	Short wires

design reflects the requirement of operation with a high track density (~ 100 tracks per event). These chambers are characterized by a large number of short sense wires (Table 4.5). Wires 10 mm long are stretched parallel to the beam direction at a distance of 2.54 mm from each other, thus forming numerous (30–40) rows. The reconstruction of an event, typical for experiments with relativistic heavy ions registered in a TPC, is demonstrated in Fig. 4.25.

TPCs have also become popular in branches of science, other than high energy physics, for instance, in double-beta decay experiments (Bellotti et al. 1983; Iqbal et al. 1986; Elliott et al. 1988; Povinec et al. 1990) or in neutron radiography (Janik et al. 1980).

The spatial resolution σ_x (the x coordinate parallel to the field $\mathcal{E}$ is determined from the electron drift time) may be of the order of 0.1 mm in the case of a small drift length, and it increases with the square root of drift length x (2.28), reaching several mm when $x \sim 1$ m (Table 4.5). The y coordinate along the wire is determined in diverse ways, to be discussed below, and therefore its r.m.s. error σ_y varies within wide limits. The resolution of adjacent tracks x_2 amounts to 5–10 mm and depends primarily on the length of the signal.

Electrodeless Drift Chambers. Recently a new interesting sort of DC has been developed, the "electrodeless drift chamber" (Allison et al. 1982, Becker et al. 1982). The term "electrodeless" is related to there being no field-shaping electrodes in the chamber, while only two essential electrodes remain: the anode

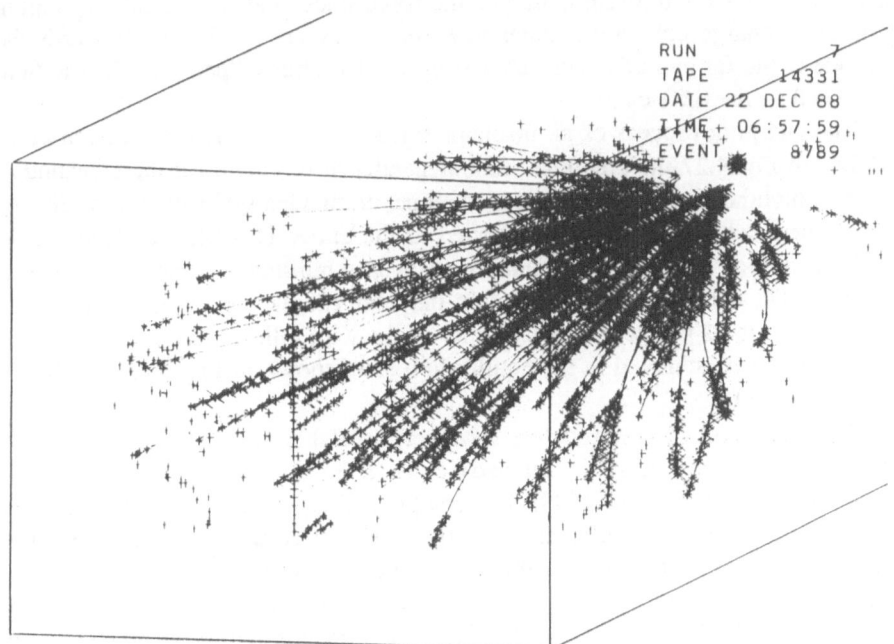

Fig. 4.25. Reconstruction of a heavy ion interaction in TPC (Etkin et al. 1989)

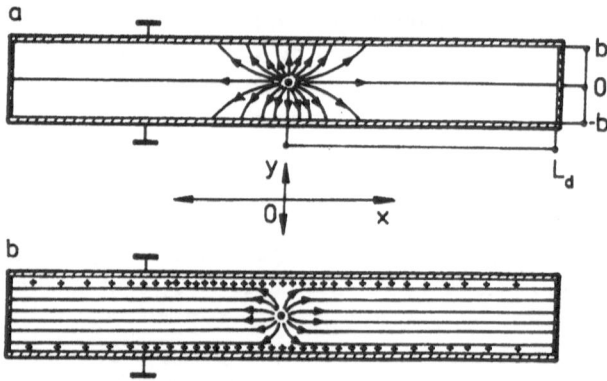

Fig. 4.26. The layout of an electrodeless DC with the force lines depicted: (a) at the switch-on time of the voltage U_a; (b) in an equilibrium state with the charge collected on the surface of dielectric

in the form of a thin wire at the middle of the chamber and the cathode at the far end of the chamber (Fig. 4.26). The walls of the chamber are made of a dielectric (usually fiberglass), metal-coated on the outside and grounded.

A positive voltage U_a is applied to the anode, the cathode being grounded. When the voltage is switched on the force lines are closed via the dielectric onto an external grounded screen (Fig. 4.26a). The positive ions produced as a result of gas amplification in the vicinity of the anode drift along the field lines and land on the surface of the dielectric. The collection of positive charge on the surface of the walls results in deformation of the force lines, and, as the self-regulating process of charge collection comes to an end, they are all directed towards the cathode at the far end of the chamber (Fig. 4.26b). Thus a "perfect" electric field is created in the drift chamber.

The charging process of an insulator surface was simulated in Allison et al. (1982). In Fig. 4.27 the computed time dependence is shown for the distribution of the potential in an electrodeless chamber with ideally insulating walls. At the moment when the voltage is applied to the anode ($t = 0$), the electric field in the chamber is essentially inhomogeneous, exhibiting a large gradient near the wire. Then, the electric field within the drift volume increases and becomes uniform, while its gradient in the vicinity of the wire falls. A perfect field for the drift chamber is achieved in the equilibrium state shown in Fig. 4.27 ($t = 200$ in arbitrary units).

The operation of electrodeless chambers is stable and efficient (Allison et al. 1982, Becker et al. 1982, Budagov et al. 1985, Dorr et al. 1985, Budagov et al. 1987, Artykov et al. 1989). For complete charging of such a chamber the intensity of cosmic radiation at the surface of the Earth is sufficient. Various materials have been utilized for the walls: fiberglass (Allison et al. 1982, Ayres and Price 1982, Budagov et al. 1985, Artykov et al. 1989), epoxy resin (Franz and Grupen 1982), polyethylene (Becker et al. 1982), polypropylene (Becker et al. 1982, Dorr et al. 1985) and organic glass (Becker et al. 1982). An attempt to make use of window glass (Becker et al. 1982) turned out to be unsuccessful.

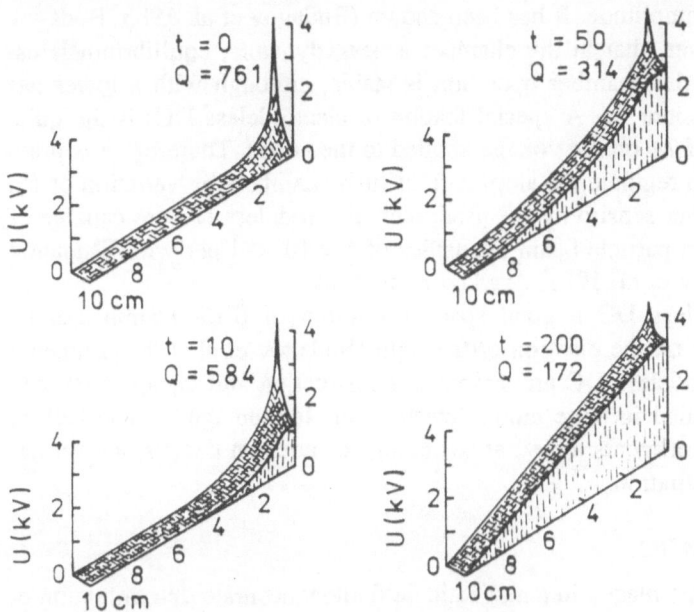

Fig. 4.27. Computed electrostatic potential as function of the distance from the signal wire x in an electrodeless DC at various times t (arbitrary units) upon switch-on of a voltage $U = 4\,\text{kV}$ (Allison et al. 1982); Q is the charge on the wire (arbitrary units)

Electrodeless chambers of various geometries have been studied: plane chambers with drift lengths up to 50 cm (Allison et al. 1982, Becker et al. 1982), cylindrical chambers (Franz and Grupen 1982, Dorr et al. 1985), as well as drift tubes, in which an effective electron drift along a distance of 1.2 m was achieved (Becker et al. 1982, Zech 1983). The principle of automatic electric field regulation in an electrodeless DC makes possible the establishment of uniform fields in volumes of complex geometry, which may turn out to be extremely attractive for the application of such chambers in X-ray radiography, nuclear spectroscopy and so on.

At the beginning electrodeless DCs were only used in beams of low intensity or for cosmic rays. They have since been shown to be capable of efficient and reliable operation in conditions typical for accelerators (Budagov et al. 1985b, Budagov et al. 1987, Artykov et al. 1989).

In electrodeless chambers interesting effects are exhibited such as current leakage through the insulating chamber wall and along its surface, the diffusion of ions, etc. The leakage current reduces the charge density on the surface of the dielectric, thus leading to the appearance of a focusing electric field in the chamber; on the other hand, the ion diffusion leads to an enhancement of the charge density and to the creation of a defocusing field. These issues are considered, for example, in (Becker et al. 1983, Budagov et al. 1987, Artykov et al. 1989).

In particle beams of high intensity the above mentioned effects play an important role: the charge density on the surface of the dielectric increases, resulting in a fall of the signal amplitude. It has been shown (Budagov et al. 1985, Budagov et al. 1987), however, that in the chamber a new, dynamic, equilibrium is established in which the chamber operation is stable, although with a lower gas amplification coefficient M. A special feature of electrodeless DCs is the quite weak dependence of M on the voltage applied to the anode. Therefore it is practically impossible to regulate the amplitude in such chambers by variation of U_a. If amplifiers of higher sensitivity are used, then electrodeless DC are capable of efficient operation in particle beam intensities of $5 \times 10^5 \, s^{-1}$ per wire (Budagov et al. 1985, Budagov et al. 1987, Artykov et al. 1989).

In an electrodeless DC a good space resolution of 0.15–0.5 mm can be achieved depending on the electron drift length (Budagov et al. 1985, Budagov et al. 1987). Electrodeless DC are simple in construction and cheap. They can be successfully applied in large muon detectors, in electromagnetic and hadron calorimeters, in experiments aimed at searching for nucleon decays, and in detectors of cosmic radiation.

4.3.2 Cylindrical MDCs

Chambers of this type placed in a magnetic field allow accurate determination of outgoing angles and measurement of charged particle momenta in experimental apparatuses in colliding beams (Sitar 1987; Sauli 1988a; Saxon 1988). They are usually termed central detectors. The requirement that detectors in colliding beams be very compact imposes strong restrictions on the radial dimensions of a central detector; and in order to achieve the required momentum resolution, it is necessary to provide high precision in the measurement of particle tracks. The principal parameters of cylindrical MDCs are given in Table 4.6.

Large **TPCs** are employed in the PEP–4 – TPC setup (TPC 1976; Barbaro-Galtieri 1982) at the PEP e^+e^- colliding beams of SLAC; in DELPHI (Delpierre 1984) and ALEPH (ALEPH Handbook 1989) in the e^+e^-–beams of LEP at CERN; and in TOPAZ (Shirahashi et al. 1988) in the e^+e^-–beams at TRIS-TAN. These chambers are characterized by a long electron drift length, which determines the measurement precision σ_y of the y coordinate along the chamber axis. The azimuthal coordinate is usually measured with a high accuracy ($\sigma_{r\varphi} \approx 0.2 \, mm$).

Chambers of the Jet type are intended to record events of large track multiplicity in particle beams of high intensity. In these chambers electrons drift short distances in azimuthal directions, which provides a good spatial resolution, $\sigma_{r\varphi} \approx 0.1$–0.25 mm. Wires are stretched along the chamber axis, so the y coordinate is determined with a low precision $\sigma_y \approx 15$–20 mm. Such chambers were used in the JADE detector (Wagner 1982) in the PETRA e^+e^-–beams of DESY; and in the AFS detector (Cockerill et al. 1981) at the ISR $p\bar{p}$–collider of CERN. They operate in the OPAL detector (OPAL Detector 1983) at the LEP e^+e^-–collider of CERN; in H1 at the HERA ep–collider of DESY (H1 Collaboration 1986); and in the KMD detector (Aluchenko et al. 1986) at the VEPP-2M e^+e^-–collider.

Table 4.6. The main parameters and space resolution of large cylindrical MDCs. This Table is a supplement to Table 3.1

Detector	Type of chamber	Inner-outer radii [cm]	Maximum drift length [cm]	Number of electronics channels	Magnetic field [Tesla]	Resolution $\sigma_{r\varphi}$ [mm]	Resolution σ_y [mm]	Double track resolution z_2 [mm]	Momentum resolution $(\sigma_p/p)^2$ [%, GeV^2/c^2]	Method of determining coordinate along sense wire
PEP-4	TPC	20–100	100	13824	1.45	0.16	0.34	10–20	3.7	Cathode pads
DELPHI	TPC	30–120	140	22000	1.2	≤0.25	< 0.8	<20	≤0.75	Cathode pads
ALEPH	TPC	31–180	220	47400	1.5	≤0.16	<1.5	15	0.15	Cathode pads
TOPAZ	TPC	36–118	150	5496	1.0	0.185	0.335		1.0	Cathode pads
JADE	Jet	21–79	7.5	3072	0.48	0.16	13	7	2.2	Charge division
AFS	Jet	20–80	2.8	3444	0.5	0.22	14	40	2.0	Charge division
OPAL	Jet	25–185	25	7680	0.4	0.12	40	2	2.0	Charge division
H1	Jet	20–79.5	5.1	5120	1.2	0.1	24	2.5	0.3	Charge division
CMD-2	Jet	2.5–29.5	2.8	1024	2.0	0.1	2		1.5	Charge division
MARK III	SSC	14–114	3.0	2000	0.4	0.25	15		1.5	Stereo layers
MARK II/SLC	SSC	19–145	3.3	5832	0.45	0.175	4	(<)5	1.5	Stereo layers
SLD	SSC	20–100	2.6	11648	1.0	0.055	9	1	0.2–0.55	Charge division
CDF	SSC	27–138	3.5	10080	1.5	<0.2	2.5	3.5	~0.1	Stereo layers
DØ	SSC	9–69		2688		0.2	2			Delay lines
ZEUS	SSC	16–79	2.5	9216	1.8	0.1–0.12	1.2–3	2	0.21 ± 0.29	Stereo layers
UCD	SSC	18–140	5.5	16000	1.5	0.05	1–2	0.5	~0.1	Stereo layers
BES	SSC	15.5–115	3.1	2808	0.45	0.2	~4.0		0.7	Stereo layers
KEDR	SSC	13–50	3		2.0	0.1		1.0	0.6	Charge division
ARGUS	CDC	15–86	9.5	5940	0.8	0.15		9	0.9	Stereo layers
CLEO II	CDC	17.5–95	7	12240	1.5	0.12	0.5		0.7	Stereo layers + Cathode pads
AMY	CDC	15–64	0.57	9796	3.0	0.14–0.45		5	0.8	Stereo layers
UA-1	PDC	10–122	18	12200	0.7	0.25	8–25	2.5		Charge division

The most progressive central detectors are the **Stereo Superlayer Chambers** (SSC) described in detail in Sect. 3.1. In addition to the SSC parameters presented in Table 3.1, Table 4.6 gives parameters of the SSC used in other detectors: D0 (Aronson 1984) at the Fermilab Tevatron.

Vector drift chambers, as SSCs are alternatively termed, are most suitable for high crossing rates and luminosity environments at the new SSC, LHC and UNK colliders (Saxon 1988).

The spatial resolution of an SSC amounts to $\sigma_{r\varphi} \sim 50\text{--}250\,\mu m$, the resolution along the sense wires to $\sigma_y \sim 1\text{--}15\,mm$, and the double track resolution to $x_2 \sim 0.5\text{--}5.0\,mm$. It is remarkable that with some detectors (SLD and UCD) attempts are being made to achieve a resolution on the level of the high resolution chambers of $\sim 50\,\mu m$ (r.m.s.). This is a very difficult task, since there are 8–12 thousand wires and SSCs are large. The high spatial resolution should allow an excellent relative momentum resolution.

Cylindrical MDCs with symmetric cells (CDC) are characterized by a short drift length and, correspondingly, by a good resolution $\sigma_{r\varphi} \approx 150\text{--}250\,\mu m$. Such chambers usually do not have many layers (36–51). They are used, for instance, in the ARGUS (ARGUS Collaboration 1989), and CLEO II (Cassel et al. 1986) detectors at the CESR e^+e^-–collider of the Cornell University and in AMY (Ueno 1988) at the TRISTAN e^+e^-–collider.

The cylindrical MDC (Calvetti et al. 1982) of the UA–1 detector at the SPS $p\bar{p}$–collider of CERN, is divided into rectangular sections (of the PDP type) by wire planes. In this chamber a moderate spatial resolution of $\sim 250\,\mu m$ has been achieved.

Table 4.6 also presents the double-track resolution x_2 in the electron drift direction and the particle momentum resolution σ_p/p^2. It must be noted that data concerning the momentum resolution have to be treated with caution. Some of the figures take no account of multiple scattering, and others take no account of the resolution improvement due to utilization of additional high precision detectors, etc.

Spectrometers operating at colliders usually contain other coordinate detectors for accurate determination of the event vertex, which aids in disentangling complex multitrack events. Such vertex detectors (Sect. 4.2) are situated in the immediate vicinity of the beam intersection point. Often, proportional chambers operating with a $\sim 50\,ns$ gate are placed near vertex detectors, which is extremely useful in performing rapid searches of real tracks in data processing.

Cylindrical drift chambers are widely used not only in high energy physics, but also in other branches of nuclear physics. They are used at meson factories: for example, the TPC for measuring $\mu \to e$ decays at the TRIUMPH accelerator (Bryman et al. 1985); the ARES proportional chamber (Baranov et al. 1986) intended for investigation of rare muon at JINR, Dubna; and the spiral drift chamber for the OBELIX detector (Bussa et al. 1986). They are used in dealing with heavy ions at the TPC of the DIOGENE detector at the SATURN synchrotron of Saclay (Alard et al. 1987). A CDC filled with hydrogen at elevated pressure

(Chapin et al. 1984) is used as an active target involving registration of recoil protons in the TRED setup at Fermilab.

4.3.3 Measurement of Coordinate Along the Sense Wires

MDCs are used for three-dimensional track reconstruction. Therefore a measurement of the coordinate along the sense wires must be performed with a good resolution, similar to the spatial resolution in the direction of electron drift. Multilayer drift chambers can be divided into several classes according to the method of coordinate measurement. In TPCs (in the PEP–4, DELPHI, ALEPH, TOPAZ detectors; Table 4.6) the azimuthal coordinate (r, φ) is determined by measuring the centroids of the signals induced in the cathode pads. The coordinate along the chamber axis (y) is determined from the electron drift time.

In chambers of the Jet and SSC types (in the JADE, AFS, OPAL, CDF, MARK II, SLD, H1 and other detectors), the (r, φ) coordinate is determined from the drift time, while the y coordinate (along the wires in these chambers) is measured in different ways, listed below. This task is certainly not a simple one. There exist several methods for determining the coordinate along a sense wire.

Charge Division. This method, proposed by Charpak, is based on the measurement of the currents I_1 and I_2 at both ends of the sense wire. The currents will be divided in accordance with the resistances R_1 and R_2 of the wire segments on one and the other side of the electron avalanche:

$$I_1/I_2 = R_2/R_1 = (L_w - y)/y \qquad (4.14)$$

where y is the distance from the beginning of the wire to the electron avalanche on the wire; L_w is the length of the wire. The y coordinate along the wire is

$$y = L_w I_2/(I_1 + I_2) \qquad (4.15)$$

The charge-division method is often applied in drift chambers, and it has been studied in detail (Radeka 1974). The measurement precision of the coordinate is determined by the signal-to-noise ratio; therefore wires with higher specific resistances ϱ_w (most often stainless steel) are used. The spatial resolution depends on the capacity of the wire $L_w C_w$, and on the charge Q collected in it (Radeka 1974):

$$\Delta y/y = 2.54\sqrt{kTL_w C_w}/Q \qquad (4.16)$$

The resolution limit of the charge-division method is not high: $\Delta y/y \approx 1\%$. Application of this method makes it necessary to use a relatively high gas amplification factor $(M \sim 10^5)$ which reduces the rate capability of the chamber and distorts the ionization measurements. The charge-division method is usually applied in chambers of the Jet type (Table 4.6).

Center-of-Gravity Method. This method is frequently applied in proportional and drift chambers (in Tables 4.5 and 4.6 reference is made to "cathode pads"). Rows of cathode pads are situated in the end caps of TPCs. The charge Q_i induced in the i-th pad, of width 4–8 mm, is measured, and the coordinate of the electron avalanche is calculated using (4.13). This method has been studied in detail and is frequently applied (Charpak, Sauli 1984; Lynch, Hadley 1982; Delpierre 1984; Sauli 1988a). The limiting resolution (Sect. 4.2.3) is not achieved in large drift chambers because of diffusion, track inclination, etc.; but in large chambers of the TPC type a resolution of 100–200 μm is obtained. A disadvantage of this method is the large number of channels (Table 4.6) and expensive analog electronics. The center-of-gravity method of the induced charge is usually applied in TPCs to determine the azimuthal coordinate. The spatial resolution $\sigma_{r\varphi}$ for TPCs, given in Table 4.6, is obtained by this method.

Delay Lines. A delay line is used as the cathode. The signal induced in a delay line propagates in opposite directions along the wire, with a known velocity. The coordinate of the electron avalanche on the wire is determined from the difference in time between the signal arriving at the anode wire and the end of the delay line. The best method is to take up the signal from both ends of the delay line, in which case several particles traversing the drift chamber can be recorded simultaneously (Budagov et al. 1980; 1989). The application of delay lines in drift chambers is described in Aronson (1984), Erwin et al. (1985) and Budagov et al. (1989). Delay lines are situated close to the sense wires so that in many cases they must contain minimum material. This results in a small delay, i.e. the signal passes through the delay line rapidly (about 10 mm/ns). For fast delay lines, 1–1.5 m long, a precision $\Delta y \sim 5$ mm is achieved. In slow delay lines (with a propagation velocity of 0.3–0.5 mm/ns) a good spatial resolution $\sigma_y \leq 0.1$ mm is achieved with a delay line length of 20 cm (Janik et al. 1980).

Stereo Layers. This method is applied in cylindrical drift chambers, particularly in SSCs and CDCs (Table 4.6), where in some of the layers the sense wires are set at a small stereo angle α_s (4–6°) to the chamber axis (Fig. 3.7) The y coordinate along the wire is determined with a precision

$$\sigma_y = \sigma_{r\varphi}/\sin\alpha_s \tag{4.17}$$

a typical value is $\sigma_y = 20 \cdot \sigma_{r\varphi}$, i.e. $\sigma_y \approx 2$–5 mm (Table 4.6)

Wires or Strips Perpendicular to the Sense Wires. This method is applied in the ASTRON detector (Babaev et al. 1978) where only some of the electrons from the particle tracks are recorded (in Table 4.5 this is denoted as "additional wires"). It yields a spatial resolution equal to the distance between the auxiliary wires (in ASTRON, $\Delta y = 15$ mm). Another similar method consists in utilization of diamond or zigzag shaped strips or pads. The amplitude of the charge induced in a strip (pad) corresponds to the position of the electron avalanche along the wire (Allison et al. 1985, Miki et al. 1985).

4.3.4 Left-right Ambiguity

A left-right ambiguity is an inconvenient property of drift chambers. In order to find out whether electrons have arrived from the right or the left side of a wire, the following measures are adopted

1) Instead of one sense wire, two are established at a small distance (less than 1 mm) from each other.
2) Two adjacent drift chambers are shifted relative to each other at half the distance between the sense wires.
3) In numerous Jet chambers the even sense wires are staggered to one side and the odd wires to the other side of the central plane. Thus, for example, in the MARK II/SLC chamber the wires are staggered by 0.38 mm (Hanson 1986), and in the H1 chamber by 0.1 mm (H1 Collaboration 1986).
4) The induced charge is measured on pick-up electrodes placed close to the sense wire. The charge turns out to be higher on the electrode situated in the part of the chamber traversed by the particle (Walenta et al. 1982).

4.3.5 MDCs in a Magnetic Field

The electron drift velocity w_M (2.31) and the diffusion coefficient D_M (2.33) are reduced in a magnetic field. The electrons travel at an angle α_M to the electric field gradient (2.32), so that their trajectories are not perpendicular to the sense wire plane, illustrated by the JADE drift chamber (Fig. 4.7). In chambers of other electrode configurations in the magnetic field an even more drastic distortion of the electron trajectories is observed (Fig. 4.28), and in this case it is difficult to reconstruct the particle track coordinate from the measured drift time. Therefore, when using drift chambers in magnetic fields, one must apply additional measures to compensate for these effects. In planar drift chambers with a distributed potential (Fig. 4.2c) the configuration of the electric field is altered in such a way that it compensates for the influence of the magnetic field

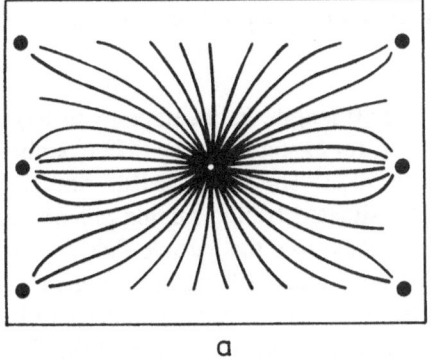 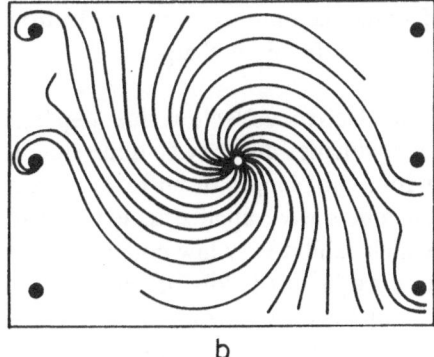

a   b

Fig. 4.28. Electron trajectories in the CELLO drift chamber: (a) – $B = 0$; (b) – $B = 1.5$ Tesla (Jaros 1980)

on the electron trajectories. Such a shift of the electric field direction is provided by applying the same potential to a pair of wires, lying at an angle α_M to the sense wire plane (Sauli 1977). The principle of compensating for the magnetic field by an electric field is also applied in large cylindrical drift chambers. The cells of such a chamber are tilted at an angle α_M, and electrons in the presence of the magnetic field of the solenoid drift in a direction perpendicular to the sense wire plane. Thus, for example, the sense wire planes in the CDF and ZEUS chambers are tilted at an angle of 45°, while in the H1 chamber the angle is 30°. Obviously, the gas mixture and the electric and magnetic field strengths must be chosen so that the electrons may drift at a given angle.

The most favorable situation arises in those drift chambers where $\mathcal{E} \parallel B$, so that the electron trajectories remain practically unchanged when the magnetic field is switched on. This condition holds in TPCs situated in a solenoid, which represents one of their best advantages.

Expression (2.32) can be rewritten in the form

$$\tan \alpha_M = eH/m_e n_a v_e \sigma_e \tag{4.18}$$

where v_e is the electron velocity, and σ_e is the cross section of elastic electron collisions with atoms of the gas. In cool gases the elastic cross section σ_e is large, so that $\tan \alpha_M \ll 1.0$ [$\alpha_M \approx 0.03$ in CO_2; $\alpha_M \approx 0.01$ in NH_3, and in C_2H_5OH at $B = 0.5$ Tesla (Drumm et al. 1980)]. The angle α_M in mixtures with a high concentration of CO_2 is smaller by nearly an order of magnitude than in "conventional" mixtures utilized in drift chambers (Becker et al. 1983).

The transverse diffusion coefficient D_T decreases in a magnetic field, especially in fast gases (Amendolia et al. 1986), but no decrease of the longitudinal diffusion coefficient D_L is observed (Sauli 1984).

4.4 Count Rate Capability of MDCs

The count rate capability of a multilayer drift chamber is limited by three main factors: the time required for electron collection, the signal length, and the accumulation of space charge.

The electron collection time depends on the drift length. In chambers with large drift volumes this may be quite significant; thus, for example, in the ISIS chamber with a drift length of up to 2 m it amounts to 50 μs. The ISIS, CRISIS and EPI chambers were applied as external identifiers situated behind bubble chambers, which exhibit very low count rate capabilities. Therefore the particle flux density in experiments involving these chambers did not exceed 10^4 particles per m^2.

In MDCs operating at colliders, the collection time is usually chosen to be smaller than the beam crossing time; for instance it is less than 4.6 μs at the SPS of CERN and less than 22.5 μs at LEP. Hence it can be seen that restrictions imposed on the collection time in these colliders are not too severe. In Jet, SSC and CDC chambers, with collection times between 1 and $2\,\mu s$, this condition is easily satisfied. More serious problems may arise in TPCs where, drift lengths being > 1 m, the collection time is between 20 and $50\,\mu s$.

The bunch spacing at colliders of the new generation is extremely small: it is 96 ns at HERA, 15 ns at LHC, 16 ns at SSC and 5 ns at the upgraded LHC. This must be taken into account in designing new central detectors. Evidently, only chambers with minimal drift lengths, SSCs and CDCs, are suitable for such small bunch spacings.

Nevertheless, it is also possible to work under such conditions, when during the electron collection time there appear in the chamber out-of-time tracks. These tracks are readily discarded in the course of track processing, since they do not pass through the interaction vertex. Such tracks are also discarded with the aid of fast coordinate detectors, for example, proportional chambers placed in front of and behind the MDC.

The length of signals occurring in drift chambers utilized for ionization measurements amounts to 50–150 ns, which imposes a limit on the particle flux of about 10^6 particles per sec per wire.

Influence of the space charge on formation of the particle track. The most serious restrictions are imposed on the admissible particle flux by accumulation of the space charge in the chamber. The number of positive ions N_+ produced in the process of gas multiplication in the vicinity of the sense wire is given by (3.6). The ion collection time τ_+ in large chambers may amount to fractions of a second. Thus, within a volume V inside the chamber a positive charge of density

$$\varrho_+ = eN_+\eta_+\tau_+/V \tag{4.19}$$

is accumulated, where η_+ is the efficiency of ion penetration from the proportional to the drift volume. A charge present in the volume V with coordinates x, y, and z alters the electric field strength, in accordance with (3.7):

$$\delta\mathcal{E}_x = \frac{\varrho_+ x}{3\varepsilon_{\mathrm{g}}} ; \quad \delta\mathcal{E}_y = \frac{\varrho_+ y}{3\varepsilon_{\mathrm{g}}} ; \quad \delta\mathcal{E}_z = \frac{\varrho_+ z}{3\varepsilon_{\mathrm{g}}} \tag{4.20}$$

where ε_{g} is the dielectric constant of the gas. This, in turn, results in changes in the drift velocity w_{d} and direction of motion of the electrons, which become noticeable if the space charge exceeds the drift electrode charge density per unit area Q_{e} (Allison 1982), i.e. if

$$\int \varrho_+(x)\,dx > 10^{-3}Q_{\mathrm{e}} . \tag{4.21}$$

If ϱ_+ increases above this limit, then serious track distortions are observed. For example, the charge density $\varrho_+ = 5\,\mathrm{nC/m^3}$ admissible in the ISIS chamber corresponds to a particle flux density not higher than $1.3 \cdot 10^3\,\mathrm{m^{-2}s^{-1}}$.

4.4.1 Methods of Enhancing the Count Rate Capability of MDCs

These methods are based on limiting the penetration of positive ions from the proportional to the drift volume or on the reduction of the ion generation time.

Screening of the drift volume is accomplished with the aid of an electric gate that does not let electrons pass from the drift volume to the proportional volume, nor let ions back in the opposite direction. Normally the gate is closed; it opens, if a trigger pulse appears, when a useful event is to be recorded. The time required for opening the gate must be short (several hundred nanoseconds), so that electrons drifting with velocities $\approx 5\,\text{cm}/\mu\text{s}$ will not be lost.

Controlling TPCs by gates is becoming popular (Nygren 1981; Lehraus et al. 1982; Nemethy et al. 1983; Bryman et al. 1985; Amendolia et al. 1985). Two types of gates are applied. In the first case the drift volume is separated from the proportional volume by two grids, at a small distance from each other, to which a potential is applied, providing for free passage of electrons through the gate in case it is open. The gate is closed by applying to the external grid (the one farthest from the proportional volume) a positive voltage relative to the internal grid. In this case the electrons cannot drift through the shutter and are collected on the external grid.

The external grid of the other type of gate (Fig. 4.29) is at a potential permitting free passage of the electrons through the gate when it is open. When the gate is closed, a positive voltage $+U_g$ is applied to the even wires and a negative voltage $-U_g$ is applied to the odd wires. The field lines between adjacent wires block the passage of electrons through the grid.

The second type of gate turns out to be more advantageous. The first type of gate requires a voltage $U_g = 100\text{--}200\,\text{V}$ to be applied, the opening time of

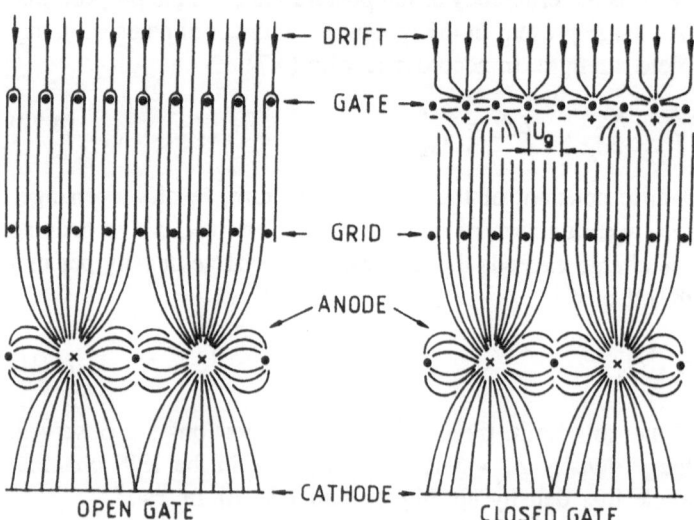

Fig. 4.29. Principle of operation of a gate in TPC ($\times$ – anode wires)

the gate being larger than 1 μs, while for the second type $U_g = 20$–50 V, and the opening time is about 200 ns (Nemethy et al. 1983; Amendolia et al. 1985). The stray pick-ups in the amplifiers for the first type of gate are also much larger. In the presence of a strong magnetic field the voltage U_g must be enhanced by 2 to 3 times (Amendolia et al. 1985). Well controlled gates operate very efficiently, and permit enhancement of the count rate capability of MDCs by approximately two orders of magnitude (Bryman et al. 1985; Amendolia et al. 1985).

The Non-Sensitive Zone in the Beam Region. A small insensitive zone can be created, from which electrons do not emerge in the region where the beam passes. This method is applied only in MDCs operating in fixed-target experiments (in cylindrical MDCs at colliders, the beam passes through a vacuum pipe). The insensitive zone is created in various ways:

a) In the EPI chamber (Aderholz et al. 1974) the sense wires in the region where the beam passes are covered with a thin layer of glass.
b) In the MDC of the European muon spectrometer at CERN the beam passes through a thin mylar pipe (Braun 1983).
c) Between the beam and the sense wires a thin metal foil can be stretched of width equal to the beam dimensions (Budagov et al. 1980).

About 10 % of the beam particles usually interact in the target; thus, when the insensitive volume around the beam is created, the particle flux can be increased by approximately an order of magnitude.

The Pulsed Anode Voltage Supply. To enhance the effective count rate, the anode voltage (usually 2 to 3 kV) is reduced by about 500 V during the time when no trigger pulses are present. This method is less effective than utilization of gates, since rapid variations of the high voltage result in stray pick-ups in the amplifier. The pulsed operation mode can be used only if the MDC operates together with a slow device, for instance, with a bubble chamber, which operates with a frequency of few Hz. The anode voltage can be switched off during data handling in the electronics. The pulsed operation mode was applied, for instance, in the ISIS chamber, where the anode voltage was switched on during only 5 % of the time, which permitted enhancement of the admissible particle flux density from 1.3×10^3 to $2.5 \times 10^4 \, \mathrm{m^{-2}s^{-1}}$ (Allison 1982).

Using Minimum Gas Amplification. The count rate capability of an MDC is usually inversely proportional to the gas gain. On the other hand, to achieve good spatial resolution, high pulse amplitudes must be used, so that optimum gas amplification must be chosen.

Reduction of the Drift Length. In choosing a suitable type of chamber it is necessary to take into account the maximum flux rate density incident upon the detector. For intense particle fluxes the best chambers are MDCs with small drift lengths (for example, SSCs, Jets and CDCs). The allowed flux density is usually inversely proportional to the drift length.

Field-Shaping Electrodes in the Drift Volume. The influence of the space charge in the drift volume can be greatly reduced by the introduction of field-shaping electrodes. To these electrodes (in the form of wires) a voltage is applied which corresponds to the potential at a given point in the chamber. They compensate for the influence of the space charge on the motion of electrons. In this way an order of magnitude enhancement of the flux density was achieved in the TPC at CERN (Dameri et al. 1985).

Application of the above mentioned measures allows enhancement of the flux density in MDCs. Usually these measures are combined, which permits operation of MDCs in particle fluxes of high intensity.

4.4.2 Calibration of Large MDCs

In large chambers with long drift lengths it is necessary to know and to control the velocity and direction of the electrons drift to a high accuracy. Thus, for a required space resolution $\sigma_{r\varphi} \approx 100\,\mu$m in a TPC with a drift length of 1 m, the indicated quantities must be known with an accuracy not worse than 10^{-4} relative to the drift length. For this purpose good calibration of the device is required.

The electron velocities and drift angles depend on the electric field strength, which may exhibit local inhomogeneities of a static or dynamic nature.

Static inhomogeneities are due to incorrect or inaccurate positioning of the field-shaping electrodes, the sense wires and the screening grids, and to an incorrect distribution of the voltage applied to the field-shaping electrodes. These setbacks can be taken into account or removed by calibration of the drift chamber before the data taking.

Dynamic variations of w_d and α_M are mainly due to an accumulation of space charge in the drift volume of the chamber, and to variations in temperature, pressure and gas composition. These variations can be taken into account only by constant calibration of the MDC throughout the experiment.

Methods of Calibration of MDCs. In track reconstruction in MDCs three methods of calibration are applied: by beam particles, automatic calibration, and by a laser beam. *Calibration by beam particles* is suitable only for MDCs involved in fixed-target experiments, in which the beam traverses the chamber. *Automatic calibration* is performed taking advantage of measured particle tracks. If the relation between w_d and α_M in each layer of the MDC is known, then, by comparing track sections in different layers, it is possible to estimate local inhomogeneities of the electric field and to introduce the appropriate corrections for track reconstruction.

The application of laser beams for the calibration of MDCs has won widespread popularity (Hilke 1980; Sadoulet 1981; Ledingham et al. 1984; Va'vra 1984; Hilke 1986). A laser beam simulates a straight track with an accuracy of $10\,\mu$m (Hilke 1986). This method is convenient for imitating straight tracks in detectors placed in a magnetic field. The main problem connected with the

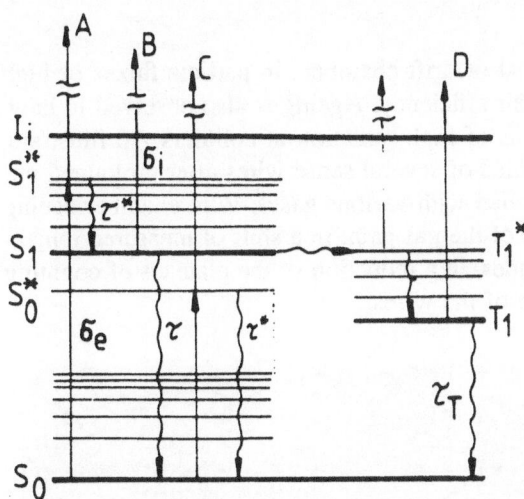

Fig. 4.30. Energy level diagram of a complex molecule and channels for 2-photon ionization (Guschin et al. 1984)

utilization of lasers working in the ultraviolet region is that their energy is insufficient not only for ionizing argon ($I_1 = 15.76 \, \mathrm{eV}$; for instance, the wavelength of a nitrogen laser is 337 nm and the energy $h\nu = 3.68 \, \mathrm{eV}$), but also for ionizing the organic gases used in proportional detectors. Ionization by laser beams can be explained by 2-photon ionization of the composite molecules. The theory of this phenomenon was presented by Guschin et al. (1984). In Fig. 4.30 the ionization mechanism of the benzene molecule is illustrated, as an example, where the most probable processes are B and D. Typical cross sections for the excitation of S_1 and S_1^* levels are $\sigma_{\mathrm{ex}} \approx 10^{-16}$–$10^{-20} \, \mathrm{cm}^2$; typical ionization cross sections are $\sigma_{\mathrm{i}} \approx 10^{-18} \, \mathrm{cm}^2$; the lifetimes of the excited states are $\tau_{\mathrm{ex}} \approx 10^{-7}$–$10^{-10} \, \mathrm{s}$, and $\tau^* \approx 10^{-11}$–$10^{-12} \, \mathrm{s}$.

Ionization by laser beams is feasible only if the gas contains an admixture of organic molecules with a low ionization potential I_1. For instance, nickelocene [Ni(C$_2$H$_5$)$_2$, $I_1 = 6.5 \, \mathrm{eV}$]; triethylamine [TEA, (C$_2$H$_5$)$_3$N, $I_1 = 7.5 \, \mathrm{eV}$]; dimethylaniline [DMA, (CH$_3$)$_2$C$_6$H$_5$N, $I_1 = 7.14 \, \mathrm{eV}$]; diethylaniline [DEA, (C$_2$H$_5$)$_2$C$_6$H$_5$N, $I_1 = 6.99 \, \mathrm{eV}$]; tetradimethylaminoethylene [TMAE: C$_{10}$H$_{24}$N$_4$, $I_1 = 5.36 \, \mathrm{eV}$] and others (Ledingham et al. 1984; Hilke 1986). These gases or vapors are introduced into proportional detectors, and the ionization density produced by the laser beam corresponds to the partial density of the gas with a low ionization potential (Sadoulet 1981; Ledingham et al. 1984; Hilke 1986).

Ionization of the gas in a proportional detector also proceeds effectively without the addition of a gas with a low ionization potential (Hilke 1980; Sadoulet 1981; Hilke 1986). This is because small admixtures (10^{-12} of the total volume) are always present in the gas. To produce an ionization corresponding to a minimum ionizing particle (in argon ≈ 100 electrons per cm), the energy density of the nitrogen laser must be $\geq 100 \, \mu\mathrm{J/mm}^2$ (Hilke 1986). To calibrate large MDCs with the aid of appropriate optical systems, laser beams longer than 10 m are created. The duration of the laser pulse is 0.5 ns; the beam divergence is $0.2 \times 0.1 \, \mathrm{mrad}$. (Hilke 1986).

4.4.3 Drift Chamber Ageing

Prolonged operation of proportional or drift chambers in particle fluxes of high intensity results in reduction of their efficiency. Ageing is also observed in large MDCs, which operate in conditions of high radiation at colliders. To illustrate, in Fig. 4.31 photographs are presented of several sense wires after prolonged operation in proportional chambers filled with various gases. Wire chamber ageing also manifests itself in a reduction of the gas gain, in a shift of measured ionization spectra towards lower amplitudes, in a reduction of the plateaus of counting rates, and, ultimately, in a damage of the wires.

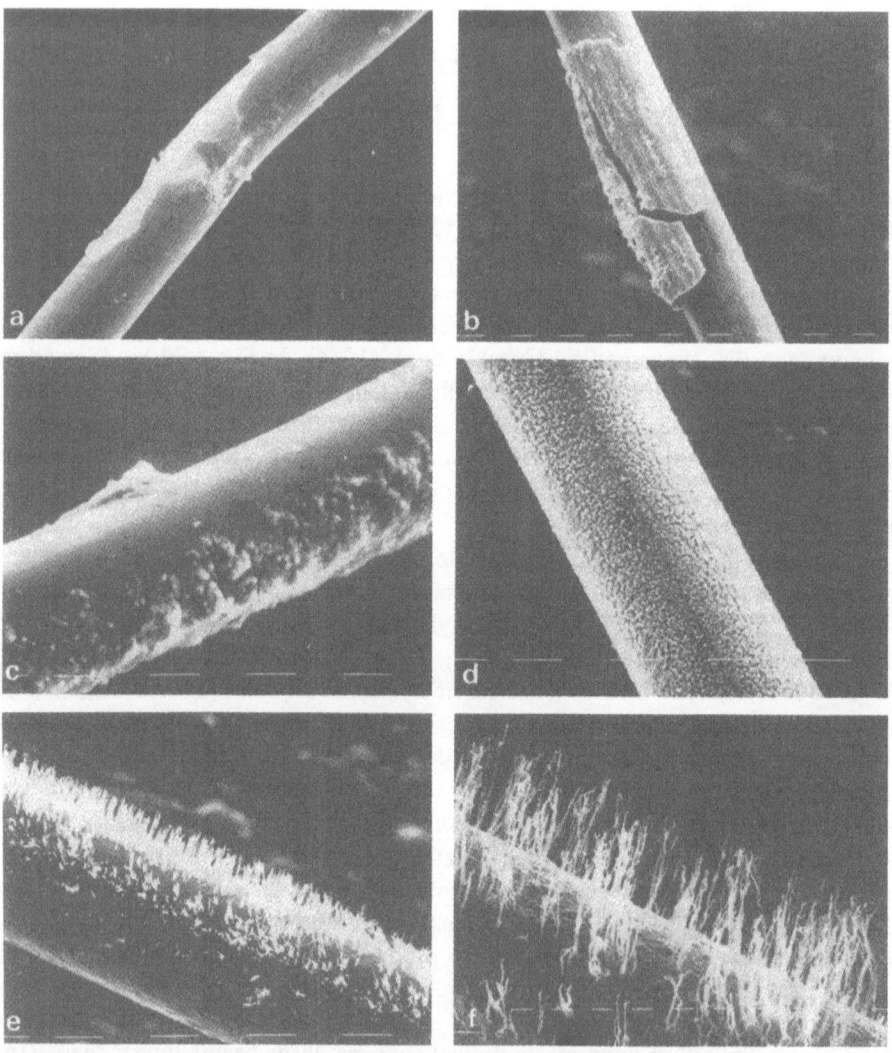

Fig. 4.31. Deposits on anode wires: (a) – Ar+C_2H_6; (b) – Ar + C_2H_6 + methylal; (c) – Ar+CO_2; (d) – perspex chamber; (e, f) – chambers with G10 fiber-glass and a cold trap (Adam 1983)

The Mechanism of the Wire Ageing. Ageing is due to the deposition and polymerization of organic molecules on the surfaces of the electrodes in the chamber. Wire chamber ageing is a very complicated process described in detail, for example, in a review by Va'vra (1986b).

The polymerization of hydrocarbons present in the gas mixture is a multistep process. Thus, for example, the following reactions take place in an $Ar + CH_4$ gas mixture:

$$CH_4 + CH_2 \rightarrow C_2H_6$$
$$C_2H_6 + CH_2 \rightarrow C_3H_8 \quad \text{and so on .}$$

(4.22)

This process continues until a solid-state chemical substance (a polymer) is produced which then sticks to the wire under the influence of electric forces. The (CH_2) radicals participating in reactions such as (4.22) are produced by dissociation of organic molecules in the course of the gas amplification.

The concentration of free radicals in a gas-discharge plasma is usually 5 to 6 orders of magnitude higher than the concentration of ions (Va'vra 1986b), so that processes such as (4.22) proceed very efficiently, and in the vicinity of the sense wires, there appears a large variety of complex molecular compounds. Numerous tests (for example, Adam 1983; Turala, Vermeulen 1983; Va'vra 1986b; Kotthaus 1986; Henderson et al. 1988; Atac 1988) have shown that ageing is observed when the total dose amounts to ≈ 0.1 C per 1 cm of wire, and that its extent depends on the composition and purity of the gas, as well as on the materials applied in the chamber constructions. Va'vra (1986b) presents a list of materials and gases not to be used for chambers with a long lifetime (≈ 1 C/cm). The following are forbidden: halogens and oil tracers in the gas, oil bubblers, rubber or silicon rubber, polyurethane adhesives, PVC, teflon or Cu tubing, soft epoxies or adhesives, aggressive solder, unknown organic materials, and large amounts of fiber-glass. It is also necessary to avoid sparking and to keep the wires clean.

Methods for decreasing ageing are:

1) Making use of minimum gas amplification.
2) Utilization of gas and vapor admixtures that are not subject to polymerization. Methylal is such a gas; using 1–3 % methylal extends the lifetime of a chamber several times, compared with an argon + hydrocarbon binary mixture.
3) Utilization of a gas mixture without hydrocarbons. Chambers filled with $Ar + CO_2$ have stable operation for count rates an order of magnitude higher than would be the case if hydrocarbons were used (Adam 1983; Turala, Vermeulen 1983).
4) Utilization of very pure gases with a minimum contamination of oil increases the lifetime by more than an order of magnitude (Kotthaus 1986).
5) The use of admixtures containing oxygen, H_2O, alcohols etc. greatly extend the lifetime of a chamber.

6) Choosing materials for the chamber construction in accordance with the list presented above.
7) Utilization of metal electrodes, instead of wires, for cathodes. Such chambers can withstand high particle fluxes (Adam et al. 1983).
8) Training the chambers. This is performed by flowing pure argon through the chamber while a voltage is carefully applied to the wires. In an argon atmosphere hydrocarbon deposits are "burnt up", which results in more efficient operation of the chamber.

4.5 Readout and Processing of Information in Multilayer Proportional Chambers

4.5.1 Multilayer Proportional Chamber (MPC) Readout

A feature characteristic of the readout of information from a multilayer proportional chamber used for ionization measurements is the utilization of analog-to-digital converters (ADC). A typical channel consists of an amplifier, an integrator and an eight-bit ADC.

In practice, the requirements of an amplifier for an MPC do not differ from the requirements of amplifiers for standard proportional chambers (Rice-Evans 1974; Cisneros 1981; Radeka 1974; Sauli 1977): an input impedance of $1-2\,k\Omega$, an input sensitivity of $1-2\,mV$, a pulse rise time of $20-50\,ns$, a noise level of $(5-10) \times 10^3$ electrons, and a signal-to-noise ratio $\approx 100:1$.

4.5.2 Multilayer Drift Chamber (MDC) Readout

In multilayer drift chambers the drift time of electrons and the amplitude (charge) of each signal are measured. In conventional drift chambers the drift time of electrons from a track is measured with the aid of a time-to-digital converter. The electronics system for "classical" drift chambers is described in many publications (Rice-Evans 1974; Verweij 1975; Sauli 1977; Guzik 1976; Pernicka 1978). In an MDC with a large drift volume several particles may traverse each layer. As a result, there may appear on each sense wire a series of pulses separated by $t \geq 100-150\,ns$. The electronics system must accurately measure the arrival time and amplitude of each signal (Fig. 4.32). These requirements are met by the multihit electronics described below. In principle, there exist two approaches to the electronics for multilayer drift chambers. In the first, drift time and signal amplitude are measured separately by a multihit time-to-digital converter (MHTDC) and an amplitude-to-digital converter (ADC), respectively (Fig. 4.32). In the second approach the entire time pattern of the arrival of a signal from the anode is stored in analog form in a charge coupled device (CCD), or the amplitude is rapidly digitized in real time in a flash ADC and stored in a fast buffer memory.

A faster amplifier is required for MDCs than for MPCs and its time jitter at the discrimination level must be of the order of 1 ns. The main parameters

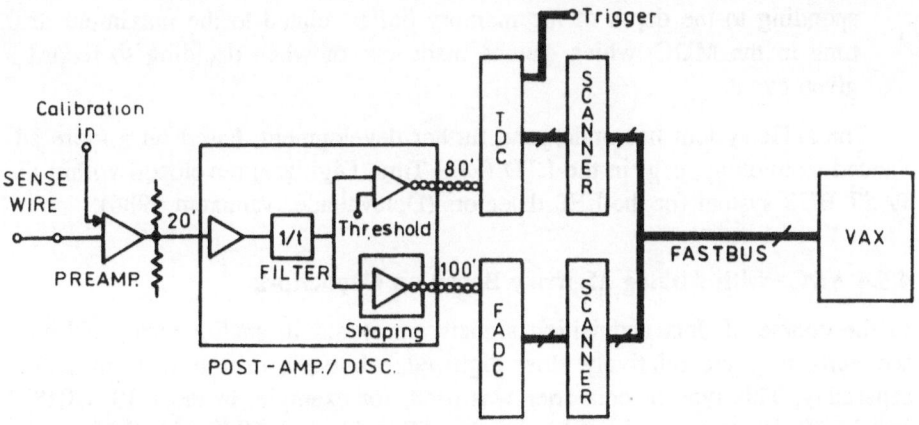

Fig. 4.32. Typical MDC read-out scheme (MARK II)

of this amplifier are the following: an input impedance of about 300 Ω, an input sensitivity lower than 2 μA, a bandwidth ∼ 100 MHz, and an amplification ∼ 0.2 V/pC.

In the next section we present the main parameters of the time-to-digital converter and of the electronics for measuring pulse heights. There exist three principal modifications of these electronics: analog memory on capacitors, an analog memory based on CCDs, and a wave-form digitizer based on a flash ADC.

4.5.3 Multihit Time-to-Digital Converter (MHTDC)

An MHTDC allows the accurate measurement of the arrival time within an interval equal to the electron collection time in the MDC. Fig. 4.32 presents an example of the scheme of a charge and time digitizer containing a multihit time-to-digital converter and a quantizer flash ADC.

Multihit TDCs, or DTRs (Drift Time Recorders), are based on direct filling of the measured interval with pulses from a standard pulse generator (100–200 MHz) which are subsequently counted and memorized in a buffer. For more accurate timing, a chronotronous time interpolator is applied, which determines the time within the main sampling period, with an accuracy of 1–4 ns (Hallgren, Verweij 1980; Festa, Sellem 1981; Ouimette et al. 1982; Budagov et al. 1985a; Delavallade, Vanuxem 1986). The dead time of a TDC is determined by the speed with which the buffer memory is filled; it amounts to 30–100 ns. Operation of the TDC memory is continuous-cyclic, and recording is terminated when a STOP signal related to the trigger appears. Such organization of a TDC has the following advantages:

- signals are recorded continuously with minimum dead time;
- almost any number of signals can be recorded;
- between the trigger and the STOP signal there is a large time delay, corre-

sponding to the depth of the memory buffer related to the maximum drift time in the MDC, which can be made use of when deciding to record a given event.

The DTR system has undergone further development, based on a more advanced technology, e. g. in the LTD (LEP Time Digitizer) developed within the FAST BUS system for the LEP detectors (Delavallade, Vanuxem 1986).

4.5.4 ADCs with Analog Memory Based on Capacitors

In the course of data acquisition signals are stored in analog form, and subsequently they are relatively slow digitized, while the drift time is measured separately. This type of converter was used, for example, in the ISIS, CRISIS and JADE detectors. In the ISIS chamber, (Brooks et al. 1978) signals from the preamplifier arrived at the amplifier, where a zero reset permitted the passage of a series of signals without distortion of their amplitudes (Fig. 4.33). Upon amplification the signals were integrated and sent to an analog memory consisting of 30 consecutively switchable capacitors C_i, each of which memorized the charge.

Upon completion of data acquisition from the chamber the potential of each capacitance was gradually transformed into a digital code by a single relatively slow (10–100 kHz) ADC. Simultaneously time information was read from the buffer memory. Then, all the information on the pulse height and drift time was subjected to further processing. The advantages of converters utilizing an array of capacitors are conversion reliability, linearity and a relatively low cost.

4.5.5 Analog Memory Based on CCDs

An analog memory can be based on charge coupled devices (CCDs). A CCD analog shifting register performs the same function as a chain of capacitors. The advantage of CCDs is a high degree of integration and similarity of each cell of the chain. This reduces the manpower required for building and adjustment and enhances the operation reliability of the electronics. CCDs can be applied in two ways: in the continuous recording mode and in the trigger mode.

Analog memory based on CCDs in the continuous recording mode is applied, for example, in handling data from TPC (Jared et al. 1982) (Fig. 4.34). When the START signal from the external trigger appears, information from the chamber starts to be read out continuously, and is recorded in the CCD. Thus, the samples are preserved, and shift along the CCD with a frequency of $f_1 = 10\,\text{MHz}$.

In this case the signal occupies several cells of a CCD. The required number of cells of a CCD register $n_b = t_d f_1$ depends on the total time of electron collection in the chamber t_d, and on the recording frequency f_1. For example, in a chamber with a drift length of 1 m, $t_d \approx 30\,\mu s$, $f_1 = 10\,\text{MHz}$ and $n_b = 300$. It must be noted that this frequency of the CCD does not meet the requirements of accuracy for the drift time measurements. It is necessary either to enhance the recording frequency f_1 or to measure the signal arrival time separately.

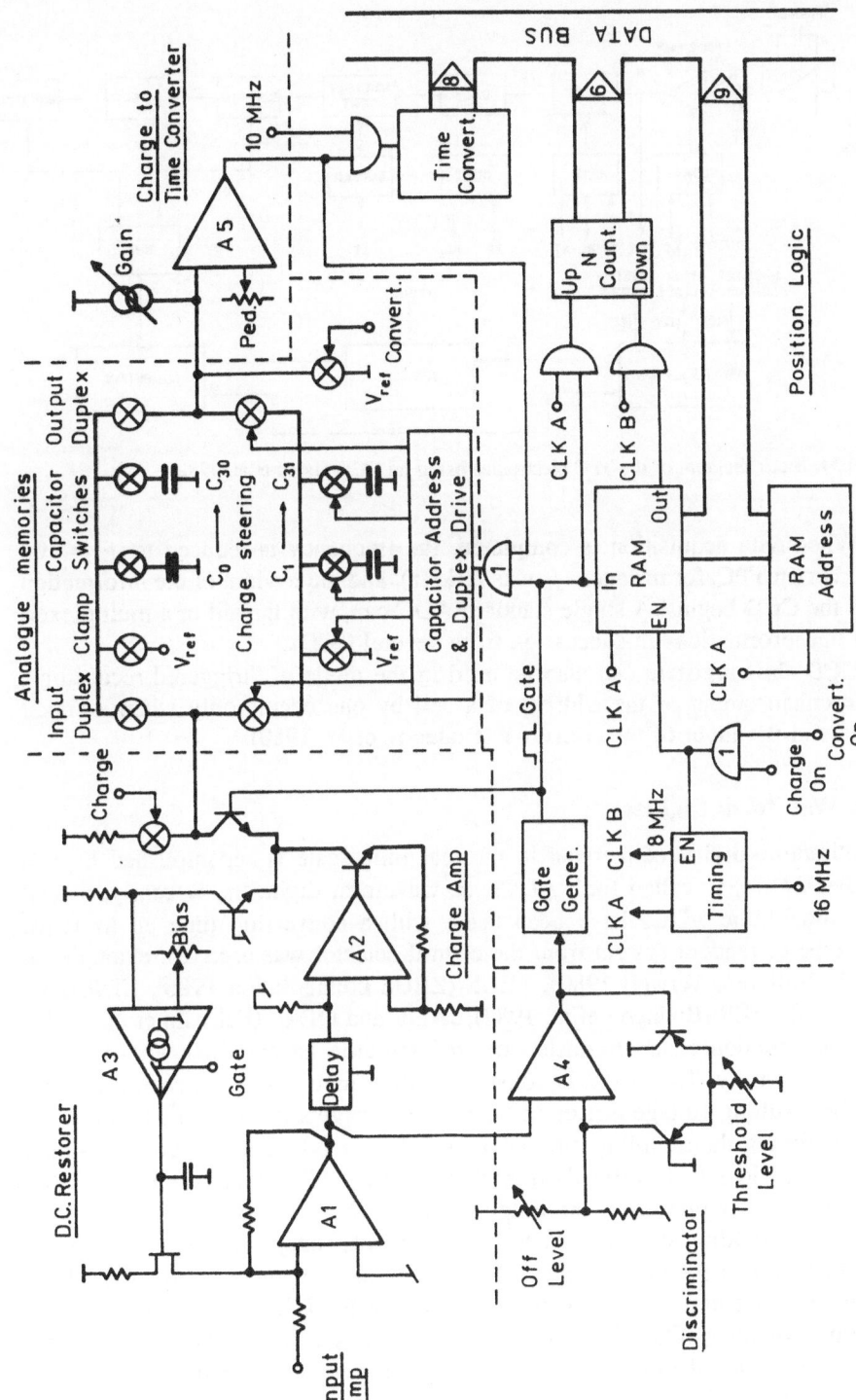

Fig. 4.33. Multi-ADC unit for ISIS

227

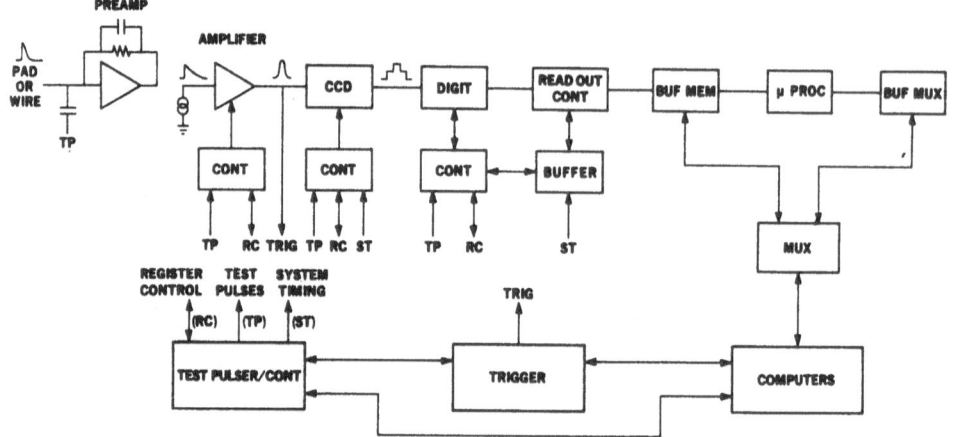

Fig. 4.34. Block diagram of the TPC electronics based on CCD (Jared et al. 1982)

When data acquisition is completed, the frequency is reduced to $f_2 = 10$–100 kHz (in TPC, for instance, f_2 was 20 kHz), and the readout of the information from the CCD begins. A single standard ADC can, with the aid of a multiplexer, read out informations in succession from several CCDs.

CCD shift registers can also be used in the mode of "triggered recording", when enhancement of the address of a cell by one occurs only when a signal appears at the input of the converter (Budagov et al. 1980).

4.5.6 Waveform Digitizer

Amplitude-to-digital conversion in the real time scale is accomplished by fast parallel ADCs, so-called flash ADCs, or waveform digitizers. Recently, 6- and 8-bit monolithic ADCs have been used, with a conversion time up to 10 ns. This type of readout system from the central detector was used, for example, in UA–1 (Hallgren, Verweij 1980), ZEUS (ZEUS collaboration 1986), CDF (Abe et al. 1987), ICS (Budagov et al. 1987), JADE and OPAL (Eckerlin et al. 1988). Each information channel consists of an integrator, an amplifier, an ADC and a buffer memory. The signal is converted in the input circuit in such a manner that the resulting voltage is proportional to the charge arriving at the input of the module during the sampling time interval (10 ns). Upon amplification the signal is quantized in a 6- to 8-bit flash ADC with a sampling interval of 10 ns. The codes of the samples from the output of the ADC are recorded in the buffer memory. Recording occurs each ten nanoseconds, independently of whether a signal appears at the input or not. Thus, at any moment time information about the signals that have arrived at the input during the full drift time is stored in the buffer memory. Quantization of the signal stops at the arrival of an external STOP signal, and then the TDC transfers information to the computer.

ADCs for ionization measurements in MDCs have to provide sufficient dynamic range [for details see, for example, Lehraus (1983)]. In Fig. 4.35a the

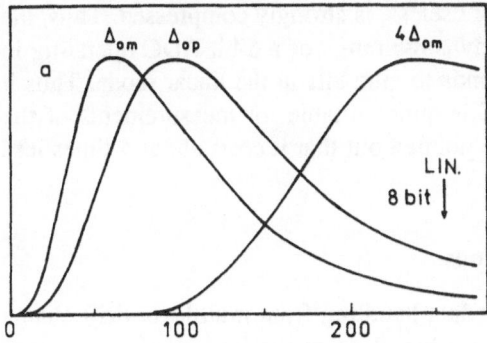

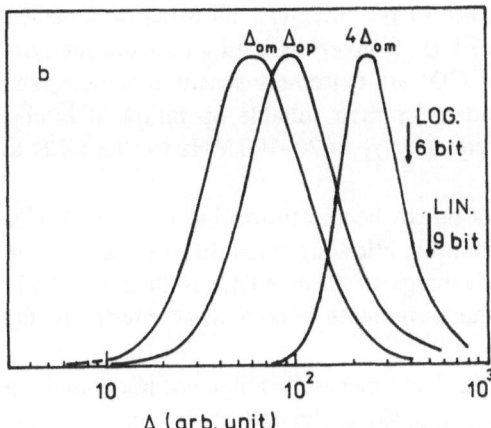

Δ (arb. unit)

Fig. 4.35. Computed ionization energy loss spectrum for particles with the most probable values Δ_{0m}, Δ_{0p}, and $4\Delta_{0m}$ for $l = 4$ mm and $P = 10^5$ Pa in argon: (a) – linear scale; (b) – logarithmic scale (Lehraus 1983)

ionization energy loss spectrum is shown for minimum ionizing particles (the most probable value of energy loss is Δ_{0m}) and the spectrum for particles of an ionization corresponding to the Fermi plateau (the most probable value being Δ_{0m}) is given. For illustration, the ionization loss spectrum, which has a most probable value equal to $4\Delta_{0m}$ is also shown. Such large losses may be observed in the case of inclined tracks and, also, when several tracks are superimposed on each other. Fig. 4.35a shows that a significant part of the latter spectrum does not fall within the dynamic range of the 8-bit ADC; its limit along the x-axis is indicated by the shaded area.

The whole spectrum may be fitted into the range of the ADC by reducing the amplification, but in this case the separation coefficient in the MDC deteriorates. The solution of this problem is based on the nonlinear conversion mode and is realized by a dynamic shift of the reference voltage at the input of the flash ADC (Hallgren, Verweij 1980). The transformation of spectra in an FADC is not linear, but logarithmic, so that the range of high amplitudes is "compressed". In Fig. 4.35b the same spectra are presented as in Fig. 4.35a, but upon transformation in the logarithmic ADC. The part of the spectrum with small amplitudes carrying the main information on the ionization energy loss is conserved, while the part with large amplitudes, which is almost useless, is strongly compressed. Thus, the ionization loss spectrum fits into the dynamic range of a 6-bit ADC operating in the logarithmic mode, which corresponds to nine bits in the linear mode. Thus it follows that a 6-bit logarithmic ADC is quite suitable for measurements of the ionization energy loss. It must also be pointed out that it costs about 4 times less than an 8-bit flash ADC.

4.5.7 Comparison of Readout Methods

Let us compare the above methods of data handling from multilayer drift chambers. Application of the analog memory based on capacitors, as in the case of the ISIS chamber (Fig. 4.33), is quite cumbersome; however, recently interest in such systems has risen again, because of the integral realization of capacitor arrays and analog switches based on FETs. Converters with an analog memory based on CCDs are progressive, but CCDs are extremely sensitive devices, and appropriate conditions must be provided for their reliable operation. It is also necessary to enhance the recording frequency f_1 to 20–100 MHz for the CCD to contain information on the drift time.

The most promising method turns out to be the utilization of flash ADCs. The electronics channel is relatively simple, allowing reliability of operation in the large system. The principle disadvantage of flash ADCs is their relatively high cost. Their accuracy in charge measurements is somewhat inferior to the accuracy of CCDs.

Flash ADCs (100 MHz) can measure drift times with a high accuracy, making use of computation of the average time centroid of the collected charge. Expressions (4.13) and (4.14) show that measurement of the center-of-gravity should be much more accurate than the leading edge timing. These issues are analyzed in detail by Bock et al. (1986) and Va'vra (1986a); they draw the following conclusions:

1) A 100 MHz FADC digitizer is not fast enough for transfer and recording of a signal without distortions; for a further improvement of the time resolution it is necessary to enhance its speed.
2) For fast gases and drift lengths smaller than 1–2 cm the leading edge timing provides a better resolution than a 100 MHz FADC.
3) In a slow gas the timing capability of a waveform digitizer and the leading edge technique are comparable.
4) In the case of large drift distances (> 2 cm) timing with a waveform digitizer is better than leading edge timing.

References

Aachen-Siegen-Zürich Vertex Group 1984: S. Miniato Meeting on Future High Energy Machines, Florence. Sect. 4.2

Abe F., Kondo K., Kurisu M., Mimashi T., Sekiguchi M., Takayama H. 1987: Nucl. Instr. Meth. A **259**, 466. Sect. 4.5

Adam J., Baird C., Cockerill D., Frandsen P.K., Hilke H.J., Hofmann H., Ludlam, T., Rosso E., Soria D., Vaughan D. 1983: Nucl. Instr. Meth. **217**, 291. Sect. 4.4

Aderholz M., Lazeyras P., Lehraus I., Matthewson R., Tejessy W. 1974: Nucl. Instr. Meth. **118**, 419; Ibid 1975: **123**, 237. Sect. 4.4

Alard J.P., Arnold J., Augerat J., Babinet R., Bastid N., Brochard F., Costilhes J.P., Crouau M., De Marco N., Drouet M., Dupieux P., Fanet J., Fodor Z., Fraysse L., Girard J., Gorodetzky P., Gosset J., Laspalles C., Lamaire M.C., Lhote D., Lucas B., Montarou G., Papineau A., Parizet M.J., Poitou J., Racca C., Schimmerling W., Taimain J.C., Terrien Y., Valero J., Valette O. 1987: Nucl. Instr. Meth. A **261**, 379. Sect. 4.3

ALEPH Handbook 1989: ALEPH 89–77 (ed. Blum W.), CERN, Note 89–03. Sect. 4.3

Alexander J., Hayes K., Hoard C., Hutchinson D., Jaros J., Odaka S., Ong R., Drell P., Fuzesy R., Harr R., Trilling G. 1986: Nucl. Inst. Meth. A **252**, 350. Sect. 4.2

Allison J., Barlow R.J., Bowdery C.K. 1982: Nucl. Instr. Meth. **201**, 341. Sect. 4.3

Allison J., Barlow R.J., Canas R., Duerdoth I.P., Loebinger F.K., Macbeth A.A., Murphy P.G., Stephens K. 1985: Nucl. Instr. Meth. A **236**, 284. Sect. 4.3

Allison W.W.M., Brooks C.B., Bunch J.N., Cobb J.H., Lloyd J.L., Pleming R.W. 1974: Nucl. Instr. Meth. **119**, 499. Sect. 4.3

Allison W.W.M. 1982: Proc. Int. Conf. on Instrumentation for Colliding Beam Physics, Stanford Linear Accelerator Report, p. 61. Sect. 4.4

Allison W.W.M., Brooks C.B., Shield P.D., Aguilar-Benitz M., Willmott C., Dumarchez J., Schouten M. 1984: Nucl. Instr. Meth. **224**, 396. Sect. 4.3

Aluchenko V.M., Khazin B.I., Solodov E.P., Snopkov I.G. 1986: Nucl. Instr. Meth. A **252**, 299. Sect. 4.3

Amendolia S.R., Blum W., Benetta R., Cherny M., Fidecaro F., Froberger J.P., Hubbard B., Jared R.C., Lehraus I., Liello F., Marrocchesi P.S., Matthewson R., May J., Meyer T.C., Milotti E., Nanni F., Peisert A., Price M.J., Ragusa F., Richstein J., Richter R., Rolandi L., Schlatter W.D., Sedgbeer J., Settles R., Stierlin U., Takashima M., Tejessy W., Tromba G., Witzeling W., Wu S.L., Wu W. 1985: Nucl. Instr. Meth. A **234**, 47. Sect. 4.4

Amendolia S.R., Binder M., Blum M., Cherny M., Farilla A., Fidecaro F., Izen J.M., Jared R.C., Lehraus I., Marrocchesi P.S., Matthewson R., May J., Milloti E., Muller D., Peisert A., Price M.J., Richstein J., Richter R., Roland L., Schlatter W.D., Settles R., Sinnis C., Stefanini G., Stierlin U., Takashima M., Tejessy W., Vayaki A., Wicklund E., Witzeling W., Wu W.M. 1986: Nucl. Instr. Meth. A **244**, 516. Sect. 4.3

Anderhub H., Anders H., Ansari S., Boehm A., Bourquin M., Bucher A., Burger J., Chen M., Commichau V., Deutschmann M., Ohina M., Fehlmann J., Fong D., Friebel W., Hangarter K., Hausammann R., Hofer H., Ilyas M.M., Kessler G., Krause B., Krenz W., Leiste R., Li Q.Z., Linnhoefer D., Lue X., Masson S., Nusbaumer M., Nierobisch H., Nowak W.D., Overlack C., Pohl M., Rau R.R., Ren D., Roderburg E., Rohde M., Schreiber H.J., Schug J., Schulte R., Stoehr B., Ting S.C.C., Tonutti M., Viertel G., Wu S.X., Wyne M.F., Wyslouch B. 1988: Nucl. Instr. Meth. A **265**, 50. Sect. 4.2

Aronson S.H. 1984: Brookhaven National Lab, Prep. BNL–34525, OG 772. Sect. 4.3

ARGUS Collaboration 1989: Nucl. Instr. Meth. A **275**, 1. Sects. 4.2, 3

Artukh A.G., Budagov Y.A., Hlinka, V., Holy K., Kladiva E., Nikitin V.A., Omelyanenko A.A., Povinec P., Seman M., Semenov A.A., Sitar B., Spalek J., Teterev Y.G. 1991: J. Phys. G. **17**, Supplement, S 477. Sect. 4.3

Artykov A.M., Glagolev V. V., Hlinka V., Kladiva E., Sitar B. 1989: Prib. Tech. Exp. **1**, 66. (English transl.: Instr. Exp. R **32**, 72). Sect. 4.3

Atac M. 1988: in *Vertex Detectors* (ed. Villa F.), Plenum, New York. Sect. 4.4

Atrazhev V. M., Yakubov I. T. 1977: J. Phys. D **10**, 2155. Sect. 4.2

Ayres D. S., Price L. E. 1982: Prep. Argon National Lab. ANL – HEB–CP–82–34. Sect. 4.3

Babaev A. I., Barski D. A., Boris S. D., Brachman E. V., Varlamov L. I., Galaktinov Yu. V., Gorodkov Yu. V., Danilov M. V., Eliseev G. P., Zeldovich O. J., Iljin M. A., Kamyshkov Yu. A., Kravcov A. I., Laptin L. P., Lubimov V. A., Nagovicyn V. V., Nozik V. Z., Popov V. P., Semenov Yu. A., Sopov V. S., Tichomirov I. N., Cvetkova T. N., Chudakov V. N., Shevchenko V. G., Shumilov E. V. 1978: Prep. ITEP–103, Moscow. Sect. 4.3

Baranov V. A., Evtukhovich P. G., Filippov A. I., Fursov A. P., Korenchenko A. S., Korenchenko S. M., Kostin B. F., Krabchuk N. P., Khomutov N. V., Kuchinsky N. A., Moiseenko A. S., Mzhavia D. A., Nekrasov K. G., Povinec P., Szarka J., Smirnov V. S., Vanko J., Zyazyulya F. E. 1986: Nucl. Instr. Meth. B **17**, 438. Sect. 4.3

Barbaro-Galtieri 1982: Proc. Int. Conf. on Instrumentation for Colliding Beam Physics, Stanford Linear Accelerator Report, p. 46. Sect. 4.3

Bari G., Basile M., Bonvicini G., Cara Romeo G., Casaccia R., Cifarelli L., Cindolo F., Contin A., D'Ali G., Del Papa C., Focardi S., Iacobucci G., Maccarrone G., Massam T., Motta F., Nania R., Palmonari F., Prisco G., Sartorelli G., Susinno G., Votano L., Zichichi A. 1986: Nucl. Instr. Meth. A **251**, 292. Sect. 4.2

Basile M., Bonvicini G., Cara Romeo G., Cifarelli L., Contin A., D'Ali G., Del Papa C., Maccarrone G., Massam T., Motta F., Nania R., Palmonari F., Rinaldi G., Sartorelli G., Spinetti M., Susinno G., Villa F., Votano L., Zichichi A. 1985: Nucl. Instr. Meth. A **239**, 497. Sect. 4.2

Baskakov V. I., Cherniatin V. K., Dolgoshein B. A., Lebedenko V. N., Romanjuk A. S., Fedorov V. M., Gavrilenko I. L., Konovalov S. P., Majburov S. N., Muravjev S. V., Pustovetov V. P., Shmeleva A. P., Vasiljev P. S. 1979: Nucl. Instr. Meth. **158**, 129. Sect. 4.2

Becker Ch., Weihs W., Zech G. 1982: Nucl. Instr. Meth. **200**, 335. Sect. 4.3.1

Becker Ch., Weihs W., Zech G. 1983: Nucl. Instr. Meth. **213**, 243. Sect. 4.3.1

Becker U., Cappel M., Chen M., White M., Ye C. H., Yee K., Fehlman J., Seiler P. G., 1983: Nucl. Instr. Meth. **214**, 525. Sect. 4.3

Belau E. R., Blum W., Hajduk Z., Sanford R. W. L. 1982: Nucl. Instr. Meth. **192**, 217. Sect. 4.2

Bellazzini R., Betti C., Brez A., Carboni E., Massai M. M., Torquati M. R. 1986: Nucl. Instr. Meth. A **252**, 453. Sect. 4.2

Bellotti E., Cremonesi O., Fiorini E., Liguori C., Raguzzi S., Rossi L., Traspedini L., Zanotti L. 1983: Prep. CERN/EP 83–144. Sect. 4.3

Benso S., Darbo G., Rossi L., Sannino M., Traspedini L., Vitle S. 1983: Nucl. Instr. Meth. **217**, 194. Sect. 4.3

Bobkov S., Cherniatin V., Dolgoshein B., Evgrafov G., Kalinovsky A., Kantserov V., Nevsky P., Sosnovtsev V., Sumarokov A., Zelenkov A. 1984: Nucl. Instr. Meth. **226**, 376. Sect. 4.2

Bock, P., Heintze J., Kunst T., Schmidt B., Smolik L. 1986: Nucl. Instr. Meth. A **242**, 237. Sects. 4.2, 5

Bondar A. E., Onuchin A. P., Panin V. S., Lelnov V. I. 1983: Nucl. Instr. Meth. **207**, 379. Sect. 4.2

Bouclier R., Charpak G., Erskine G. A., Guerard B., Santiard J. C., Sauli F., Solomey N. 1988: Nucl. Instr. Meth. A **265**, 78. Sect. 4.2

Braun H. M. 1983: Prep. CERN 83–02. Sects. 4.3, 4

Breskin A., Zwang N. 1977: Nucl. Instr. Meth. **144**, 609. Sect. 4.2

Brooks C. B., Shield P. D., Allison W. W. M. 1978: Nucl. Instr. Meth. **156**, 297. Sect. 4.5

Bryman D. A., Leitch M., Navon I., Numao T., Schlatter P., Dixit M. S., Hargrove C. K., Mes H., MacDonald J. A., Skegg R., Spuller J., Burnham R. A., Hasinoff M., Poutissou J. M., Azuclos G., Depommier P., Martin J. P., Poutissou R., Blecher M., Hotow K., Carter A. L. 1985: Nucl. Instr. Meth. A **234**, 42. Sects. 4.3, 4

Budagov Yu. A., Vinogradov V. B., Hlinka V., Dzhelepov V. P., Pisut J., Pisutova N., Povinec P., Semenov A. A., Sitar B., Strmen P., Usacev S., Flyagin V. B., Chudy M., Cerny V., Stefunko I., Janik R. 1980: Prep. UKJF 80–30, Bratislava. Sects. 4.3–5

Budagov Yu. A., Seman M., Semenov A. A., Sitar B., Spalek J. 1985a: Nucl. Instr. Meth. A **234**, 302. Sect. 4.5

Budagov Yu. A., Nagajcev A. P., Omelyanenko A. A., Semenev A. A., Sergeev S. V., Hlinka V., Povinec P., Sitar B., Spalek J., Artykov A. M. 1985b: Nucl. Instr. Meth. A **238**, 245. Sect. 4.3

Budagov Yu. A., Zinov V. G., Seman M., Semenov A. A., Sitar B., Spalek J. 1987: Prib Tech. Exper. **1**, 59. (English transl.: Instr. Exp. R **30**, 62). Sect. 4.5

Budagov Yu. A., Glagolev V. V., Korolev V. M., Omelyanenko A. A., Semenov A. A., Sergeev S. V., Hlinka V., Povinec P., Sitar B., Janik R., Kladiva E., Seman M., Spalek J., Jordanov A. B., Artykov A. M., Omelyanenko M. N. 1987: Nucl. Instr. Meth. A **255**, 493. Sect. 4.3

Budagov Yu. A., Glagolev V. V., Omelyanenko A. A., Semenov A. A., Hlinka V., Povinec P., Sitar B., Kladiva E., Seman M., Spalek J., Artykov A. M. 1989: Nucl. Instr. Meth. A **284**, 433. Sect. 4.3

Bussa M. P., Busso L., Erdman K., Gastaldi U., Grasso A., Maggiora A., Marcello S., Masoni A., Panzieri D., Sabev C., Tosello F. 1986: Nucl. Instr. Meth. A **252**, 321. Sect. 4.3

Calvetti M., Cennini P., Centro S., Chesi E., Cittolin S., Cnops A. M., Dumps L., DiBitonto D., Geer S., Haynes B., Jank W., Jorat G., Karimaki V., Kowalski H., Kinnunen R., Lacara F., Maurin G., Norton A., Piano Mortari G., Pimia M., Placci A., Queru P., Rijssenbeek M., Rubbia C., Sadoulet B., Sumorok K. S., Tao C., Vuillemin V., Vialle J. P., Verweij H., Zurfluh E. 1982: Proc. Int. Conf. on Instrumentation for Colliding Beam Physics, Stanford Linear Accelerator Report. Sect. 4.3

Carter J. R., Elcombe P. A., Hill J. C., Carter A. A., Lasota M. M. B., Kyberd P., Lloyd S. L., Newman-Coburn D., Pritchard T. W., Wyatt T. R. 1989: Nucl. Instr. Meth. A **278**, 725. Sect. 4.2

Cassel D. G., Desalvo R., Dobbins J., Gilchriese M. G. D., Gray S., Hartill D., Mueller J., Peterson D., Pisharody M., Riley D., Kinoshita K. 1986: Nucl. Instr. Meth. A **252**, 325. Sect. 4.3

Chapin T. J., Cool R. L., Goulianos K., Jenkins K. A., Silverman J. P., Snow G. R., Sticker H., White S. N. 1984: Nucl. Instr. Meth. **225**, 550. Sect. 4.3

Charpak G., Majewski S., Sauli F. 1975: Nucl. Instr. Meth. **126**, 381. Sect. 4.2

Charpak G., Sauli F. 1978: Phys. Lett. **78**, 523. Sect. 4.2

Charpak G., Sauli F. 1984: Ann. Rev. Nucl. Part. Sci. **34**, 285. Sect. 4.2

Chirikov-Zorin I. E., Davydov Yu. I., Feshchenko A. A., Flyagin V. B., Sergeev S. V., Spalek J., Strmen P. 1987: Nucl. Instr. Meth. A **260**, 142. Sect. 4.2

Cisneros E., Hutchinson D., McShurley D., Richter R., Shapiro S. 1981: IEEE Trans. Nucl. Sci. NS-**28**, 465. Sect. 4.5

Cockerill D., Fabjan C. W., Frandsen P., Hallgren A., Heck B., Hilke H. J., Hogue R., Jeffreys P., Jensen H. B., Killian T., Kreisler M., Lindsay J., Ludlam T., Lissauer D., Molzon W., Nielson B. S., Oren Y., Queru P., Rosselet L., Rosso E., Rudge A., Scire M., Wang D. W., Wang Ch. J., Willis W. J., Botner O., Bøggild H., Dahl-Jensen E., Dahl-Jensen I., Dam Ph., Damgaard G., Hansen K. H., Hooper J., Møller R., Nielsen S. Ø., Schistad B., Akesson T., Almehed S., von Dardel G., Henning S., Jarlskog G., Lorstad B., Melin A., Mjornmark U., Nilsson A., Albrow M. G., McCubbin N. A., Evans W. M. 1981: Physica Scripta **23**, 649. Sect. 4.3

Christophorou L. G. 1970: *Atomic and molecular radiation physics*, Wiley, New York. Sect. 4.2

Crawford O. H., Delgarno A., Hays P. B. 1967: Molec. Phys. **13**, 181. Sect. 4.2

Crompton R. W., Elford M. T., Jory R. L. 1967: J. Phys. **20**, 369. Sect. 4.2

Damerell C. J. S. 1987: Vertex Detectors, in *Techniques and Concepts of High Energy Physics*. (ed. Ferbel T.), Plenum, New York, NATO ser. B 164. Sect. 4.2

Dameri M., Darbo G., Lamanna E., Osculati B., Petrera S., Rossi L., Vitale S. 1985: Nucl. Instr. Meth. A **235**, 279. Sect. 4.4

Delavallade G., Vanuxem J. P. 1986: Nucl. Instr. Meth. A **252**, 596. Sect. 4.5

Delpierre P. 1984: Nucl. Instr. Meth. **225**, 566. Sect. 4.3

Dolgoshein B. A. et al. 1984: in *High Energy Physics*, Energoatomizdat, Moscow, p. 50. Sect. 4.2

Dorr R., Grupen C., Noll A. 1985: Nucl. Instr. Meth. A **238**, 238. Sect. 4.3

Drumm H., Eichler R., Granz B., Heinize J., Heinzelmann G., Heuer R. D., von Krogh J., Lennert P., Nozaki T., Rieseberg H., Wagner A., Warming P. 1980: Nucl. Instr. Meth. **176**, 333. Sects. 4.1–3

Duinker P., Guo J. C., Harting D., Hartjes F., Hertzberger L. O., Hoekstra J., Konjin J., Massaro

G.Ġ.G., Walenta A.H. 1982: Nucl. Instr. Meth. **201**, 351. Sect.4.2

Eckerlin G., Elsen E., v.d. Schmitt H., Wagner A., Walter P.V., Zimmer M. 1988: Nucl. Instr. Meth. A **263**, 206. Sect.4.5

Elliott S.R., Hahn A.A., Moe M.K. 1988: Nucl. Instr. Meth. A **273**, 226. Sect.4.3

Erwin A.R., Hongfang Chen, Hasan A., Hollenhorst R., Kuhr G., Kuehn C.E., Nelson K.S., Thompson M.A. 1985: Nucl. Instr. Meth. A **237**, 493. Sect.4.3

Etkin A., Eiseman S.E., Foley K.J., Hakenburg R.W., Longrace R.S., Love W.A., Morris T.W., Platner E.D., Saulys A.C., Lindenbaum S.J., Chan C.S., Kramer M.A., Hallman T.J., Madansky L., Bonner B.E., Buchanan J.A., Chiou C.N., Clement J.M., Corcoran M.D., Krishna N., Kruk J.W., Miettinen H.E., Mutchler G.S., Nessi-Tedaldi F., Nessi M., Phillips G.C., Roberts J.B. 1989: Nucl. Instr. Meth. Sect.4.3

Fehlmann J., Viertel G. 1983a: *Compilation of Data for Drift Chamber Operation*, ETH Zürich–IHP. Sect.4.1

Fehlmann J., Paradiso J.A., Viertel G. 1983b: "WIRCHA" Program Package to Simulate Drift Chambers, ETH Zürich. Sect.4.1

Festa E., Sellem R. 1981: Nucl. Instr. Meth. **188**, 99. Sect.4.5

Filatova N.A., Nigmanov T.S., Pugachevich V.P., Riabtsov V.D., Shafranov M.D., Tsyganov E.N., Uralsky D.V., Vodopianov A.S. 1977: Nucl. Instr. Meth. **143**, 17. Sect.4.1

Fischer J., Radeka V., Smith G.C. 1986: Nucl. Instr. Meth. A **252**, 239. Sect.4.2

Franz A., Grupen C. 1982: Nucl. Instr. Meth. **200**, 331. Sect.4.3

Garabatos C. representing the NA36 Collaboration 1989: Nucl. Instr. Meth. Sect.4.3

Gaukler G., Schmidt-Boching H., Schuch R., Schule R., Specht H.J., Tserruya J. 1977: Nucl. Instr. Meth. **141**, 115. Sect.4.2

Guschin E.M., Lebedev A.N., Somov S.V. 1984: Nucl. Instr. Meth. **228**, 94. Sect.4.4

Guzik Z. 1976: Fermilab Rep. FN–301. Sect.4.5

H1 Collaboration 1986: Technical Report for the H1 Detector, DESY. Sect.4.3

Hallgren B., Verweij H. 1980: IEEE Trans. NS–27, 333. Sect.4.5

Hanson G.G. 1986: Nucl. Instr. Meth. A **252**, 343. Sect.4.3

Hayes K.G. 1988: Nucl. Instr. Meth. A **265**, 60. Sect.4.2

Henderson R., Openshaw R., Fasrer W., Salomon W., Salomons G., Sheffer G. 1988: IEEE Trans. NS–35, 477. Sect.4.4

Hendrix J., Lentfer A. 1986: Nucl. Instr. Meth. A **252**, 246. Sect.4.2

Hilke H.J. 1980: Nucl. Instr. Meth. **174**, 145. Sect.4.4

Hilke H.J. 1986: Nucl. Instr. Meth. A **252**, 169. Sect.4.4

Huxley L.G., Crompton R.W. 1974: *The Diffusion and Drift of Electrons in Gases*, Wiley, New York. Sect.4.2

Huth J., Nygren D. 1985: Nucl. Instr. Meth. A **241**, 375. Sect.4.2

Iqbal M.A., O'Callaghan B., Henrikson H., Boehm F. 1986: Nucl. Instr. Meth. A **243**, 459. Sect.4.3

Janik R., Moucka L., Peshekhonov V.D., Sitar B. 1980: Nucl. Instr. Meth. **178**, 71. Sect.4.3

Jared R.C., Landis D.A., Goulding F.S. 1982: IEEE Trans. NS–29, 57. Sect.4.5

Jaros J.A. 1980: Stanford Linear Accelerator Report, PUB–2647. Sects.4.1, 3

Jaros J.A. 1988: in *Vertex Detectors*, ed. Villa F., Plenum, New York, p. 37. Sect.4.2

Karpenko G.V., Krumshtein Z.V., Tokmenin V.V., Khazins D.M., Khovanski N.N. 1987: Proc. Int. Symp. on Position Detectors in High Energy Physics, Dubna, p. 237 (in Russian). Sect.4.2

Khazins D.M. 1989: Prep. JINR P13–89–361, Dubna (in Russian). Sect.4.2

Khrapak A.G., Yakubov I.T. 1981: *Electrons in Dense Gases and Plasma*, Nauka, Moscow (in Russian). Sect.4.2

Kirkby J. 1988: in *Vertex Detectors*, ed. Villa F. (Plenum, New York) p. 225. Sect.4.2

Kotthaus R. 1986: Nucl. Instr. Meth. A **252**, 531. Sect.4.4

Ledingham K.W.D., Raine C., Smith K.M., Campbell A.M., Towrie M., Trager C., Houston C.M. 1984: Nucl. Instr. Meth. **225**, 319. Sect.4.4

Lehraus I., Matthewson R., Tejessy W. 1982: Nucl. Instr. Meth. **196**, 361. Sect.4.4

Lehraus I. 1983: Nucl. Instr. Meth. **217**, 43. Sect. 4.5

Lynch G. R., Hadley N. J. 1982: Proc. Int. Conf. on Instrumentation for Colliding Beam Physics, Stanford Linear Accelerator Report, p. 85. Sect. 4.3

Manfredi P. F., Ragusa F. 1986: Nucl. Instr. Meth. A **252**, 208. Sect. 4.2

Michel E., Schmidt-Parzefall W., Appuhn R. D., Buchmuller J., Kolanoski H., Kreismeier B., Lange A., Siegmund T., Walther A., Edwards K. W., Fernholz R. C., Kapitza H., MacFarlane D. B., O'Neill M. O., Parsons J. A., Prentice J. D., Seidel S. C., Tsipolitis G., Ball S., Babaev A., Danilov M., Tichomirov I. 1989: Nucl. Instr. Meth. A **283**, 292. Sect. 4.2

Miki T., Itoh R., Kamae T. 1985: Nucl. Instr. Meth. A **236**, 64. Sect. 4.3

Muller Th. 1986: Nucl. Instr. Meth. A **252**, 387. Sect. 4.2

Nelson H. N. 1988: in *Vertex Detectors*, ed. Villa F., Plenum, New York, p. 115. Sect. 4.2

Nemethy P., Oddone P. J., Toge N., Ishibashi A. 1983: Nucl. Instr. Meth. **212**, 273. Sect. 4.4

Nygren D. 1981: Phys. Scripta **23**, 584. Sect. 4.4

OPAL Detector Technical Proposal 1983: CERN/LEPC/83–4. Sects. 4.2, 3

Ouimette D., Porat D., Tilgham A., Wojcicki S., Young C. 1982: IEEE Trans. NS–**29**, 290. Sect. 4.5

Palladino V., Sadoulet B. 1975: Nucl. Instr. Meth. **128**, 323; Prep LBL–3013 (1974). Sect. 4.1

Peisert A., Sauli F. 1984: Prep. CERN 84–08. Sect. 4.1

Peisert A., Sauli F. 1986: Nucl. Instr. Meth. A **247**, 453. Sect. 4.2

Pernicka M. 1978: Nucl. Instr. Meth. **156**, 311. Sect. 4.5

Polyakov V. A., Rykalin V. I. 1987: Proc. Int. Symp. on Position Detectors in High Energy Physics, Dubna. Sect. 4.2

Potter D. M. 1984: Nucl. Instr. Meth. **228**, 56. Sect. 4.2

Povinec P., Sitar B., Kladiva E., Seman M., Budagov Yu. A., Semenov A. A. 1990: Nucl. Instr. Meth. A **293**, 562. Sect. 4.3

Radeka V. 1974: IEEE Trans. NS–**21**, 51. Sects. 4.3, 5

Rehak P., Gatti E., Longoni A., Kemmer J., Holl P., Klanner R., Lutz G., Wylie A. 1985: Nucl. Instr. Meth. A **235**, 224. Sect. 4.2

Rice-Evans P. 1974: *Spark, Streamer, Proportional and Drift Chambers*, Richelieu, London. Sect. 4.5

Ritson D. M. 1988: Nucl. Instr. Meth. A **265**, 88. Sect. 4.2

Roderburg E., Backer A., Krieger H., Mattern D., Walenta A. H., Zech G. 1986: Nucl. Instr. Meth. A **252**, 285. Sect. 4.2

Sadoulet B. 1981: Phys. Scripta **23**, 434. Sect. 4.4

Sauli F. 1977: Prep. CERN 77–09. Sects. 4.3, 5

Sauli F. 1984: in *The Time Projection Chambers* (ed. MacDonald J. A.), Amer. Inst. Phys., New York, p. 171. Sect. 4.3

Sauli F. 1988a: Nucl. Instr. Meth. A **273**, 805. Sect. 4.3

Sauli F. 1988b: Z. Phys. C **38**, 339. Sect. 4.3

Saxon D. H. 1988: Nucl. Instr. Meth. A **265**, 20. Sect. 4.3

Schultz G., Gresser J. 1978: Nucl. Instr. Meth. **151**, 413. Sect. 4.1

Shirahashi A., Aihara H., Itoh R., Kamae T., Kusuki N., Tonaka M., Fujii H., Fujii K., Ikeda H., Iwasaki H., Kobayashi M., Matsuda T., Miyamoto A., Nakamura K., Ochiai F., Tsukamoto T., Ymauchi M. 1988: IEEE Trans. NS–**35**, 414. Sect. 4.3

Simon M., Braun T. 1983: Nucl. Instr. Meth. **204**, 371. Sect. 4.2

Sitar B. 1985: Proc. 8th Conf. Czechoslov. Phys., Bratislava. Sect. 4.3

Sitar B. 1987: Fiz. Elem. Chastits At. Yadra **18**, 1080 (English transl: Sov. J. Part. Nucl. **18**, 460). Sect. 4.3

Sosnovtsev V. V. 1987: Thesis, Institute of Physics and Engineering (MIFI), Moscow (in Russian). Sect. 4.2

Toothacker W. S., Trumpinski S., Whitmore J., Lewis R. A., Gress J., Ammar R., Coppage D., Davis R., Kankel S., Kwak N., Elcombe P. A., Hill J. C., Neale W. W., Kowald W., Robertson W. J., Walker W. D., Lucas P., Voyvodic L., Bishop J. M., Biswas N. N., Cacon N. M., Kenney V. P., Mattingly M. C. K., Ruchti R. C., Shepard W. D., Ting S. J. Y. 1988: Nucl. Instr. Meth. A **273**, 97. Sect. 4.3

TPC 1976: Proposal for PEP Facility Based on the Time Projection Chamber, Stanford Linear Accelerator Report, PUB–5012. Sect. 4.3

Turala M., Vermuelen J.C. 1983: Nucl. Instr. Meth. **205**, 141. Sect. 4.4

Ueno K. 1988: Prep. Univ. of Rochester, UR–1050, New York. Sect. 4.3

Va'vra J. 1984: Nucl. Instr. Meth. **225**, 13. Sect. 4.4

Va'vra J. 1986a: Nucl. Instr. Meth. A **244**, 391. Sects. 4.2, 5

Va'vra J. 1986b: Nucl. Instr. Meth. A **252**, 547. Sect. 4.4

Verweij H. 1975: IEEE Trans. NS–**22**, 437. Sect. 4.5

Villa F. 1983: Nucl. Instr. Meth. **217**, 273. Sect. 4.2

Villa F. (ed.) 1988: *Vertex Detectors*, Plenum, New York. Sect. 4.2

Volkov A.D. et al. 1985: Prep. JINR 13–85–417; 13–85–418, Dubna. Sect. 4.2

Wagner E.B., Davis F.J., Hurst G.S. 1967: J. Chem. Phys. **47**, 3138. Sect. 4.2

Wagner A. 1981: Phys. Scripta **23**, 446. Sect. 4.3

Wagner A. 1982: Proc. Int. Conf. on Instrumentation for Colliding Beam Physics, Stanford Linear Accelerator Report, p. 76. Sect. 4.3

Wagner R.L. 1988. Nucl. Instr. Meth. A **265**, 1. Sect. 4.2

Walenta A.H. 1979: IEEE Trans. NS–**26**, 73. Sect. 4.2

Walenta A.H. 1982: Proc. Int. Conf. on Instrumentation for Colliding Beam Physics, Stanford Linear Accelerator Report, p. 34. Sects. 4.2, 3

Walenta A.H., Kapitza H., Krämer M., Jönsson L., Roderburg E., Zech G. 1988: Nucl. Instr. Meth. A **265**, 69. Sect. 4.2

Warren R.W., Parker J.H. 1962: Phys. Rev. **128**, 2661. Sect. 4.2

Zech G. 1983: Nucl. Instr. Meth. **217**, 209. Sect. 4.3

ZEUS Collaboration 1986: The ZEUS Detector: Technical Proposal. Sect. 4.5

5. Ionization Measurements in Gas Track Detectors

In addition to the pictorial clarity of the recorded phenomena and high space-angular resolution in track (visual) detectors, a remarkable property is their sensitivity to the ionizing power of particles. These properties of track detectors have always ensured the high information content of physical results. Thus, for example, track detectors enabled the discovery of nuclear fission, the detection and investigation of the principal decay modes of many elementary particles, the study of the most basic characteristics of their interaction with matter, etc..

Certain disadvantages of track detectors must be acknowledged: e. g. the necessity of photographing images; and significant expenditure of manual labor for film scanning and primary processing of film information. At present, however, such problems are gradually being overcome by transition to filmless image recording and automation of information analysis.

Not all known gas track detectors comply with the requirements of modern experiments in high energy physics. For instance, cloud chambers (the Wilson and diffusion chambers), which played such an outstanding role in the development of experimental nuclear physics (Rochester and Wilson 1952), have hardly been used since the middle of the 1960s; this is because they are complicated in construction and have a long duty cycle (more than 10 s). The application of spark chambers is also restricted, mainly to investigations of cosmic radiation; this is because of the long recovery time (the duty cycle is longer than 0.1 s), and the dependence of the particle recording efficiency upon the number of particles and their directions (Rice-Evans 1974). The gas track detector that has turned out to be the most long-lived is the streamer chamber, which even now is used in a number of large-scale experimental installations aimed at searching for rare processes and detailed studies of multiparticle events at super-high energies of particle accelerators (Schroeder 1979; Sandweiss 1982; Rohrbach 1988).

5.1 Ionization Measurements in Cloud Chambers

The Wilson cloud chamber and the diffusion cloud chamber are classical representatives of condensation detectors; their operation is based on the condensation of supersaturated vapor (most often a mixture of H_2O and C_2H_5OH) on ions produced in the gas by charged particles. These detectors were utilized, in their time, for studies of the ionizing powers of fast charged particles (Ghosh et. al.

237

1954; Kepler et al. 1958; Ballario et al. 1961) and for the identification of the decay products of non-stable particles, as illustrated by Rochester and Wilson (1952); they were also used to identify the products of nuclear interactions at high energies ($E = 10^{12}-10^{14}$ eV) (Fretter and Hansen 1960), and to determine the π/p ratio for cosmic ray particles at mountain altitudes (Puchkov 1973) using measurements of the particle momenta and ionization of the gas.

The vapor supersaturation corresponding to the threshold of droplet condensation decreases with enhancement of the gas ionization, since then the average number of like electric charges within the volume of a droplet increases. This serves as the basis for methods of relativistic electron background suppression developed by Gorbunov et al. (1957), of discrimination between particles of different ionizing powers used by Nikitin et al. (1965), and of the operation of discharge-condensation detectors, suggested by Mandjavidze and Roynishvili (1967), and developed by Grigalashvili et al. (1975) and Barnaveli et al. (1975). In the latter, droplets are not formed on individual ions or clusters of several ions, but on short Townsend avalanches which are developed on the ionization electrons in a strong pulsed electric field applied to the detector.

A Wilson cloud chamber permits the measurement either of the total or of the primary ionization, depending on whether the expansion of the chamber takes place either *after* or *before* the passage of the particle of interest through it. In the first case, typical of experiments with cosmic rays, the droplets land on the separate ions produced by the particle in the gas, (Ghosh et al. 1954; Kepler et al. 1958), and which have had time to diffuse a distance of $l_d \gtrsim 1$ mm. In the second case, they land on ionization clusters, which have had no time to split up into separate ions because of rapid condensation (Ballario et al. 1961).

When ionization measurements are performed by counting droplets, the total ionization along a track is proportional to the restricted energy loss (1.14): clusters consisting of a number of droplets $N_d > (N_d)_{max} = 30$–40 are discarded, since it is impossible to resolve individual droplets. This corresponds to restrictions being imposed on the energy transfers in individual collisions $w \leq T_0$, where

$$T_0 = w(N_d)_{max} \simeq 1 \text{ keV}$$

($w \simeq 30$ eV; Table 1.3). For light gases ($Z \leq 10$) the quantity $(-dE/dx)_{T_0}$ is computed in accordance with (1.50). In the case of heavy gases ($I_{K, L, \dots} \gg 1$ keV), calculations are in agreement with experimental data only when (1.50) is modified to the form (1.64), taking into account the fact that electrons belonging to deep atomic shells take no part in collisions with $\omega < I_{K, L, \dots}$ (Merson 1972; Asoskov et al. 1982) (Fig. 1.10). The typical accuracy of dN_t/dx measurement for a track 40 cm long ($N_d \simeq 10^3$) amounts to $\sigma_J = 5$–10% (Ghosh et al. 1954).

The diffuse nature of tracks, which is necessary to determine the ionization by droplet counting, excludes simultaneous precise measurements of momenta in the Wilson cloud chamber. This obstacle is overcome with the aid of *relative* measurements of the cluster density along a track, limiting ion diffusion by reduction of the time delay of the chamber expansion. Here particle tracks

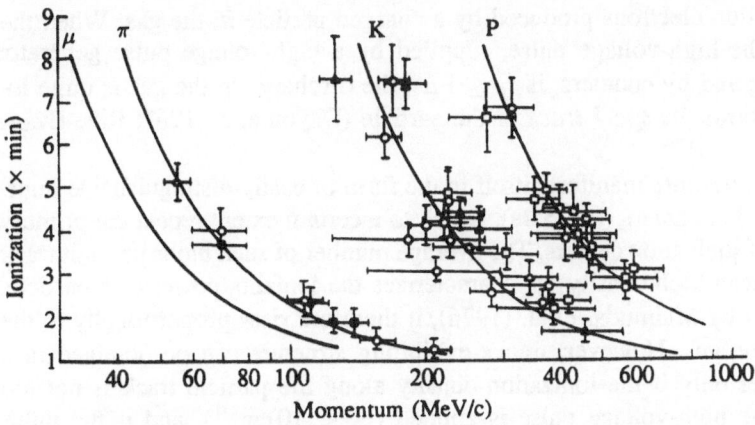

Fig. 5.1. Identification of non-relativistic mesons and protons in a Wilson cloud chamber according to momentum and photometric ionization measurements (Bjornerud 1956): o – positive particles; • – negative particles

of known ionizing powers are chosen for calibration, e. g. tracks of relativistic electrons or those of the beam particles (Cowan 1954).

Measurement of the primary ionization is characteristic of a diffusion cloud chamber with constant supersaturation and of discharge-condensation and discharge-diffusion chambers, in which the number of Townsend avalanches corresponds to the number of primary ionization clusters (Grigalashvili et al. 1975).

In the case of small-scale photography, and when strongly ionizing particles are involved, or in condensation chambers of high pressure, tracks are continuous and structureless. In this case, relative ionization measurements and the identification of particles can be performed by estimating the width and densities of their track images (Schluter 1954; Bjornerud 1956) (Fig. 5.1, Sect. 5.4).

5.2 Determining the Ionizing Powers of Particles with Spark Chambers

The localization of a pulse discharge in a gas along the trajectory of a charged particle is utilized in high energy physics for visualization of particle tracks and for coordinate measurements in spark chambers. It was noticed long ago that the structure, brightness and width of a track in a spark chamber depend on the ionizing power of the particle. These properties of a spark chamber served as the basis for developing the methods of ionization measurements dealt with below. The method, common to all gas-discharge detectors, of determining the ionization from the efficiency of the spark chamber, is described in Sect. 6.2.

The Structure, Brightness and Width of a Track in a Spark Chamber. A discharge in a spark chamber is due to a breakdown of the conductive plasma channel resulting from the development of avalanches and streamers on the clusters of ionization electrons produced by a charged particle in the gas. When the duration of the high-voltage pulse, supplied by a high-voltage pulse generator (HVPG) triggered by counters, is $\tau_p \leq 1\,\mu s$, the discharge in the gas is quite localized and forms the *spark track* of the particle (Dayon et al. 1967; Rice-Evans 1974).

The track structure manifests itself in the form of easily distinguishable luminous blobs and streamers (Fig. 5.2a), which to a certain extent repeat the primary distribution of ionization centers. The average number of such blobs per unit track length (the mean blob density) g_b characterizes the ionizing power of a particle, and, as shown by Maimudar et al. (1976), it therefore rises proportionally to the pressure of the gas. However, tracks exhibiting structure can be obtained in a spark chamber only if the ionization density along the particle track is not too high when the high-voltage pulse is applied ($g_b < 10\,\mathrm{cm}^{-1}$), and if the pulse duration does not exceed 10^{-7} s.

To obtain tracks of intermittent structure, low pressure gases which have a small specific ionization are used, or electro-negative admixtures (air, SO_2) are introduced and the high-voltage pulse delay is increased to $\tau_d = 0.7\text{--}1.5\,\mu s$. To reveal the track structure, it is necessary to slow down the growth rate of streamers. To this end, organic admixtures such as C_3H_8 are utilized, the high-voltage pulse is shortened, and the HVPG output power is lowered.

Estimations of ionization via the density of luminous blobs along a track are quite rough, since g_b is several times smaller than the actual ionization density and

(a)   (b)   (c)   (d)

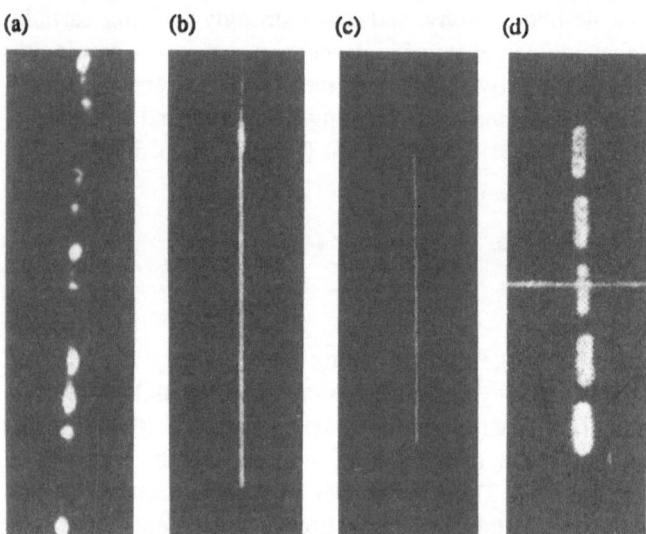

Fig. 5.2. Particle tracks in a spark chamber: (a) – structured track of a relativistic muon in a four-gap argon spark chamber, $l = 1.9$ cm: $P = 0.1\text{--}0.4 \cdot 10^5$ Pa, $30 < RC < 120$ ns; $U_0 = 15$ kV (Maimudar et al. 1976); (b) and (c) – luminosities of proton and pion tracks with total ion densities g_p and g_π, respectively, $(g_p/g_\pi = 3.5)$ in a neon spark chamber with a 0.03 % C_3H_8 admixture, $l = 15$ cm (Dayon et al. 1967); (d) – broadening of a proton track slowing down in the plates of a 18-gap neon spark chamber, $l = 1$ cm (Aganianz et al. 1969)

in addition exhibits large fluctuations. This is due to the fact that for the development on *each individual* ionization electron of a discharge limited in space, and for its subsequent photo-recording, quite rigorous conditions (Sect. 5.3), which are usually not fulfilled in spark chambers, must be satisfied.

The *track brightness* is measured by means of photometric methods by the blackening of the image on the photographic film (Sect. 5.4). Not only tracks of genetically related particles with differing ionizing powers differ in brightness (which can be explained by the corresponding division of the high-voltage pulse energy), but also tracks of separate, non-simultaneous particles differing in time (Fig. 5.2b, c).

The brightness H_s of a spark is proportional to the energy W_R released in it. If the electric resistance of a spark channel $R_s \ll R$, where R is the resistance of a spark chamber terminator, then

$$H_s = k_s W_s = k_s(W_g - W_R) = k_s \left[CU_0^2/2 - \int_0^{\tau_f} (U^2(t)/R)\, dt \right] . \qquad (5.1)$$

Here U_0 is the output voltage of the HVPG, while C is its output capacitance; $W_g = CU_0^2/2$, and W_R are the respective energies stored in the HVPG and released in the resistance R during the *time τ_f required for formation of the discharge*; k_s is a constant factor. In the case of a sharp rise of the pulse from the HVPG, $U(t) = U_0 \exp(-t/RC)$, and in accordance with (5.1), the track brightness is

$$H_s = k_s W_g \exp(-2\tau_f/RC) . \qquad (5.2)$$

Therefore the track brightnesses H_{s1} and H_{s2} ($H_{s1} < H_{s2}$) of two particles with differing ionizing powers, g_1 and g_2, respectively, are related by the following expression:

$$\ln(H_{s2}/H_{s1}) = (2/RC)(\tau_{f1} - \tau_{f2}) \equiv (2/RC)\,\delta\tau_f . \qquad (5.3)$$

The discharge formation time τ_f depends on the average time required for the development up to the streamer stage, of avalanches initiated on ionization blobs of size l_d, with a mean number of electrons in each blob of $\overline{N}_e \simeq g l_d$. According to the streamer theory of breakdown in a gas, transition to a spark discharge occurs when the number of charges in an avalanche reaches the critical value N_{cr}:

$$N_{cr} = \overline{N}_e \exp \left[\int_0^{\tau_f} \alpha_T(t) w_d(t)\, dt \right] . \qquad (5.4)$$

Here $\alpha_T(t)$ is the first Townsend ionization coefficient, and w_d is the electron drift velocity (Chap. 2). Within a broad range of the electric field strengths $\mathcal{E}(t) = U(t)/l$,

$$\alpha_T(t)w_d(t) = a[\mathcal{E}(t)]^b ,\qquad(5.5)$$

where a and b are constant coefficients for the given gas composition and pressure (Chap. 2). Applying (5.2) to (5.5) we obtain the following equation:

$$\ln(\overline{N}_{e2}/\overline{N}_{e1}) = \ln(g_2/g_1) = k[\exp(-2\tau_{f1}/RC)]^{b/2}[(H_{s2}/H_{s1})^{b/2} - 1] ,$$

where $k = a(U_0/l)^b RC/b$. Consequently, the ratio H_{s2}/H_{s1} of the track brightnesses of two particles depends not only on their relative ionizing powers g_2/g_1, but also on the relation between τ_{f1} and the time constant RC of the HVPG. The same parameters, including U_0, determine the track brightness H_{s1} of minimum ionizing particles (5.2) for the chosen gas composition and pressure, and electric field strength. Since the track brightness has an upper limit $H_s = k_s W_g$ [for $\tau_f \ll RC$, see (5.2)], it varies within a range that is larger when the track brightness of minimum ionizing particles is weaker. For example, in a wide-gap neon spark chamber ($l = 20$ cm), when $g_2/g_1 = 15$ and $H_{s1} = 20$ arbitrary units, the ratio of brightnesses $H_{s2}/H_{s1} = 12$, while for $H_{s1} = 40$ arbitrary units $H_{s2}/H_{s1} \simeq 4$ (Bolotov et al. 1970).

In measuring the dependence $H_s(g)$ (Fig. 5.3) one may pass from the distribution of track brightness to the distribution of ionization. For $lP \gtrsim 20 \cdot 10^5$ cm $\cdot$ Pa the latter is close to the Landau distribution (Galaktionov et al. 1965; Bolotov et al. 1970).

The track brightness is particularly affected by oscillations of the high voltage, which influence both τ_f and W_g. According to Pavlovsky (1966) the high voltage must be subjected to rigorous stabilization, since a variation only of 15 % results

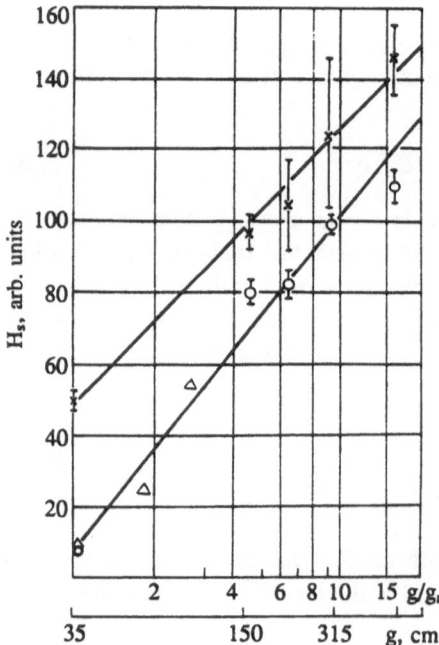

Fig. 5.3. Spark brightness versus specific ionization produced by relativistic particles in a wide-gap spark chamber ($l = 20$ cm) (Bolotov et al. 1970): *Circles* – $RC = 100$ ns; *Crosses* – $RC = 115$ ns; *Triangles* – data obtained by Galaktionov et al. (1965); solid lines – approximation of experimental points

in changing the track brightness by a factor of about 100. The brightness of a track in a spark chamber also depends on the angle between the track and the electric field. Both of the above problems impose serious restrictions on the application of this method.

Control of the track brightness by varying the operation conditions of the spark chamber permits to discriminate particles according to their ionizing powers. Thus, simply by regulating the shunting resistance of the HVPG, the track brightness of particles of minimum ionizing powers can be reduced to such an extent that the spark chamber remains sensitive only to the slow, strongly ionizing products of nuclear reactions. This makes it possible, for instance, to eliminate the background of beam particles and of secondary relativistic particles present in experiments involving recoil nucleus detection, which are performed with spark chambers at accelerators. This operation mode was realized, for example, in studies of π^0-meson photo-production on helium nuclei in a high-intensity $(10^8$–$10^9\,s^{-1})$ photon beam of an energy up to 3.5 GeV (Alexanian et al. 1977).

The dependence of the track brightness in a spark chamber on the ionization density was also utilized in the search for particles of fractional charge $z = 1/3$ in cosmic radiation (Krider et al. 1970). Tracks of such particles were simulated by introducing an electro-negative admixture of SF_6 $(10^{-2}\,Pa)$ into the gas, so that when the time delay of the high-voltage pulse was $\tau_d = 4\,\mu s$, 8/9 of the ionization electrons were captured by the admixture. In the case of small time delays $(\tau_d \ll 0.2\,\mu s)$ the presence of the admixture did not affect the track brightness of particles with $z = 1$.

The track width, like its brightness, correlates with the specific ionization. That can be seen when particles are slowed down in a multigap spark chamber (Fig. 5.2d). The track width of such particles increases linearly with their ionizing power (Fig. 5.4). The slope of this dependence is determined by the energy released in the spark, i. e. by the HVPG parameters, and by the number of spark gaps through which the particle has passed.

A track broadens with the increase of specific ionization for the same reasons that its brightness is enhanced. This is related to the decrease of τ_f because

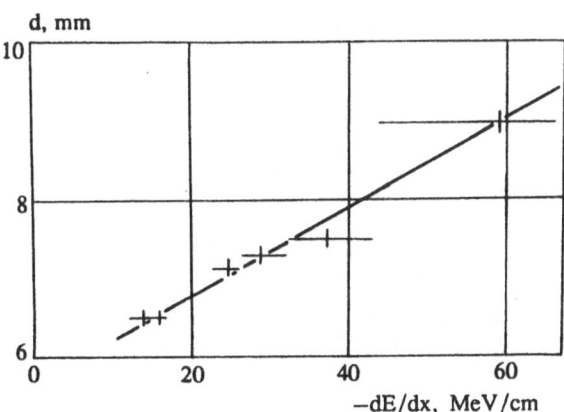

Fig. 5.4. Dependence of the average width of a spark track upon the ionizing power of particles decelerated in the plates of a spark chamber (Aganianz et al. 1969)

of which, in accordance with (5.1), the part of the HVPG energy released in the spark increases, as does, consequently, the number of light emitting atoms excited in the spark channel. Absorption of the short-wave part of this emission leads to photo-excitation of the gas at the periphery of the discharge, where the formation of molecular ions proceeds intensively (Lozansky, Firsov 1976) (Chap. 2). As a result, within a very short time ($t < 10^{-9}$ s $\ll \tau_f$) there arises in that region an additional ionization leading to a broadening of the spark channel. This phenomenon is particularly noticeable in a pure inert gas at low pressures ($P \lesssim 10^4$ Pa), where, owing to an enhancement of the photon absorption length, tracks are not only broadened, but are also accompanied by a wide luminous halo (Asoskov et al. 1977). In a gas containing admixtures, their photo-ionization can serve as an additional cause of track broadening.

The Spark Discharge Formation Time. The development of avalanches and streamers on ionization electrons in a gas up to the spark discharge stage proceeds the faster the smaller the mean distance $l_c = g_b^{-1}$ between the centers of primary ionization blobs, i.e. the higher the ionizing power of the particle (Dayon et al. 1967; Allkofer 1969) (Fig. 5.5). Therefore the average formation time τ_f of a discharge in a spark chamber, other conditions being equal, serves as a measure of the ionization produced by a particle (Gebauer 1977; Ananyin, Stibunov 1983). It is expected that in the case of weak diffusion of electrons away from the primary blob, during the delay time τ_d of the high-voltage pulse, when the diffusion displacement $l_d \ll l_c$, τ_f will depend on the primary ionization; while if $l_d \gg l_c$, it depends on the total ionization in the gas.

The discharge formation time is determined by the interval between the moment the high-voltage pulse is applied to the spark gap, where free electrons produced by the charged particle are present, and the formation along the particle track of a stable spark breakdown. The latter is characterized by a sharp

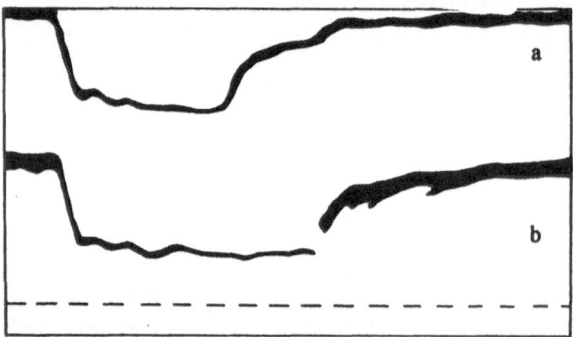

Fig. 5.5. Oscillograms of voltage pulses in a helium spark chamber (Ananyin et al. 1979): (a) – α-particles, $\varepsilon_\alpha = 5.3$ MeV; (b) – β-particles, $\varepsilon_\beta \simeq 2$ MeV. The pulse duration corresponds to the discharge formation time; the trailing edge corresponds to the development of the spark breakdown; the marks indicate 10 ns; $l = 1.5$ cm; $\mathcal{E}_0 = 8$ kV/cm

increase of current and a corresponding fall of voltage across the gap. It follows that there are two feasible methods of measuring τ_f. The first consists in integration of the voltage pulse $U(t)$, read from the electrodes of the spark chamber, so that the amplitude of the resultant signal is proportional to τ_f (Gebauer 1977; Ananyin, Stibunov 1983). The second consists in direct measurement of τ_f, i.e. of the duration of the voltage pulse. To this end the signal is differentiated and applied to a start-stop electronic circuit, which allows time measurements to be performed (Ananyin et al. 1979).

An analysis of the factors determining the spark formation time reveals that $\tau_f = \tau_0 \cdot \ln N_c^{-1}$, where τ_0 is a function of P, U_0/l and RC, and also depends on the gas composition; and N_c is the average number of electron clusters in the spark gap at the moment the high-voltage pulse is applied. Typical values of τ_f are 40–90 ns. Since τ_f depends on the composition and pressure of the gas and the amplitude and shape of the high-voltage pulse, it is reasonable to deal with the difference $\delta\tau_f$ between the track formation times of two particles, one of which is usually chosen to be a particle of minimum ionization ($\tau_f \equiv \tau_{fm}$). Thus, for example, for a helium spark chamber ($P = 10^5\,\mathrm{Pa}$, $\mathcal{E}_0 = 8\,\mathrm{kV/cm}$, $l = 3\,\mathrm{cm}$), when $\tau_d < 1\,\mu\mathrm{s}$ and the electron diffusion from the primary cluster can be neglected (Fig. 2.1),

$$\delta\tau_f = \tau_0 \ln(N_1/N_{1m}) \qquad (5.6)$$

(Fig. 5.6). Here $\delta\tau_f = (\tau_{fm} - \tau_f)$; $\tau_0 = 6.22\,\mathrm{ns}$; N_1 and N_{1m} represent the numbers of primary electrons produced by the particle under investigation and by minimum ionizing particles, respectively. This dependence of $\delta\tau_f$ on (N_1/N_{1m})

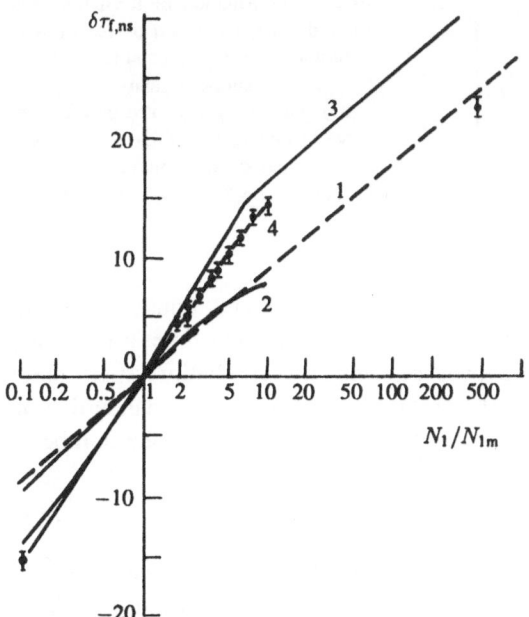

Fig. 5.6. Dependence of discharge formation time in a helium spark chamber on the primary ionization (Ananyin, Stibunov 1983); $l = 3\,\mathrm{cm}$, $\mathcal{E}_0 = 8\,\mathrm{kV/cm}$: the computed curves correspond to the avalanche (*1*), streamer (*2*) and combined avalanche and streamer (*3*) stages of discharge development; (*4*) – approximation of experimental data according to (5.6)

245

follows from our understanding of the durations of the avalanche and streamer stages of a discharge when $N_1/N_{1m} \lesssim 10$. In the case of a high primary ionization density, when the longitudinal dimensions of an avalanche $l_a \gtrsim l_b$ overlap the distances between adjacent blobs, a conductive channel is formed before the streamer stage is initiated, and the dependence in (5.6) becomes flatter.

The fluctuations of the spark formation time $f(\delta\tau_f)$ computed with the aid of (5.6), assuming the number of primary ionizing collisions of a particle in the gas to exhibit a Poisson distribution, are described by the relation

$$ f(\delta\tau_f) = \tau_0^{-1}[\exp(-N_1)] \left\{ N_1^{N_{1m}\exp(\delta\tau_f/\tau_0)} / \Gamma[N_{1m}\exp(\delta\tau_f/\tau_0)] \right\}, \qquad (5.7) $$

where Γ is the gamma-function (Fig. 5.7). The mean value $dN_{1m}/dx = 3.73 \pm 0.30\,\text{cm}^{-1}$ for this distribution corresponds to the pressure $P = 10^5\,\text{Pa}$ in the helium spark chamber under consideration. This result is close to the data obtained by other methods and to theoretical predictions (Chap. 1). As the specific ionization increases, distribution (5.7) becomes narrower and more symmetric, so that at $N_1 \simeq 30$ its asymmetry practically vanishes.

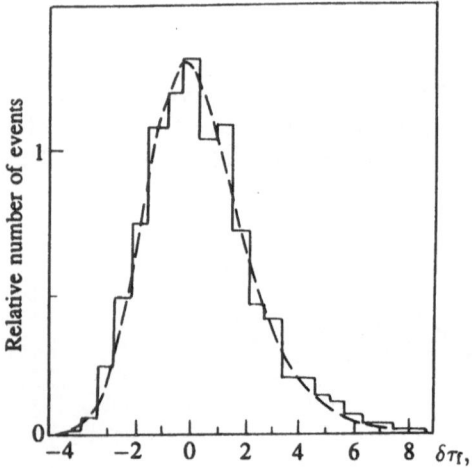

Fig. 5.7. Distribution of formation time for a discharge initiated by electrons of minimum ionizing capability in a helium spark chamber (Ananyin, Stibunov 1983). The histogram corresponds to the experimental data; the dashed curve is the computed distribution for $dN_{1m}/dx = 3.73\,\text{cm}^{-1}$; $l = 3\,\text{cm}$; $\mathcal{E}_0 = 8\,\text{kV/cm}$

To enhance the precision of ionization measurements and, consequently the reliability of particle identification by the discharge formation time, a multi-gap spark detector is required. Studies carried out with a 4-gap helium spark spectrometer ($l = 3\,\text{cm}$) (Ananyin et al. 1979) have revealed that the standard root-mean-square deviation σ_n of ionization measurements depends on the number n of gaps as follows

$$ \sigma_n = \sigma_1 n^{-1/2}. \qquad (5.8) $$

Here $\sigma_1 \approx 0.4$ is the resolution of a single gap spark chamber. When summing the information signals from all spark gaps there must appear in (5.8) an additional

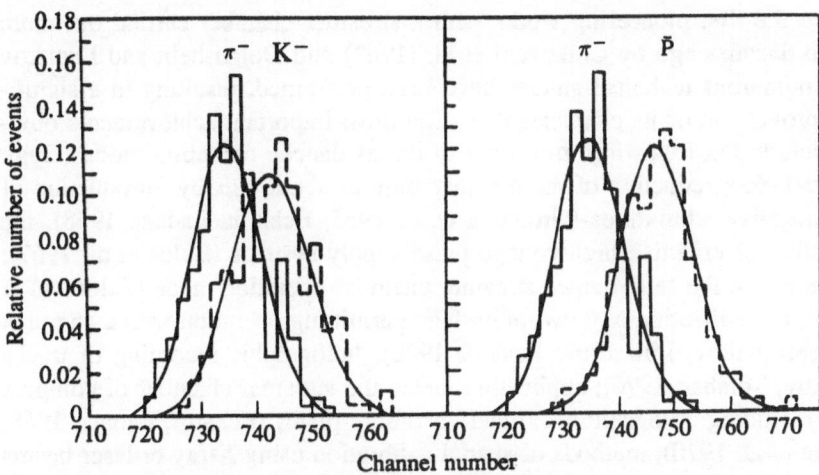

Fig. 5.8. Distribution of the average discharge formation time in a 30-gap spark chamber exposed to a 6 GeV/c particle beam (Gebauer 1977): the chamber is filled with a 30 % He + 70 % Ne mixture; $l = 1$ cm. The experimental histograms are approximated by Gaussian distributions

factor $(1 + \nu_f^2)^{1/2}$, where ν_f is the average relative deviation $\delta\tau_f$ within individual gaps.

The feasibility of identifying relativistic particles using the discharge formation time in a 30-gap spark chamber is illustrated in Fig. 5.8, where total amplitudes of the current signals in all gaps are presented.

Using the methods involving measurement of the spark discharge formation time requires both the high voltage pulse amplitude to be stabilized to better than 0.5 % and constant gas composition and pressure. The difficulties of measuring the specific ionization, when several particles pass through the detector, and the dependence of τ_f on the angle between the track and the electric field, imposes significant restrictions on the application of this method.

5.3 Ionization Measurements in a Streamer Chamber

The Main Properties of a Streamer Chamber. A modification of the spark chamber developed in the early 1960s, the streamer chamber[1], is used in experimental nuclear physics and in high energy physics as a universal track detector. Unlike the spark chamber, the streamer chamber exhibits almost isotropic properties. It records, with a 100 % efficiency, charged particles, independently of their number and direction, thus providing a reliable space image of an event.

---

[1] The term "streamer chamber" is also extended below to avalanche chambers in which the gas discharge is terminated at the avalanche, instead of the streamer stage, in accordance with conventions adopted in the russian physical literature.

Since the first pioneering works with a streamer chamber carried out more than two decades ago by Chikovani et al. (1963) and Dolgoshein and Luchkov (1964), numerous technical studies have been performed, resulting in a significant improvement of its characteristics. The most important achievements obviously include the following: transition to the avalanche operation mode (Gygi, Shneider 1966); reduction of the memory time to about 1 μs by introduction of electro-negative admixtures (Gromova et al. 1965; Eckardt, Ladage 1968); the construction of effective high-voltage pulse supply systems (Bulos et al. 1967); development of the self-shunted streamer chamber operation mode (Falomkin et al. 1967); the utilization of a two-pulse feed permitting operation with a complex trigger (Mikhailov, Iljin 1969; Eckardt 1970); holographic recording of tracks (Stabnikov, Tombak 1970); application inside the streamer chamber of complex targets (including cryogenic ones) and absorbing plates (Eckardt, Ladage 1968; Vardenga et al. 1970); methods of spatial calibration using X-ray or laser beams (Guschin et al. 1984); and so on. The space resolution of large streamer chambers has now achieved a value of 150 μm, while that of small high pressure chambers is as good as 20 μm (Sandweiss 1982).

At present, streamer chambers are utilized as vertex detectors or coordinate visual detectors in large-scale experimental installations at high-energy accelerators like RISC and E–128 (Russia); NA–5, NA–9, NA–28, NA–35, WA–44, UA–5, UA–5/2, PS–179 (CERN) and others [see review articles by Somov (1975) and Rohrbach (1988)]. There exist well known examples of the application of streamer chambers in experiments on hadron production by photons and hyperon production by kaons at SLAC, in searching for quarks both in cosmic rays and accelerator beams (FRG, CERN), in investigations of the decays of K^0- and K^+-mesons, of the polarization of direct muons, of nucleus-nucleus collisions (Russia, CERN), and so on. The high reliability in recording complex multiparticle events, good spatial resolution, isotropy of detection, large counting rates, and simplicity of construction and operation, all have contributed to establishing the streamer chamber firmly in high energy experimental physics. We predict that there will be a wide range of applications for this device for a long time to come.

Particle tracks in a streamer chamber represent chains of short luminous avalanches or streamers growing on the ionization electrons under the influence of a strong pulsed electric field (Fig. 5.9). Streamer chambers exhibit excellent time and space characteristics, are not subject to rigorous size restrictions, are reliable in operation, and are relatively inexpensive. A streamer chamber is capable of operation in particle beams of high intensity (up to $10^6 \, \mathrm{m}^{-2} \, \mathrm{s}^{-1}$), is not sensitive to secondary interactions or to the background of γ-quanta and neutrons. Because a streamer chamber is a triggered detector, it records events of a given class corresponding to the chosen trigger. Owing to weak multiple scattering of particles in a gas and to the small diffusion track width a streamer chamber placed in a magnetic field permits high-precision measurements of particle momenta and, under certain conditions, of the primary ionization (Davidenko et al. 1970a; Morrison 1972; Rohrbach 1988; Merson 1990). Thus, a streamer cham-

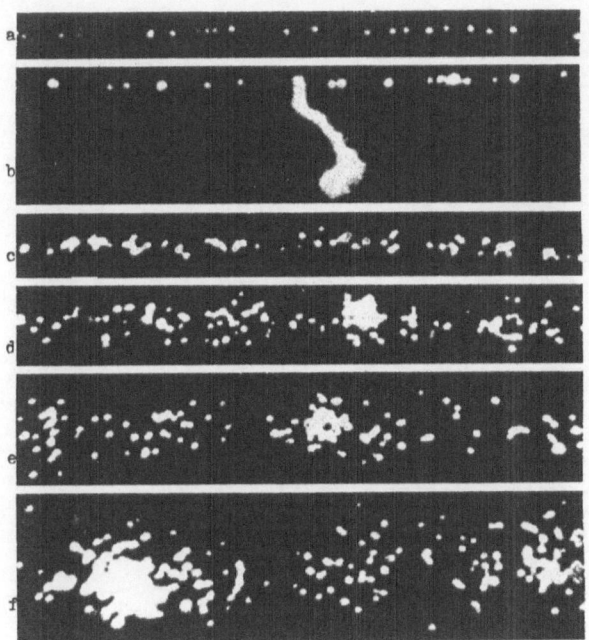

Fig. 5.9. Tracks of relativistic particles in a streamer chamber (Davidenko et al. 1968a): (a), (b) – He, $P = 6 \cdot 10^4$ Pa; (c–f) – Ne, $P = 8 \cdot 10^4$ Pa. The time delay of the high-voltage pulse is: (a), (b) – 0.2 μs; (c) – 1 μs; (d) – 2.5 μs; (e) – 5.0 μs; (f) – 10.0 μs

ber can serve as an excellent ionization identifier of particles without imposition of any restrictions on their multiplicity. Shortcomings of the streamer chamber exist, such as low track luminosity, which creates difficulties in recording, and the small rate of accumulation of events (the triggering frequency is $\lesssim 10$ Hz). Film recording only permits off-line analysis of the results of measurements. It must also be stressed that the construction and operation of large streamer chambers capable of serving as ionization identifiers of relativistic particles represents quite a complicated technical problem.

The Avalanche and Streamer Modes. Two operation modes of the streamer chamber can be distinguished, the avalanche and the streamer mode. In the avalanche mode, the gas amplification M is relatively low ($M \leq 10^8$), and on each ionization blob corresponding to a single primary electron, avalanches elongated along the electric field develop with approximately the same geometrical dimensions (Fig. 5.10a). The number of electrons in each avalanche is $N_e \simeq M\omega/w$, its length is $l_a \simeq 1$ mm, and its diameter d_a is determined by the electron diffusion $d_a \simeq 2(2\tau_d D)^{1/2} = 0.5\text{--}1.0$ mm, where the electron diffusion coefficient D is a function of $\mathcal{E}_0/P$. The avalanches emit about 10^8 photons per 1 mm^2, so that in order to photograph them, light amplification using image intensifiers is required. The avalanche operation mode of the streamer chamber is characterized by a high spatial resolution, and permits measurement of the primary ionization.

249

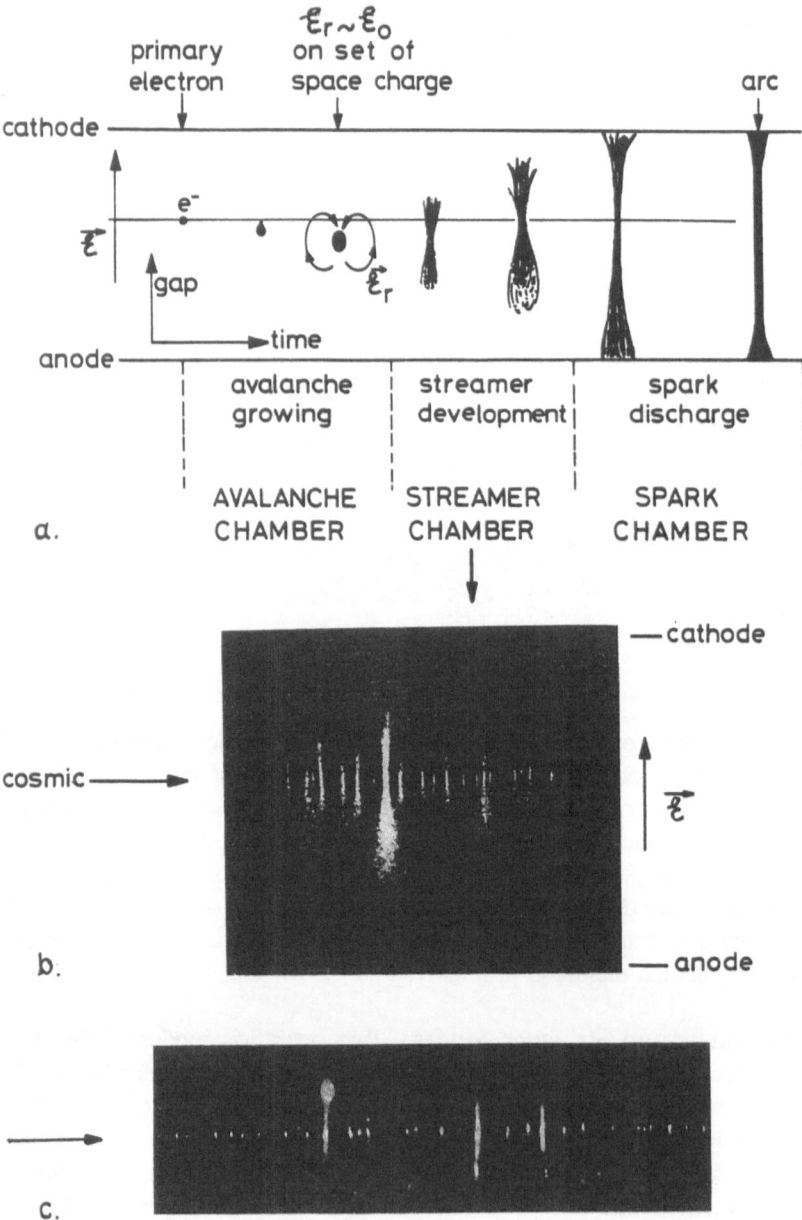

Fig. 5.10. (a) Development of a gas discharge in a streamer chamber (Allkofer 1969, Lecoq et al. 1979): $\mathcal{E}_0$ – field strength of the external electric field; $\mathcal{E}_r$ – field strength of the additional space charge field arising in the avalanche-to-streamer transition. (b) Picture of a cosmic ray track in a hydrogen streamer chamber (Rohrbach 1988). (c) Picture of a relativistic charged particle track in the self-shunted streamer chamber (Falomkin et al. 1975b)

As the avalanche develops, tending towards its critical size determined by the condition $(N_e)_{cr} \simeq 10^8$, there appears in it a space charge, caused by the difference in the mobilities of electrons and positive ions. This charge generates an additional field $\mathcal{E}_r$ (Fig. 5.10a), which contributes to a further enhancement of the acceleration of electrons in the front part of the avalanche. There occurs an avalanche-to-streamer transition: an avalanche transforms into streamers, the development of which is strongly promoted by photo-ionization of the gas and by associative ionization (Sect. 2.4). Streamers initiated on the avalanche develop in both directions toward the positive and the negative electrodes (Fig. 5.10a). The growth of streamers is much more rapid than that of avalanches, but is irregular and exhibits large fluctuations (Davidenko et al. 1968b). The diameter

$$d_s = 0.4(l_s d_a)^{1/2}$$

of streamers of length $l_s = 1\text{–}2\,\text{cm}$ amounts to $1.5\text{–}2.0\,\text{mm}$ and is determined mainly by the electro-static repulsion of electrons (Davidenko et al. 1969b). Such streamers emit $10^{10}\text{–}10^{11}$ light photons per mm^2 and they can be photographed directly by applying wide aperture optics and high-sensitive photographic films.

The growth of the avalanches and streamers can be suspended, resulting in localization of the gas discharge, if the accelerating electric field is switched off sharply. Otherwise the streamers reach the electrodes of the chamber leading first to the projection and then to the spark modes of operation (Fig. 5.10a). Thus, the operation mode of a track spark chamber, and in particular of a streamer chamber, depends on the amplitude and duration of the applied high voltage pulse, and on the direction of motion of the particles.

The Parameters of the High Voltage Pulse. To produce an avalanche of $l_a \simeq 1\,\text{mm}$, with a characteristic growth velocity of $v_g \gtrsim 10^7\,\text{cm/s}$, in a field $\mathcal{E}_0 = 10\text{–}20\,\text{kV/cm}$, an extremely short pulse of duration $\tau_p = l_a/v_g \lesssim 10\,\text{ns}$ is required. The growth velocity of a streamer is much higher ($v_g \simeq 10^8\,\text{cm/s}$) so that a pulse duration of $\tau_p = 20\text{–}40\,\text{ns}$ corresponds to the streamer length of $l_s \simeq 1\,\text{cm}$. In the avalanche mode, the choice of $\mathcal{E}_0$, τ_p and of the gas pressure P corresponds to a mean number of electrons in the avalanche $\overline{N}_e$, such that a sufficient number of photons $\overline{N}_p \propto \overline{N}_e$ are emitted to record the track. According to Tan et al. (1970), for avalanches starting from a single electron, the mean number of electrons is determined from the equation

$$\ln \overline{N}_e = \alpha_T w_d \tau_p \propto \mathcal{E}_0^3 \tau_p P^{-2}\,,$$

where α_T is the first Townsend coefficient. Therefore the avalanche operation mode is retained if

$$\mathcal{E}_0^3 \tau_p P^{-2} = \text{const} \qquad\qquad (5.9)$$

(Fig. 5.11). The constant in the right-hand side of this relation depends on the gas composition and is equal to $(2.9 \pm 0.4) \cdot 10^{-4}$ for H_2, $(4.8 \pm 0.5) \cdot 10^{-5}$ for He, and $(1.7 \pm 0.2) \cdot 10^{-5}\,(\text{kV/cm})^3 \cdot \text{ns/Pa}^2$ for a 75 % Ne + 25 % He mixture (Rohrbach et

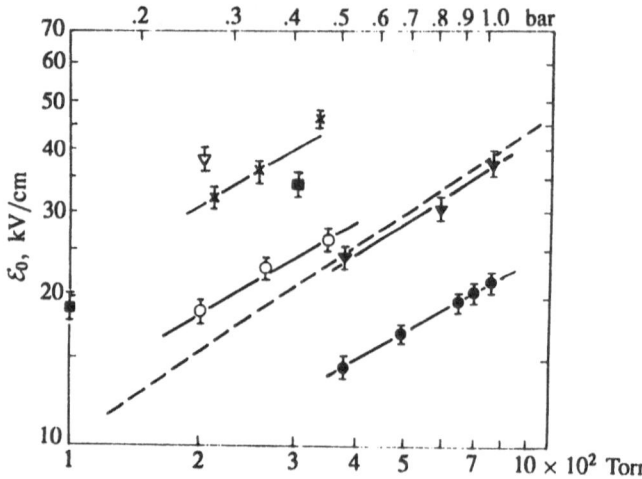

Fig. 5.11. Parameters of the high-voltage power supply of a streamer chamber for different gas mixtures (Rohrbach et al. 1973): ($\triangledown$) CH_4, $\tau_p = 9\,ns$; ($\times$) $70\% H_2 + 30\% CH_4$, $\tau_p = 9\,ns$; (o) $H_2 + 0.5\% CH_4$, $\tau_p = 9\,ns$, ($\square$) $H_2 + 5 \cdot 10^{-5} SF_6$, $\tau_p = 6\,ns$; ($\star$) He, $\tau_p = 6\,ns$, ($\blacktriangledown$) He, $\tau_p = 9\,ns$; (•) $25\% He + 75\% Ne$, $\tau_p = 16\,ns$. The slope of the dashed straight line is calculated from (5.9)

al. 1973). In the case of higher values of the product (5.9), the projection mode is observed; while in the case of lower values no visible tracks are present. In streamer chambers filled with He, Ne and their mixtures at $P \simeq 10^5\,Pa$, typical values are $\mathcal{E}_0 = 20\text{--}30\,kV/cm$, while in hydrogen chamber, $\mathcal{E}_0 = 30\text{--}60\,kV/cm$. The avalanche mode corresponds to $\tau_p = 3\text{--}10\,ns$, while the streamer mode has values of $\tau_p = 20\text{--}40\,ns$.

The time required for generation, formation and transmission of the high-voltage pulse determines the delay of the discharge development with respect to the passage of the particle. In most streamer chambers this delay time amounts to $\tau_d = 0.8\text{--}1.0\,\mu s$. Because of effects related to transport processes in the gas (Sect. 2.2) the delay time may affect the precision of coordinate and ionization measurements in the streamer chamber.

High-voltage pulses with amplitudes up to 1 MV are created using multistage Arkadiev-Marx generators, while formation of a short pulse and its transmission to the streamer chamber is performed by means of a coaxial Blumlein line, coupled to the electrodes of the streamer chamber by a strip line (Bulos et al. 1967) (Fig. 5.12). Active resistances matched with the characteristic impedance of the line serve as the terminators. The difficulties in producing pulses with amplitudes $U_0 > 1\,MV$ limit the depth of a streamer chamber to less than about 1 m, while its length and width follow the requirements and conditions of the particular experiment.

Self-Shunted Streamer Chamber. Streamers initiated on the ionization electrons, produced by a charged particle in the gas, can be localized not only by deliberate abrupt termination of the high-voltage pulse, but also by letting the

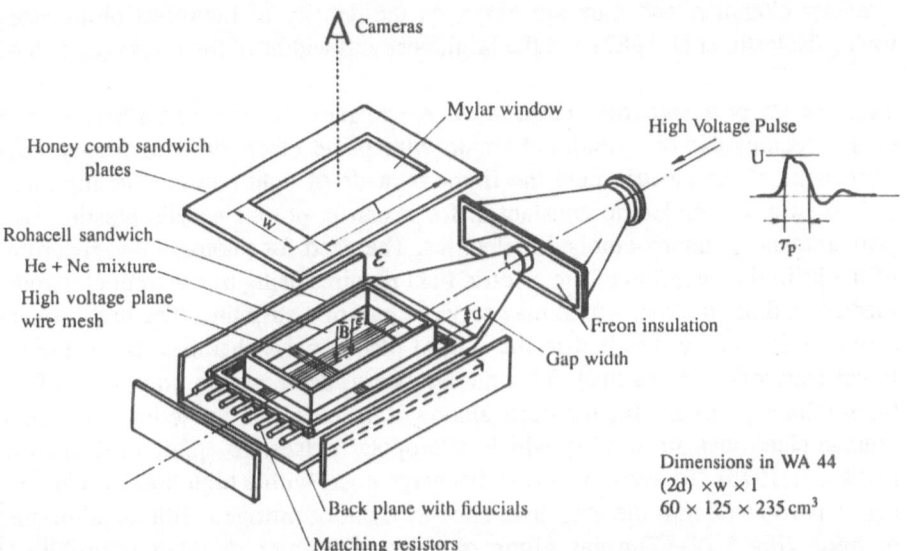

Cameras

Mylar window

High Voltage Pulse

Honey comb sandwich
plates

Rohacell sandwich

He + Ne mixture

High voltage plane
wire mesh

$\mathcal{E}$

B

d

Freon insulation

Gap width

U

τ_p

Dimensions in WA 44
(2d) ×w × l
$60 \times 125 \times 235\,cm^3$

Back plane with fiducials

Matching resistors

Fig. 5.12. The $0.6 \times 1.25 \times 2.35\,m^3$ AC–235 streamer chamber filled with a Ne + He mixture and operated in avalanche mode; $P = 10^5\,Pa$, $\tau_p = 12\,ns$ (WA–44 experiment) (Lecoq et al. 1979): B and $\mathcal{E}$ are the magnetic and electric fields, respectively

discharge current within the chamber itself shunt the high voltage pulse. Such a situation generally occurs in every spark chamber, where the moment of shunting when the current rises sharply, is determined by the formation time of the discharge (Sect. 5.2). If the gas filling of a spark chamber operating in the projection mode includes small ($\leq 0.1\,\%$) admixtures with low excitation potentials, then the brightness of the discharge channels becomes non-uniform: the central regions corresponding to the passage of a charged particle moving along the electrodes are much more luminous. This results in the track of a particle being localized to the same extent as in a conventional streamer chamber, while its brightness is significantly higher (Falomkin et al. 1967) (Fig. 5.10b).

The non-uniformity of the discharge luminosity is caused by admixtures in the gas which lower the mean electron energy in the discharge channel significantly; this causes excitation of the atoms of the main gas to take place intensively only in the "neck" of the channel corresponding to the highest electron density, i. e. precisely on the trajectory of the particle (Falomkin et al. 1975a). When helium is used as the main gas, the "cooling" admixtures are usually hydrocarbons, while in the case of neon the admixture is nitrogen. The best results with a streamer chamber operating in a magnetic field are obtained with inorganic admixtures, N_2 and Xe, at a level of $\sim 10^{-2}\,\%$. Self-shunting streamer modes can also be realized in a hydrogen chamber (Falomkin et al. 1975b). Since in this operating mode streamers merge with each other, the apparent ionization density along particle tracks turns out to be significantly lower than the primary ionization. Therefore only *relative* ionization measurements are feasible in a self-shunted

253

streamer chamber, and they are based on the density of luminous blobs along tracks (Balestra et al. 1987) and the brightness and widths of the tracks (Sect. 5.4).

The Design of a Streamer Chamber. A streamer chamber has a hermetically closed rectangular or cylindrical frame with plane electrodes. To reduce edge distortions of the electric field the frame is made of light non-conducting materials with a low dielectric constant ε: for instance, porous acrylic plastic, foam polyurethane or honey-combed dielectrics. The need for photographic recording of tracks in the direction of the electric field requires using transparent electrodes made of a thin wire net, which may serve as part of a strip line. The high-voltage electrode is usually installed in the central plane of the chamber, while the external electrodes are earthed. This allows one to decrease the amplitude of the high-voltage pulse and facilitates insulation of the chamber. Sometimes two high-voltage electrodes are used to which heteropolar pulses are applied (Mikhailov, Lomtadze 1975). To avoid a corona discharge a gas with a high breakdown voltage is introduced into the strip line, such as freon or nitrogen with an admixture of freon (Fig. 5.12). The gas filling of a large streamer chamber is usually at atmospheric pressure, for convenience of operation.

Targets, including ones with liquid H_2 or D_2 (Eckardt, Ladage 1968), scintillators used for formation of the trigger, and γ-ray converters made of a heavy dielectric (Vardenga et al. 1970) are often installed inside streamer chambers. They are also equipped with auxiliary devices for flushing or recirculating the gas, and for controlling its composition and purity.

Choice of Gas. The choice of the gas filling of a streamer chamber depends on whether it serves only for particle detection or as an "active target", i.e. simultaneously as a target and a detector. In the first case Ne and He are chosen as well as mixtures of the two, while in the second, He, H_2, D_2 are used. Pure neon has a low breakdown voltage and yields bright tracks. However, because of a high electron diffusion coefficient, such tracks cannot be used for simultaneous measurements of momentum and ionization. When neon is mixed with helium, the diffusion coefficient in the mixture turns out to be lower than in pure helium at the same partial pressure (Chap. 2), while the track is as bright as that of tracks in pure neon. The constant composition of the gas is controlled by means of mass-spectrometry or by the current of an auxiliary ionization chamber connected to the streamer chamber. The introduction of admixtures to the gas filling of streamer chambers is widely practiced for extremely diverse purposes, but is used mainly for improving spatial, time and ionization resolutions (Davidenko et al. 1969a) (Table 5.1).

A streamer chamber is capable of operating at pressures lower and higher than atmospheric pressure: the actual pressure is chosen in accordance with the particular experimental requirements. Enhancement of the gas pressure improves spatial resolution, owing to reductions of the diffusion track width and of the streamer (avalanche) radii. Thus, in a small high-pressure ($P \simeq 20 \cdot 10^5\,\text{Pa}$) streamer chamber 50 mm in diameter, tracks were achieved with a streamer den-

Table 5.1. Utilization of admixtures in the gas filling of a streamer chamber

Main gas	Admixture; concentration	Purpose of admixture
Ne	He; 10–50 %	Reduction of thermal diffusion of electrons
Ne, He, Ne + He	N_2, N_2O, H_2O; $\sim 10^2$ Pa	Reduction of electron thermalization time and length
Ne, He, Ne + He	Saturated vapor of H_2O or C_2H_5OH	Reduction of thermal diffusion of electrons
Ne	Xe; 10–20 %	Absorption of photons of X-ray transition radiation
Ne + He	CH_4; 0.2–0.6 %	Reduction of the high voltage pulse amplitude by about 15 %
Ne + He	SF_6, CCl_4, CF_3Cl; 10^{-5}–10^{-3} %	Reduction of the memory time to $\lesssim 1\,\mu s$
H_2	CH_4; 0.5–5 %	Reduction of the high voltage. Absorption of the UV radiation emitted by streamers in order to avoid breakdown
He	CH_4; 5–30 %	Enhancement of the refraction of the gas for holographic or shadowgraphic recording of streamers

sity of $60\,\text{cm}^{-1}$ and $d_s \simeq 25\,\mu m$ (Eckardt et al. 1983). Such detectors are very promising, for example, for investigating decays of short-lived non-stable mesons and baryons containing heavy quarks (Sandweiss 1982).

Time Characteristics of a Streamer Chamber. These are the sensitivity time, the dead time and the duty cycle.

The sensitivity time (memory time) τ_s corresponds to a delay of the high-voltage pulse for which the recording efficiency of streamers (the density g_s of streamers along a track) decreases by a factor of e. τ_s is determined by the lifetime of free electrons in the gas and depends on its composition and pressure. In pure neon, for instance, $\tau_s \gtrsim 100\,\mu s$, which imposes significant limits on the admissible beam load; this is because of "old" tracks of particles which have passed through the chamber during the time τ_s before the trigger pulse is recorded. To reduce the memory time, molecular admixtures with large cross-sections for the capture of thermal electrons, such as SF_6, CCl_4, $CClF_3$ (Table 2.9) are introduced into the gas filling of a streamer chamber at a level of 10^{-3}–10^{-5} %. The sensitivity time then falls to $\tau_s \simeq 1\,\mu s$, while the counting rate increases to $10^6\,\text{s}^{-1}$ (Fig. 5.13).

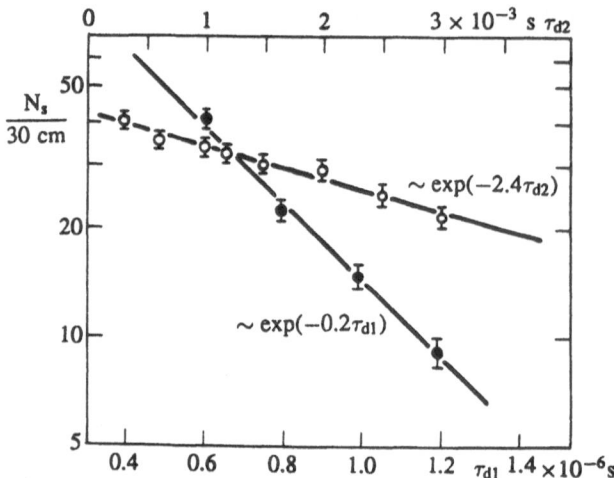

Fig. 5.13. Time characteristics of the streamer chamber filled with Ne + He mixture with admixtures of 10^{-5} % SF_6 (Eckardt 1970): ($\bullet$) conventional operation mode (*lower scale*), $\tau_p = 25$ ns; $\mathcal{E}_{01} = 10-15$ kV/cm; (o) two-pulsed mode (*upper scale*): the second (recording) pulse, $\tau_p = 7$ ns, $\mathcal{E}_{02} = 25$ kV/cm, is applied at a time τ_{d2} later than the first (preliminary) pulse which is delayed for $\tau_{d1} = 0.3$ μs. N_s is the number of streamers; τ_{d1}, τ_{d2} – delay times

The dead time (recovery time) τ_m is the sensitivity time relative to the preceding *recorded* event. It depends on the rates of ion recombination and of relaxation of excited atoms in the gas. The density of charges in the plasma of streamers, or in avalanches, is higher by several orders of magnitude than in the primary ionization blobs, but is much lower than in the spark channel, so that $\tau_m \gg \tau_s$. At the same time, τ_m is significantly shorter than the recovery time of a spark chamber. In a pure gas with small admixtures $\tau_m \lesssim 1$ ms (Anisimova et al. 1971). In gas mixtures long-lived metastable excited atoms may give rise to additional ionization through the Penning reaction (Sect. 2.4). As a result τ_m increases to roughly 10 ms. Since the diffusion coefficient for atoms in a gas is small, this effect is used for memorizing track information in the *two-pulse operation mode* of a streamer chamber, without loss of spatial resolution (see below).

The duty cycle (repetition time) τ_r of a streamer chamber is limited: by the time required for its preparation for recording the next event, e. g. by the HVPG recharging, by the photographic film rewinding etc.. That is, the limitations are purely technical. τ_r is about 0.1 s. Obviously $\tau_r \geq \tau_m$ always. The duty cycle determines the recording frequency f_s of useful events: $f_s \leq \tau_r^{-1} \lesssim 10$ Hz.

Two-Pulse High-Voltage Supply of a Streamer Chamber. The capability of spark and streamer chambers to memorize a recorded event for a time τ_m is utilized when working with a complex trigger, the preparation of which requires 0.1 to 1 ms, for instance when reference is made to a digital processor. To this end, a two-pulse supply is used (Mikhailov, Iljin 1969; Eckardt 1970). The amplitude of the first (preliminary) high-voltage pulse is chosen to be somewhat lower than required for the formation of visible streamers. In this way a considerable number of metastable atoms are produced which retain the ionization along a track during a time $\sim \tau_m$, due to the Penning effect (Fig. 5.13). Here a two-component mixture is used, in which the ionization potential of one of the components, for example, Ne or Kr, is lower than the metastable potentials of the second component (He and Ne, respectively; Table 2.5). In a time $t = 0.1$–2.0 ms $< \tau_m$ after the trigger signal appears, a second (recording) high-voltage pulse is applied with an amplitude sufficient for recording tracks. Owing to the small diffusion coefficient of atoms ($D_a = 0.13$ cm^2/s in He, and $D_a = 0.03$ cm^2/s in Ne) the two-pulse supply does not result in a deterioration of the spatial resolution of the streamer chamber. This method was applied successfully, for example, for recording $\mu - e$-decays in the polyethylene plates inserted into a streamer chamber, with the aim of determining the polarization of direct muons produced in nucleon-nucleon interactions at an energy of 70 GeV (Somov 1975).

Spatial Resolution and Precision of Momentum Measurement. When dealing with the spatial resolution of a streamer chamber it is necessary to distinguish between the resolution of individual streamer (avalanches), the resolution of two adjacent tracks and the accuracy of space track reconstruction (Lecoq et al. 1979).

The spatial resolution σ_s of streamers (avalanches) is determined by their mean diameter d_r reconstructed from images: $\sigma_s = d_r/2$, where

$$d_r^2 = d_s^2 + \frac{(\delta z)^2}{(F_p/k_p)^2} + (\lambda F_p/k_p)^2 + \frac{1}{4k_p^2} \sum_j \sigma_j. \qquad (5.10)$$

Here d_s is the actual diameter of an avalanche or streamer; δz is the depth of the photographic field (depth of the streamer chamber); $F_p > 1$ is the lens aperture (the relative aperture is $1/F_p$); $k_p < 1$ is the magnification factor; σ_j is the resolution of the j-th element of the optical system (the lens, the image amplifier, the photographic film etc.); and λ is the wavelength of the emitted light. The value of d_r depends on the composition and pressure of the gas, and on the amplitude, duration and time delay of the high-voltage pulse. For example, in the streamer chamber AC–235, $\sigma_s = 0.85$ mm for $k_p = 1/60$, while $\sigma_s = 0.5$ mm for $k_p = 1/30$ (Basile et al. 1981b). σ_s is smaller in the avalanche mode than in the streamer mode. The resolution is also improved in gases with a small diffusion coefficient (He, He + Ne mixtures with "cooling" admixtures, e. g. with CH$_4$, iso-C$_4$H$_{10}$, etc.), when the amplitude of the high-voltage pulse is increased and its duration reduced, when the optical resolution and the luminosity of streamers are enhanced, and when the magnification factor of the lens is increased. It is also improved when the magnetic field strength is increased due to a decrease in the field of the diffusion displacement of thermal electrons, during the delay time of the high-voltage pulse (Sect. 2.2).

The spatial resolution of two tracks $\sigma_{12} \simeq \sigma_s$ corresponds to the minimum distance at which two parallel adjacent tracks can be distinguished.

When the number of streamers along a track is large, it is reconstructed with an accuracy σ_{tr} depending on the precision with which a line in space is reconstructed from its stereo-image. This accuracy mainly depends on the photographic film graininess, which in the image scale amounts to about $5\,\mu m$, and for $k_p = 1/50$ gives $\sigma_{tr} = 0.25\,mm$. If a fine-grained photographic film is used, with a streamer chamber operating at atmospheric pressure in the avalanche mode, then $\sigma_{tr} \simeq 0.1\,mm$, while in the case of high-pressure chambers $\sigma_{tr} \simeq 0.01\,mm$.

The trajectory of a high-energy particle in a streamer chamber in a magnetic field $B \parallel \mathcal{E}$, because ionization losses in a gas are small, has the form of a cylindrical spiral of constant curvature with a radius R_t, related to the momentum of the particle p by the well-known relation

$$p = 0.3 R_t H / \sin \theta_H$$

where R_t is expressed in meters and p is expressed in GeV/c; H is the magnetic field strength in Tesla, and θ_H is the angle between the particle track and the magnetic field direction.

At small p the accuracy of momentum measurement is limited by multiple Coulomb scattering, while at high p by the error σ_{tr} of track reconstruction in space (Fig. 5.14)

$$(\delta p/p)^2 = (\delta p/p)^2_{\text{Coulomb}} + (\delta p/p)^2_{\text{reconstr.}} \, .$$

For tracks of length L [cm]. σ_{tr} [mm] and $\theta_H = \pi/2$ (Lecoq et al. 1979),

$$(\delta p/p)^2 = 577^2/(\beta^2 H^2 X_0 L) + 0.12^2 \sigma_{tr}^2/(H^2 L^5) \, ,$$

where X_0 [cm] is the radiation length of the gas (Table 5.2).

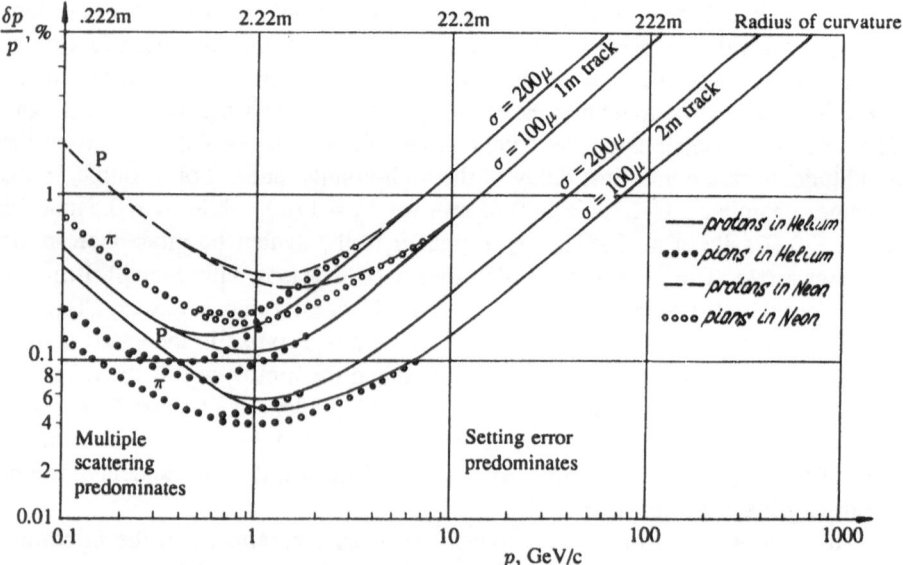

Fig. 5.14. Computed accuracy of momentum measurements in a streamer chamber installed in a magnetic field of 1.5 T for track lengths $L = 1$ and $L = 2\,m$ (Lecoq et al. 1979): (——) protons in He; (...) pions in He; (- - -) protons in Ne; (o o o) pions in Ne. The rise of the curves at low momenta is due to the enhancement of multiple Coulomb scattering in the gas

Table 5.2. Radiation length X_0 in gases at normal pressure and 20°C

Gas	H_2	He	Ne	Ar	N_2	CO_2	CH_4	iso-C_4H_{10}	0.3 He + 0.7 Ne
X_0 [m]	7533	5640	346	117	326	198	696	187	373

Spatial Calibration of a Streamer Chamber. In the photographic recording of tracks there unavoidably arise optical distortions, which until recently were corrected by photographing fiducial marks. This method, however, does not solve the problem of space positioning of tracks and of correcting distortions caused by the drift of ionization electrons during the time interval between the passage of a particle and the formation of a visible track. In large streamer chambers, for example, the error related to displacement of the track significantly exceeds σ_s or σ_{tr}, and its fluctuations may amount to 100 %.

For a solution of this problem, fiducial tracks with precise positioning of trajectories are required of particles of infinite momentum. It turns out to be possible to simulate such tracks using a narrow beam of directed electro-magnetic radiation (ultraviolet, X-ray or laser) not affected by the magnetic field (Guschin et al. 1984). Thus, when a narrow (about 1×1 mm) light beam from an N_2-laser ($\lambda = 337$ nm) is used, a fiducial track is formed as a result of two-step ionization of the vapors of complex organic admixtures, with low ($I_1 \simeq 7$ eV) ionization potentials. The saturated vapor pressure of these admixtures amounts to about 10^{-4} Pa. By appropriate choice of intensity of the laser radiation it is possible to obtain straight tracks in a streamer chamber placed in a magnetic field, which are practically identical with the tracks of relativistic charged particles.

The simulation of particle tracks in a streamer chamber, and in other gas detectors, with the aid of lasers, represents a unique means of calibration. It makes possible the simultaneous solution of a series of diverse problems, including the positioning of tracks in space, with an error not exceeding 10 μm, measurement of the spatial resolution σ_{12} of two adjacent tracks, and determination of the instrumental energy loss resolution FWHM (full width at half maximum) at a 3 % level (about 0.1 of the FWHM of the Landau distribution).

The Detection Efficiency of Multiparticle Events. When multiple particle production occurs in the gas filling a streamer chamber, the development of avalanches and streamers at distances greater than σ_s in the registration plane and at distances greater than l_s in the plane of the electric field proceeds independently, while the energy expended by the electric field upon the formation of tracks is negligible. Therefore, if it exists, the mutual influence of adjacent tracks on each other can only manifest itself at the interaction vertex. Far from the vertex the streamer chamber records particle tracks within a 4π solid angle with 100 % efficiency, independently of their multiplicity. The luminosity of a track, however, depends on its orientation. Because streamers can merge with each other, tracks directed towards the electrodes of a streamer chamber are brighter than tracks which are parallel to the electrodes.

Factors Affecting Ionization Measurement in a Streamer Chamber. Early publications on the streamer chamber noted the possibility of determining the ionizing power of charged particles from the number of streamers per unit track length (the track density, g_s), and from the mean luminosity $\bar{h}_s$ and mean length $\bar{l}_s$ of the streamers. It soon became apparent, however, that the measurement accuracy of $\bar{h}_s$ and $\bar{l}_s$ are inadequate for particle identification. The luminosity and length of streamers depend on the characteristics of the high-voltage pulse, and the angle between the track and the electric field; their fluctuations are extremely large. Thus, for instance, the spread of the streamer brightness approaches two orders of magnitude.

The problem of determining the ionization from the streamer density also turned out to be very much more complicated than was initially expected. Bright tracks suitable for direct photographing are obtained when $l_s \geq 1$ cm. In this case $g_s = 2$–$3\,\mathrm{cm}^{-1}$, which is several times lower than the primary specific ionization (Table 1.3), and it varies weakly as the latter is enhanced. Only if $l_s \simeq 1\,\mathrm{mm}$, when the track consists of very short weakly luminous streamers, or, to be more precise, of avalanches that have not yet developed into streamers, and at low time delays of the high-voltage pulse, their number is close to the number of primary electrons (Gygi, Schneider 1966; Davidenko et al. 1969b).

Detailed investigations performed mainly at the end of the 1960s in Moscow and at CERN revealed that measurements of the primary specific ionization are feasible in a streamer chamber, only if the following three mandatory conditions are fulfilled:

1) Each primary ionization cluster gives rise to the formation of a single avalanche (streamer), and the number of avalanches is conserved in the gas amplification process. This means that the characteristic times of transport phenomena in the gas (diffusion, electron attachment, the Penning effect, etc.) must exceed significantly the delay time of the high-voltage pulse. The interaction of adjacent streamers must also be totally excluded and all the external factors affecting track formation must be stabilized.
2) The recording efficiency of avalanches (streamers) equals to 100%. For this condition to be fulfilled the gas multiplication must be sufficiently high, the avalanche (streamer) luminosity must be as homogeneous as possible, and the sensitivity threshold of the recording system must be low enough.
3) The procedure of track information analysis takes into account correctly the overlapping of the avalanche (streamer) images.

If these conditions are satisfied, the results of ionization measurements are stable with respect to variations of the operating mode of the streamer chamber; that is they are independent of fluctuations of the shape of the high-voltage pulse, and of fluctuations of τ_p, τ_d, U_0, l_s, h_s, etc. (Davidenko et al. 1968a). We shall now proceed to analyze the above requirements in greater detail.

1a) *The influence of transport processes in the gas.* In the case of a streamer chamber these processes, which were dealt with in detail in Sect. 2.4, exhibit the following qualities. If, during a time τ_d, the diffusion displacement of electrons is smaller than the radius of a streamer ($l_d < r_s$), then the number of avalanches initiated corresponds to the primary ionization, and each group of electrons produced in an individual ionizing collision is recorded as a single streamer (a single avalanche). When a large energy transfer occurs, the track of the δ-electron departing from the main track (Fig. 5.9b) is discarded in the track analysis procedure. Thus, the measured values of g_s correspond to the primary ionization when $\tau_d < r_s^2/2D$, where D is the electron diffusion coefficient (Table 2.1). Thus, for $P = 10^5\,\mathrm{Pa}$, $T = 293\,\mathrm{K}$ and $r_s = 0.5\,\mathrm{mm}$, this requirement is met by $\tau_d < 4.9\,\mu\mathrm{s}$ in He and by $\tau_d < 0.44\,\mu\mathrm{s}$ in Ne. Therefore, when the time delay $\tau_d \simeq 1\,\mu\mathrm{s}$, which is the case for large streamer chambers, measurements of dN_1/dx are feasible in He and in He + Ne mixtures, but not in pure Ne (Davidenko et al. 1968a; Davidenko et al. 1969a, 1969b; Basile et al. 1981a). The diffusion of electrons decreases as molecular admixtures with large transport cross-sections are introduced into the gas (Table 2.2), and when the streamer chamber is placed in a magnetic field (Dayon et al. 1970, Anisimova et al. 1972) (Sect. 2.2).

The capture of electrons in the gas, and collisions between excited atoms and molecules of an admixture exhibiting low ionization potential, reduce or enhance, respectively, the apparent measured ionization. The contribution of these processes is significant when their characteristic times are of the order or less than τ_d. Thus, for example, in dealing with He and He + Ne mixtures one must take into account associative ionization reactions (Sect. 2.4) occurring with a characteristic time $\tau_A \lesssim 10^{-9}\,\mathrm{s} \ll \tau_d$. They take place nearly instantaneously, and the additional ionization produced in such reactions (about 8 % for He) cannot be distinguished experimentally from the primary ionization. The contribution of transport processes actually only enhances or reduces slightly the threshold particle energy transfer ω_m corresponding to ionization of the gas and, in accordance with Fig. 1.14, almost does not alter the relativistic rise of primary ionization; this was confirmed in experiments with a streamer chamber containing admixtures of H_2O (Davidenko et al. 1970a) and SF_6 (Eckardt et al. 1977).

1b) *The interaction of streamers.* When a chamber is operated in the streamer mode, the reduction of the track density is caused by the electromagnetic interaction between adjacent streamers. It leads to the merging of streamers, or to the slowing down of the growth, and even to the suppression of retarded streamers developing in the electric field which is distorted by the space charge of adjacent streamers. This effect is stronger the greater the streamer length and diameter. It gives rise to large fluctuations of the dimensions and luminosities of streamers (Fig. 5.15). The interaction of adjacent streamers manifests itself at distances of the order of 0.1 mm. It can be excluded by reducing the streamer length to $l_s \leq 1\,\mathrm{mm}$ (i.e. by transition to the avalanche mode), or by increasing the mean distance $\bar{l}_g = g_s^{-1}$ between the streamers by lowering the gas pressure and utilizing light gases (such as He or H_2).

1c) *The role of technical factors.* Ionization measurements are extremely sensitive to the composition, pressure and temperature of the gas which all must

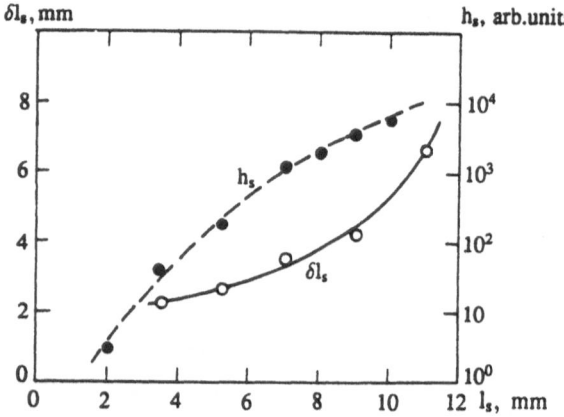

Fig. 5.15. Dependence of the length spread, δl_s, and brightness, h_s, of streamers on their mean length $\bar{l}_s$ (Davidenko et al. 1969a)

be well stabilized. To this end, it is reasonable to put the streamer chamber inside a thermostatically controlled metal container filled with a neutral gas, for instance, nitrogen with an admixture of freon, to provide additional protection from corona discharges. In addition, such a container screens the other detectors from pickups related to the high-frequency discharge in the streamer chamber. The composition, pressure, temperature and flow of the working gas through the chamber must be under permanent control. Thus, for example, in the CERN WA–44 experiment each trigger was accompanied by the recording of 32 parameters on a magnetic tape, as a result almost all (99,9 %) of the photographs taken were useful (Basile et al. 1981a).

Another important factor is the stability of the high-voltage pulse amplitude; this allows the measured values of dN_1/dx to remain constant for particles of identical γ (the voltage plateau of the dN_1/dx). In the WA–44 experiment, a reduction of U_0 by 1 %, to a point lower than the saturation level, led to an uncertainty of 12 % in the measurement of dN_1/dx (Fig. 5.16), which represents a violation of (5.9). Extremely important, also, is the homogeneity of the electric field within the chamber volume. This is provided by an appropriate choice of material, with a small dielectric constant for the casing; and by the choice of resistive terminators matched to the waveguide impedance of the chamber (Fig. 5.17)

The main requirement for the optical track recording system is the homogeneity of its characteristics throughout the entire depth of field: the stability of the photographic image magnification, of the relative aperture and of the image brightness and resolution in each optical channel. If image amplifiers are used they must be chosen to exhibit identical characteristics and to maintain the same amplification during their operation time. It is important to use contrasting photographic film (with a contrast coefficient $\gamma_f = 3.5$–4.0), and to stabilize the conditions of its development. All these optical parameters must also be checked during the experiment.

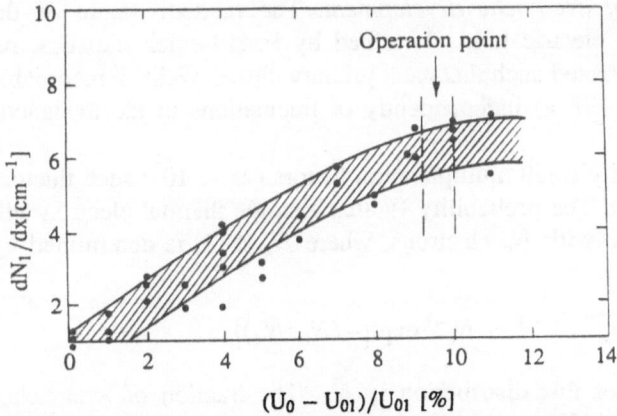

Fig. 5.16. Dependence of the measured primary specific ionization on the amplitude U_0 of the high-voltage pulse in the AC–235 streamer chamber (Basile et al. 1981a). U_0 is measured relative to an arbitrary chosen level U_{01}. The *dashed lines* define the fiducial interval of U_0

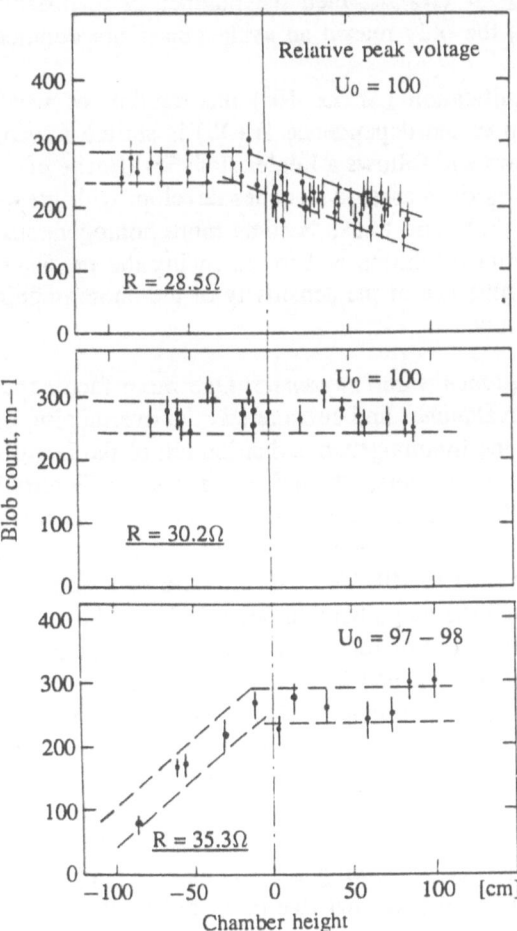

Fig. 5.17. Influence of the non-uniformity of the electric field on the results of ionization measurements in a streamer chamber (Basile et al. 1981a): R is the terminating resistance

2a) *The efficiency of avalanche development.* The random nature of the avalanche growth in an electric field, described by Farri-Legler statistics, requires conditions to be created such that each primary cluster yields a recordable avalanche (Basile et al. 1981a) independently of fluctuations in the avalanche development.

In the case of relatively small multiplication factors ($M < 10^7$) such fluctuations are quite significant. The probability W that a single thermal electron will give rise to an avalanche with N_e electrons, where $N_e \gg 1$, is determined by the relation

$$W(N_e) = \overline{N}_e^{-1}(1 - \overline{N}_e^{-1})^{N_e - 1} \simeq N_e^{-1} \exp\left[-(N_e/\overline{N}_e)\right] .$$

The standard deviation of this distribution is $\overline{N}_e$. The fraction of avalanches that are recorded is $\exp[-(N_e)_{min}/\overline{N}_e]$, where $(N_e)_{min}$ is the minimum number of electrons in an avalanche required for it to be recorded. In accordance with Farri statistics, 63 % of the avalanches contain a number of electrons $N_e < \overline{N}_e$. Therefore, when $\overline{N}_e \simeq (N_e)_{min}$, a significant number of avalanches will not be recorded. If, on the contrary, $\overline{N}_e \gg (N_e)_{min}$, then the number of recordable avalanches increases. In this case, the only reason an avalanche is not counted is geometrical overlapping.

In the case of high gas amplification ($M \simeq 10^8$) the fraction of weak avalanches falls and the maximum of the dependence $W(N_e)$ is shifted toward $\overline{N}_e$. The distribution $W(N_e)$ narrows and follows a Legler statistics (Basile et al. 1981a). Thus, as $\mathcal{E}_0$ increases and τ_p decreases, avalanches develop, with growing efficiency in accordance with (5.9), and tracks become more homogeneous. This contributes to improving spatial resolution and to enhancing the precision of ionization measurements. The influence of the sensitivity of the photographic recording system is considered below.

3a) *Overlapping images of avalanches and streamers.* One more factor that reduces the measured number of avalanches and streamers is the overlapping of their images. This effect is due to the inhomogeneous distribution of the lengths l_g of the intervals between the streamer centers, which due to a random (Poisson) distribution of ionizing collisions along the track, is exponential and favors small l_g. In addition, the overlapping effect depends on the diameter of a streamer and on the resolution of the optical system (5.10). Thus, in the case of a streamer chamber of 0.3–0.5 m in depth, when the magnification factor is $k_p = 1/40$–$1/30$, the streamer image is determined by an out-of-focus circle, the size of which may exceed d_s significantly. The minimum distance l_m at which adjacent streamers are resolved reliably [the mean length of overlapping (Davidenko et al. 1969b)],

$$l_m = 2\left[\frac{\sigma_{opt}}{k_p}\left(d_s - \frac{\sigma_{opt}}{k_p}\right)\right]^{1/2}$$

is always greater than the radius of a streamer, owing to the finite resolving power σ_{opt} of the optics ($\sigma_{opt}/k_p = 0.1$–0.15 mm). Knowledge of l_m makes it possible to calculate the influence of the geometrical overlapping of streamer images on the results of the measurement of g_s. Several methods of film analysis have also been developed that permit determination of g_s from the measured parameters of track structure, without use of information on d_s and l_m, but at the expense of the loss of $\simeq 50\%$ of the statistics (Sect. 5.4).

Recording of Particle Tracks in a Streamer Chamber. In experiments with streamer chambers two methods of track recording are applied:

1) Registration of light emission from streamers and avalanches. In the streamer mode tracks are photographed directly onto a highly sensitive photographic film using a high-aperture lens. In the avalanche mode the track brightness is enhanced by means of light amplifiers. Filmless methods of photographic recording are also applied; these utilize transmission TV tubes and solid-state matrices sensitive to light (Sect. 5.4).
2) Photographing gas inhomogeneities (which are caused by the development of streamers) in laser light using shadow or holographic methods. In this case track images are obtained in the form of shadowgrams or holograms (which is the reason the latter method was termed holographic).

Recording the Light Emission from Streamers and Avalanches. The luminosity h_s of a streamer rises sharply with its length l_s, so that even small fluctuations of l_s give rise to a large spread of h_s (Fig. 5.15). So a significant number of streamers may be unobserved, since although it is quite simple to photograph the most luminous ones, the recording of all streamers is much more complicated (Fig. 5.18). Hence it is clear that the requirements to be met by the system of track photo-registration are much more severe in ionization measurements than in coordinate measurements, when the streamer density along the track plays no decisive role.

To solve the above problem a recording system possessing the following properties is required:

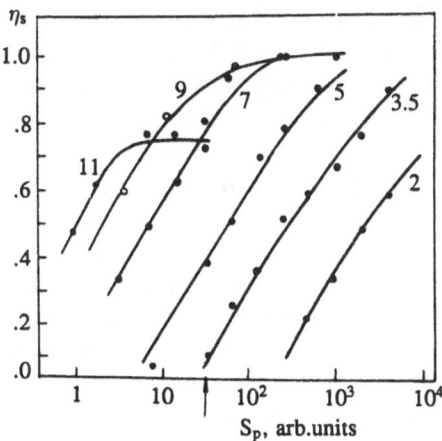

Fig. 5.18. Dependence of the streamer detection efficiency η_s on the sensitivity S_p of the photographic recording system ($P = 8 \cdot 10^4$ Pa, Ne) (Davidenko et al. 1969a): the figures indicate the mean streamer length $\bar{l}_s$ (in mm) for the given curves; the *arrow* indicates the sensitivity threshold of the recording system in the case of direct photographing with the relative aperture of 1 : 1.5

a) it must be quite sensitive to record streamers with $l_s \ll 1$ cm. That corresponds to the avalanche mode, when no merging occurs, the length fluctuations are minimal, and the luminosity is weak but more homogeneous (Fig. 5.15);

b) the dynamic range of its sensitivity must exceed the range of fluctuations of the streamer brightness.

Figure 5.18 shows that in the case of direct photographing, streamers with $l_s > 0.9$ cm are recorded with an efficiency lower than 100 %, because they merge with each other. At the same time, weak streamers and avalanches that are beyond the recording threshold are also lost. This happens because for $k_p \simeq 1/50$ and $F_p = 8$, the luminous flux density incident on the photographic film from an avalanche is $\simeq 10^5$ photons per mm^2, and from a streamer 10^7–10^8 photons per mm^2, while the sensitivity threshold of a standard photographic film is close to 10^8 photons per mm^2.

It follows that ionization measurements in large streamer chambers are feasible only when the chambers are operated in the avalanche mode and light amplifiers are used. The successful results obtained in early studies of the primary ionization by direct photographing were due to the small dimensions and depth (3.5 cm) of the streamer chambers, which allowed recording of extremely weak streamers with an aperture of 1 : 1.5 and a magnification factor of $k_p = 1/5$ (Davidenko et al. 1968a, 1969a).

An avalanche luminosity deficit of the order of 10^3 requires using light amplifiers with the proper light amplification factor. Two types of light amplifier are known involving either electrostatic or electromagnetic focussing. The latter exhibits better resolution and more homogeneous amplification. Taking pictures of the image from the screen of the light amplifier also involves significant losses of light ($\simeq 10^3$). This makes it necessary to employ light amplifiers with an amplification factor of 10^5–10^6. Four-stage light amplifiers exhibit such amplification; however, their resolution does not exceed 30 μm, which limits the spatial resolution of a streamer chamber with $k_p = 1/50$, to the value $\sigma_s \simeq 2$ mm (Lecoq et al. 1979). If the image at the output of the light amplifier is transmitted with the aid of optical fiber links, an amplification of the order of 10^3 is sufficient, so that a two-stage light amplifier can be used with a resolution of $\simeq 15$ μm. In this case, if the optical contact of the fiber link with the photographic film is sufficiently good, the resolution of the image may amount to 20 μm, which for $k_p = 1/50$ yields $\sigma_s \simeq 1$ mm.

The resolution of a light amplifier is mainly limited by scattering and refraction of light and by back-scattering of electrons on the grains of the screen luminophor of the light amplifier. This difficulty can be overcome by direct recording of the electron image, instead of its transformation into an optical image. Devices which do this, electrographic chambers, are utilized, for example, in astronomy. They have a resolution of about 10 μm within a circle up to 130 mm in diameter and exhibit significantly higher homogeneity than standard light amplifiers. This possibility has not been applied yet in experiments with streamer chambers.

From the point of view of photographic registration of avalanches, of great interest is the possibility of enhancing the sensitivity of the photographic emulsion by imposing on it a high voltage pulsed electric field , $\mathcal{E}_0 > 1\,\text{MV/cm}$ (Artamonov et al. 1984). Certain technical difficulties arise when trying to use this method, but if they could be overcome, ionization measurements along particle tracks in a streamer chamber could be performed without making use of light amplifiers.

Laser Registration of Tracks. As a streamer develops, there arises in its vicinity a local optical inhomogeneity related to an explosive gas-dynamical phenomenon: the shock wave caused by fast heating of the gas at the center of the streamer. The change of the gas refractive index inside these inhomogeneities perturbs the passing light wave, such that they can be photographed in the laser light (Kalimov et al. 1978, 1979). The contrast in the resulting track images is higher the greater the refractive index n_r, the higher the gas pressure, and the lower the gas temperature. The streamer luminosity is irrelevant in this case. Thus, for example, the quality of track images in a streamer chamber filled with H_2, CH_4 and their mixtures, i. e. in gases with a relatively high refractive index (Table 5.3), is much better than in He or Ne, in which streamers are much more luminous.

Table 5.3. Refractivity $(n_r - 1) \cdot 10^4$ of gases for wavelengths of ruby and nitrogen lasers (Tombak 1979)

λ [nm]	He	Ne	H_2	Ar	CH_4	70 % He + 30 % CH_4
694	0.350	0.670	1.380	2.830	4.390	1.562
337	0.355	0.680	1.450	2.930	4.800	1.689

A broad laser beam is used to illuminate the inhomogeneities produced by streamers in the gas to provide their registration. For this purpose, pulsed optical lasers are used with an energy of 5–6 mJ and pulse duration from 10 to 20 ns; for example, N_2 lasers ($\lambda = 337.1$ nm involving transformation of the wavelength to $\lambda = 560$ nm in a dye solution), ruby lasers ($\lambda = 694.1$ nm), etc.. The laser is synchronized with a trigger. In the case of a laser with a large time lag of pulse (exceeding 1 μs) a two-pulse supply of the streamer chamber is utilized. In order to obtain clear images, the light flash is delayed by 50–300 ns with respect to the high voltage pulse, i. e. it is delayed by the time required for the formation of a shock wave in the vicinity of the streamer.

Laser registration of tracks in streamer chambers is performed, as in bubble chambers, by applying both shadow (near-field) and holographic (far-field) methods. In the first case the track projections are photographed onto a diffuse-scattering or a directed-scattering screen or directly on the photographic film. To obtain stereo-images, light beams are used that differ either in color or in polarization, while the photocamera is supplied with appropriate light filters. Then, in

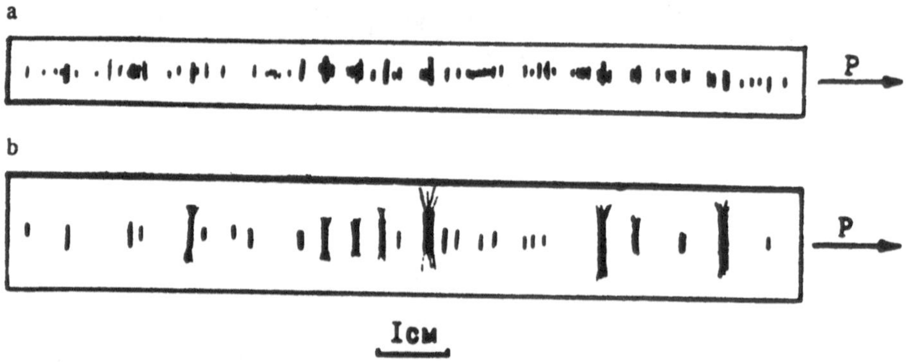

a

b

$$\underset{\text{1 CM}}{\rule{1.5cm}{0.4pt}}$$

Fig. 5.19. Pictures of 1 GeV proton track in a streamer chamber filled with 70 % He + 30 % CH_4, $P = 10^5$ Pa, recorded simultaneously applying the method of laser shadowgrams (a) and by direct photographing (b) at angles to the electric field of 40° and 50°, respectively (Kalimov et al. 1978)

the plane normal to $\mathcal{E}_0$ the streamer images are round dots surrounded by dark and light diffraction rings. When the far-field method is applied, as a result of interference between the light scattered and diffracted by the track elements and the coherent background, a hologram is formed which can be reconstructed by the means of coherent optics.

The laser method of registration completely removes the problem of weak avalanche and streamer luminosity, sharply enhances their spatial resolution and also their recording efficiency (Fig. 5.19), and consequently, enhances the precision of ionization measurements (Table 5.4). The requirements of the high-voltage pulse amplitude are also greatly reduced. Thus, for the direct photography of tracks in a hydrogen streamer chamber at $P = 10^5$ Pa, field strengths up to $\mathcal{E}_0 = 70$–80 kV/cm are required (in the presence of admixtures of CH_4 or SF_6, $\mathcal{E}_0 = 60$ kV/cm) (Fig. 5.11). On the other hand, for laser recording it is sufficient to have $\mathcal{E}_0 = 30$ kV/cm (Kalimov et al. 1981). With this method it turns out to be possible to record extremely weak tracks, even those which cannot be distinguished by eye after adaptation for a long time in the dark. Enhancement of refraction in the gas and of the spatial resolution with the increase of pressure,

Table 5.4. Parameters of proton (1 GeV) and electron (approximately 2 MeV) tracks in a streamer chamber filled with a mixture of 70 % He + 30 % CH_4 (Kalimov et al. 1978, 1979)

Method of track recording	d_s [mm]	l_s [mm]	l_d [mm]	g_s [cm^{-1}]	
				p	e$^-$
Direct photography	1.53 ± 0.04	6.33 ± 0.25	0.240 ± 0.022	2.50 ± 0.05	
Laser registration	0.38 ± 0.01	2.40 ± 0.09	0.195 ± 0.011	8.20 ± 0.06	7.80 ± 0.10
Computation			0.186	8.10	7.4

makes this method especially promising for high-pressure streamer chambers applied as vertex detectors, and in studies of the decays of short-lived relativistic charmed particles with lifetimes of 10^{-12}–10^{-14} s (Eckardt et al. 1983).

Identification of Relativistic Particles by Measurement of the Primary Ionization and Momentum in a Streamer Chamber. The possibility of measuring the primary ionization in a streamer chamber is an important advantage in relation to other gas detectors. First of all, the fluctuations of primary ionization follow the Poisson distribution, which is more narrow and symmetric than the distribution of total ionization (Fig. 1.20). The width of a track in the case of small delays of the high-voltage pulse ($\tau_d < 1\,\mu s$), corresponding to the measurement of dN_1/dx (Fig. 5.9a, b), is significantly smaller than for $\tau_d \gg 1\,\mu s$, when the number of streamers on the track corresponds to the total ionization (Figs. 5.9c–f). Finally, the relativistic rise R_{N1} of primary ionization in the light noble gases used to fillings in streamer chambers, is not less than that of the ionization energy losses (Fig. 1.11). All these factors favor simultaneous precise measuring both momentum and ionization in a streamer chamber, which is necessary for identification of charged relativistic particles.

The possibility of determining the specific primary ionization in a streamer chamber was demonstrated experimentally at the end of the 1960s and at the beginning of the 1970s (Gygi, Schneider 1966; Davidenko et al. 1968a, 1969a, 1970a; Kanofsky, Shoen 1969). The results of measurements in He and in Ne + He mixtures (Fig. 5.20) are in good agreement with theoretical predictions (Ermilova et al. 1969; Asoskov et al. 1982) (Sect. 1.4) concerning both the absolute values of dN_1/dx and its relativistic rise. Therefore, calculations of the primary ionization can serve as the basis of realistic estimations of the measure-

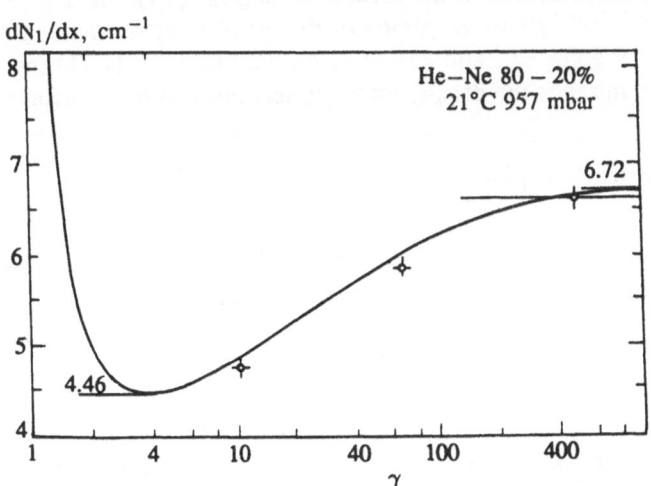

Fig. 5.20. Results of measurements of the specific primary ionization γ produced by charged relativistic particles in the AC–235 streamer chamber of volume $2.35 \times 1.25 \times 0.6\,m^3$ filled with 80 % He+20 % Ne at $P = 0.95 \cdot 10^5$ Pa and at a temperature of 21°C (Basile et al. 1981a). The *smooth curve* is computed using (1.71)

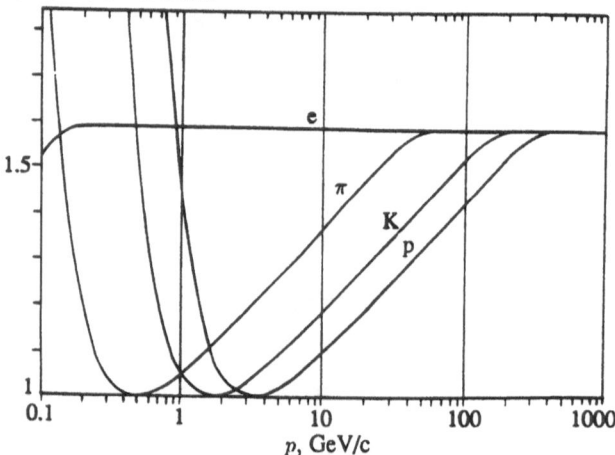

Fig. 5.21. Dependence of computed relative specific primary ionization of e, π, K, p in neon under normal conditions upon the particle momentum

ment precision required for separation of secondary relativistic hadrons, when products of high energy interactions are recorded in a streamer chamber.

Estimation of the reliability of particle identification at high energies by primary ionization measurement in a streamer chamber have much in common with an estimation based on the probable ionization measurement in proportional detectors. This follows from the similarity between the behavior of dN_1/dx and $\Delta_0(\gamma)$ in the relativistic energy region (a minimum at $\gamma \simeq 4$, logarithmic rise, saturation, Fermi plateau) (Fig. 1.11). In the momentum scale the dN_1/dx curves are shifted according to the logarithm of the particle mass (Fig. 5.21). The resulting relative difference for dN_1/dx in the region of the logarithmic rise amounts to about 25 % for π/p, 15 % for π/K and 8 % for K/p. We find from (1.71) (neglecting in this momentum range the density effect correction) that the measured mass of a particle is

$$m = p/(\beta\gamma c) = (p/c)\exp\{-(1/2)[A_1^{-1}\beta^2(dN_1/dx) - A_2 + \beta^2]\} ,$$

where A_1 and A_2 are constants of (1.71). Accordingly, the relative standard uncertainty σ_{ma} in determining the mass from momentum and primary ionization measurements is related to their respective relative uncertainties σ_p and σ_I by the relation

$$\sigma_{ma}^2 = \sigma_p^2 + (\ln\beta\gamma - \beta^2/2 + A_2/2)^2\sigma_I^2 \simeq \sigma_p^2 + (8 \pm 1)^2\sigma_I^2 ,$$

since for all inert gases $A_2 \simeq 11$ (Table 1.6). Hence it follows that for an uncertainty in the primary ionization measurement of $\sigma_I = 3\%$, the reliability of hadron separation in the momentum range 10 to 100 GeV/c is 99.9 % for π/p, 98 % for π/K and about 85 % for K/p (Fortune 1969).

It may seem that such measurement accuracy can be achieved, in the case of an optimum track density of $g_s = 6\text{--}8\,\text{cm}^{-1}$, for track lengths of $L = 1.5\,\text{m}$.

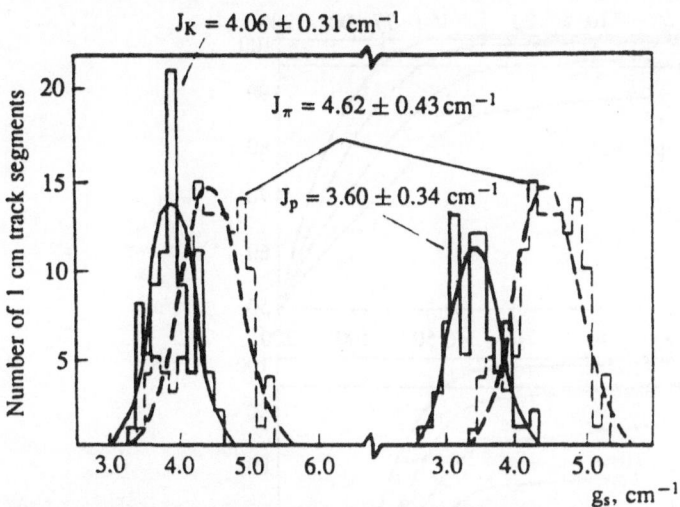

Fig. 5.22. Example of separation of pions, kaons and protons in a streamer chamber for track lengths of 0.9 m (Eckardt et al. 1977). The *smooth curves* are results of approximation of the histograms by Gaussian distributions

Actually, because images of individual avalanches overlap, a significant part of the information is lost. Therefore, such high precision is achieved only when $L > 4$ m. In a streamer chamber with a complex filling of 70 % Ne + 30 % He + 0.3 % iso-C_4H_{10} + 10^{-7} SF_6 $\sigma_J = 8$ % for the track length $L = 0.9$ m (the statistical and the systematic uncertainties are 7 % and 4 % respectively) (Fig. 5.22). A decrease in gas pressure in the streamer chamber, which seems an attractive way to extend the logarithmic region and to enhance the relativistic increase of the primary ionization, actually yields no gain, because of deterioration of the statistical measurement accuracy (Davidenko et al. 1970a). The spectrometric capabilities of a streamer chamber are demonstrated in Fig. 5.23.

Using electro-negative admixtures for reducing the memory time of a streamer chamber leads, owing to the capture of electrons, to a decrease in g_s which becomes several times less than dN_1/dx (Eckardt et al. 1977); this makes it necessary to perform relative, instead of absolute, ionization measurements. Comparisons of g_s using different photographs is possible only if they are obtained under strictly identical conditions (i. e. identical characteristics of the high-voltage pulse, of the optical channels; identical properties of the film and conditions of its development, etc.). It has been shown by the RISC collaboration that, because the measured track density and the mean diameter of a streamer image are directly correlated, the results of ionization measurements can be normalized to a certain streamer diameter (Bannikov et al. 1986). Taking it into account increases significantly the precision of calibration measurements of g_s. In the conditions of the RISC experiment the value of $\sigma_J \simeq 3$ %, corresponding to good separation of relativistic hadrons, is achieved for $L \gtrsim 8$ m.

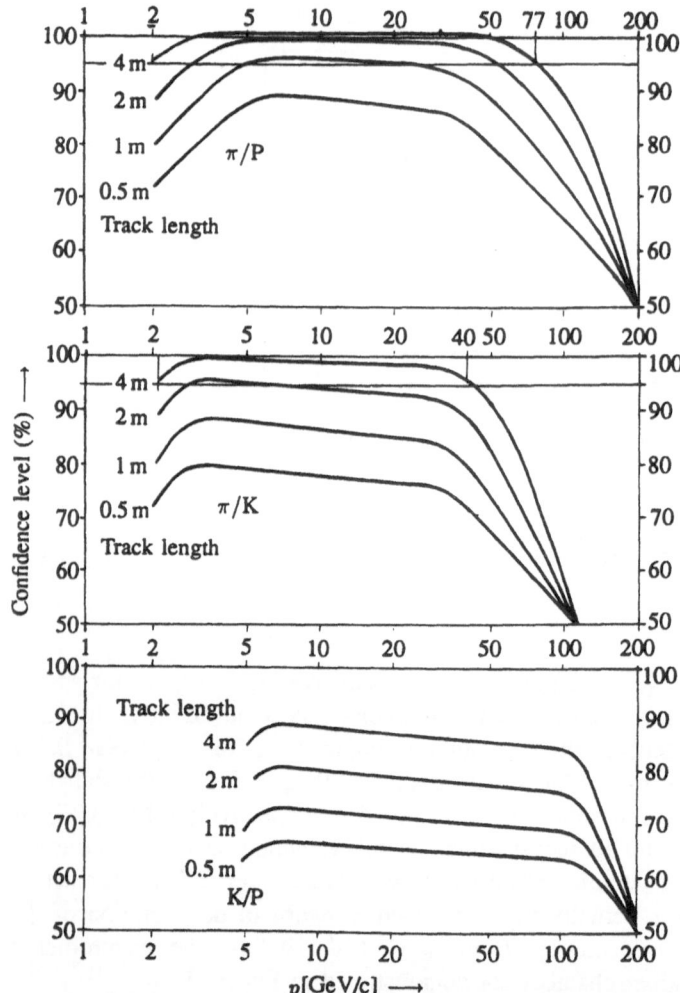

Fig. 5.23. Dependence of the calculated confidence level for hadron separation in a streamer chamber filled with a 95 % He + 5 % Ne mixture upon the particle momentum (Eggert 1974)

Examples of streamer chambers used for the identification of relativistic particles based on primary ionization measurements are presented in Table 5.5. Most of these experiments are related to the search for free quarks of fractional charge at accelerators (Fig. 5.24) and in cosmic radiation. As a visual detector, a streamer chamber requires much time for processing the photographs of tracks, which hinders its utilization. Therefore the future of this method depends on the development of filmless recording methods and on the automation of the particle track analysis.

Table 5.5. Application of streamer chambers for identification of high energy particles

Volume of streamer chamber V [cm³]	Filling	Gas pressure P [10^5 Pa]	Amplitude and duration of high-voltage pulse U_0 [kV], τ_P [ns]	Particles, energy (momentum)	Experiment
$11 \times 10 \times 10$	30 % He + 70 % Ne	1.0	300; 5	Secondary beams from the 27 GeV/c CERN proton synchrotron	Search for free quarks of fractional charge (Allaby et al. 1969)
$100 \times 30 \times 16$	95 % He + 5 % Ne; 50 % He + 50 % Ne	0.9; 1.0	400; 7	Cosmic radiation	Search for free quarks in extensive atmospheric showers (Eggert et al. 1973)
$270 \times 125 \times (2 \times 25)$	30 % He + 70 % Ne + 0.3 % i-C_4H_{10} + 10^{-5} SF$_6$	1.0	400; 15	π^-, K^-, $\bar{p}$ 27 GeV/c	NA-5 experiment. Investigation of inelastic $\bar{p}p$-interactions at 60 GeV CMS energy (Eckardt et al. 1977)
$235 \times 125 \times (2 \times 30)$	80 % He + 20 % Ne	1.0	730; 15	Charged particles produced in neutrino interactions	WA-44 experiment. Search for free quarks in high energy neutrino interactions (Basile et al. 1980)

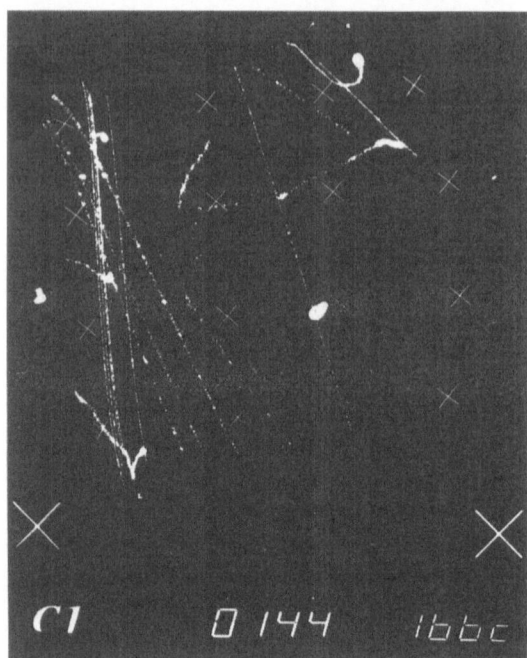

Fig. 5.24. Particle tracks in a 2.35 × 1.25 × 0.6 m³ streamer chamber exposed in the neutrino beam of the CERN proton supersynchrotron [WA–44 experiment (Rohrbach 1988)]

5.4 Methods of Track Information Analysis

The tracks of charged particles in visual detectors are composed of separate elements: e. g., of droplets, avalanches, streamers, bubbles, grains etc.. Their number per unit length (the track density g_s) serves as a measure of the ionizing power of the particles. A straightforward count of such elements is inconvenient, as a rule, because they overlap. This overlapping is due to the random nature of the distribution of track structure elements, and to the finite sizes of their images, which depend on their true diameter and on the properties of the optical recording system (5.10). Therefore, in analyzing track information one has to deal with clusters of track elements (blobs) and with gaps between them.

When the ionization density is very high, the gaps disappear, the clusters merge, and the track acquires the form of a continuous, nearly structureless, formation. In this case, information on the ionizing power of a particle is derived from measurements of the optical track density (i. e. from blackening) or of the track width (Sects. 5.1 and 5.2).

Ionization Measurement Based on the Track Structure Parameters. Although the following considerations mainly deal with a streamer chamber, the methods for determining the ionizing power of a particle from the parameters of the track structure are similar for all track detectors (Lomanov, Chirikov 1957). They are based on the determination of the number and dimensions of blobs and gaps, or on the measurement of their distribution parameters or of average values of such quantities.

In experiments with streamer chambers, several methods are applied which allow information to be obtained on the ionization, independently of the reconstructed diameter d_r of an avalanche or a streamer. The distances l_g between the streamer centers obey the distribution of random interval lengths:

$$N(l_g > l) = N_0 \exp(-g_s l) .$$ (5.11)

In conditions where avalanches or streamers overlap, it is often impossible to determine the distance between the streamer centers, so the distance between the edges of clusters must be measured. In this case, small intervals are encountered much less often than (5.11) suggests. This occurs because of finite optical resolution and fluctuations of the streamer dimensions, and also due to effects of suppression and merging of adjacent streamers when the distance between them is smaller than $l_{min} \simeq d_r/2$. The resolution length l_m of adjacent streamers can be considered as the effective diameter:

$$l_m \simeq d_r + l_{min} \simeq 1.5 \cdot d_r .$$

The methods of determining the ionization from parameters of the track structure take into account gaps between clusters of length $l_g \geq l_m$ (Glassnek, Peter 1975).

Now we shall list the track structure parameters which permit the determination of the track density. Illustrations of the accuracies of various methods are given in Fig. 5.25 and Table 5.6.

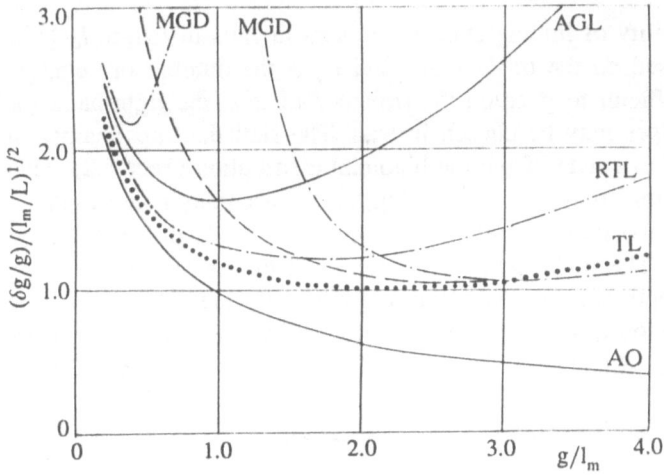

Fig. 5.25. Accuracy of the track density g_s measurement methods (l_m resolution length) taking into account of the overlapping of track structure elements (Lomanov, Chirikov 1957): MGD – mean gap density; AGL – average gap length; TL – track lacunarity; RTL – roughened track lacunarity; AO – absence of overlapping

Table 5.6. Measurement accuracy of primary ionization along tracks of relativistic particles in streamer chambers filled with the 50 % He + 50 % Ne mixture ($P = 0.9 \cdot 10^5$ Pa, $k_p = 1/30$) (Glassneck, Peter 1975)

Method of measurement	Total track length			Uncertainty of ionization density measurement ($\delta g_s / g_s$) [%]	
	L [m]	l_{min}^{**} [mm]	g_s [cm^{-1}]	total	statistical
MSC	0.11	1.5	6.26 ± 1.12	17.8	
AGL	0.22	3.0	6.48 ± 1.41	21.8	12.0
ABL	0.22	3.6	5.75 ± 1.26	21.9	13.8
MGD	0.22	3.6	6.39 ± 1.06	16.6	15.0
TL	0.22	3.6	6.49 ± 0.97	15.9	13.4
AGL + ABL	0.22	3.6	6.08 ± 1.12	18.4	12.0
AGL + TL	0.22	3.6	6.48 ± 1.02	14.9	
MGD + TL	0.22	3.6	6.45 ± 0.96	14.9	
IGLD	11*	3.0	6.5 ± 0.2	3.1	
AGL	11*	3.0	6.5 ± 0.2	3.1	

* For 50 tracks 0.22 m long each

** $l_m = 0.7$ mm + l_{min}

1. *Track lacunarity (TL)*,

$$\mathcal{L} = \exp(-g_s l_m) \, .$$

2. *Roughened track lacunarity (RTL)*,

$$\mathcal{L}_0 = 1 - \eta = \exp(-g_s l_0) = n_0 l_0 / L \, ;$$

where η is the probability of finding streamer centers in cells of length $l_0 \geq l_m$ on a scale superimposed on the track image and n_0 is the number of "empty" cells. Because it is difficult to position the streamer centers, the right-hand (or left-hand) streamer edges may be chosen instead. The statistical uncertainty in determining η may be calculated from the binomial distribution (Sect. 6.2). This method is convenient for rapid track processing, but its accuracy is not high.

3. *Integral gap length distribution (IGLD)*. In the region $l_g > l_m$ this distribution corresponds to (5.11). The track density is determined by measuring the slope of the integral gap length distribution within the range of values l_g, where it is exponential. In the region $l_g < l_m$ this distribution is distorted, so it cannot be used for determining g_s (Fig. 5.26). The related losses of the statistics are close to 60 %.

4. *Modified streamer counting (MSC)* (Schneider 1969),

$$g_s = N_s' / (L - l_m N_s') \, .$$

Here N_s' is the number of streamers on the track, each situated at a distance greater than l_m from the preceding *counted* streamer. The effective track length is $L' = L - l_m N_s'$, and the losses for $l_m = 2.5$ mm at $g_s = 5$–6 cm^{-1} also amount

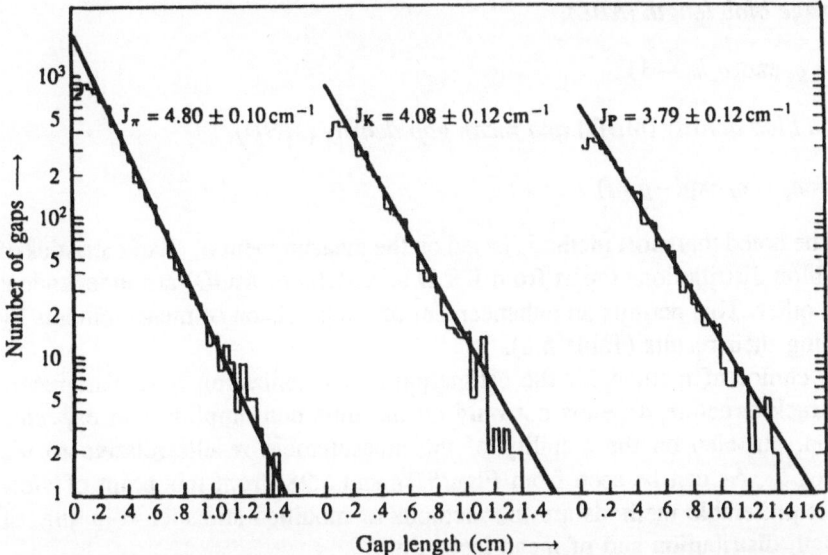

Fig. 5.26. Integral distribution of gap lengths l_g on particle tracks in a streamer chamber filled with a 30 % He + 70 % Ne + 10^{-5} SF$_6$ mixture at $P = 10^5$ Pa (Eckardt et al. 1977): the slopes of straight lines approximating histograms determine the track density g_s. Deviations from the linear dependence are observed for $l_g \leq l_m = 2.5$ mm

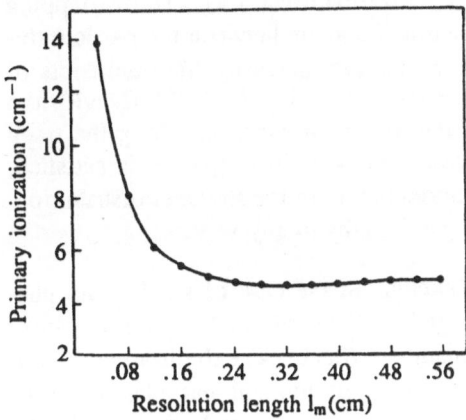

Fig. 5.27. Stability of results of track density measurements by the modified method of streamer counting when the resolution length l_m is varied (Eckardt et al. 1977). (In the region $l_m > 2.4$ mm g_s is independent of l_m)

to about 60 %. The value of l_m is chosen to lie on the plateau of the dependence depicted in Fig. 5.27. It coincides with the value of l_m at which the histogram of Fig. 5.26, obtained in the same experiment, deviates from the exponential dependence (5.11).

5. *Average gap length (AGL)*,

$$\bar{l}_g = g_s^{-1} + l_{min} \ .$$

6. *Average blob length (ABL)*,

$$\bar{l}_b = g_s \exp(g_s l_m - 1) \,.$$

7. *Mean blob density (MBD) and mean gap density (MGD)*,

$$g_b = g_g = g_s \exp(-g_s l_m) \,.$$

It must be noted that most methods based on the measurement of characteristics of gap or blob distributions (apart from IGLD and MBD or MGD) are independent of each other. This permits an enhancement of the precision of measurements by combining their results (Table 5.6).

The choice of method for the estimation of the ionization from parameters of the track structure depends not only on the time consumption and expected precision, but also on the stability of the measurement results relative to the choice of l_m. As can be seen from Figs. 5.27 and 5.28, from this point of view the most preferable methods are the methods of modified streamer counting, of gap length distribution and of mean lacunarity.

The overlapping of images of the track structure elements makes it impossible to enhance the ionization measurement precision at the expense of an g_s increase by increasing the gas pressure or by utilizing heavy gases. For $d_r \simeq 1$ mm the optimum track density is 6–8 cm^{-1}. In a streamer chamber such a density is readily obtained using He + Ne mixtures, which are characterized by bright streamers and a large relativistic rise of the primary ionization (Sect. 1.4). The overlapping of streamer images is enhanced as the inclination angle between the particle trajectory and the electrodes of the streamer chamber increases. Measurements of the primary ionization are possible, however, for angles $\theta_s \leq 30°$ (Davidenko et al. 1968a). We note that diffusion of the ionization electrons along the track does not affect the measurement precision if the condition $l_d < r_s$ is satisfied, since such diffusion does not alter the random nature of the streamer distribution. Transverse diffusion does not affect the gap lengths in any way.

Ionization Measurements on Dense Tracks. In the case of small-scale photographing the tracks of relativistic charged particles in condensation detectors (Sect. 5.1) and nearly always in spark chambers (Sect. 5.2) exhibit a continuous structure. This is also the case for tracks of strongly ionizing particles. For such tracks, relative ionization measurements are performed on the basis of the width or of the optical density of their images.

The *width of a track d* is related to the particle ionizing power, although in different detectors the track width depends on diverse processes. In condensation detectors this is ion diffusion. In accordance with Valley and Vitale (1949), the relative ionizing power g_2/g_1 of two particles of track widths d_2 and d_1, respectively is

$$g_2/g_1 = \exp[-(d_1^2 - d_2^2)/(16 D_i \tau_d)] \,.$$

where D_i is the ion diffusion coefficient, τ_d is the time delay before the beginning

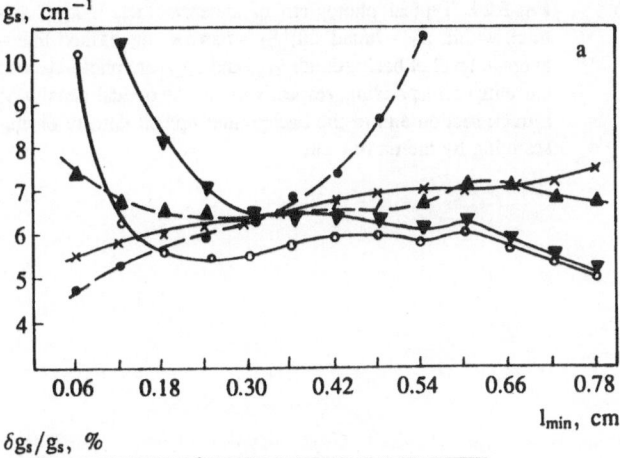

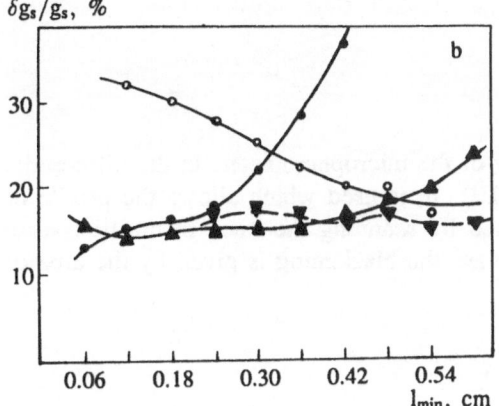

Fig. 5.28. Stability of results of track density measurements by various methods in a streamer chamber filled with a 50 % He + 50 % Ne mixture at $P = 0.9 \cdot 10^5$ Pa (Glassneck, Peter 1975): (a) track density g_s, [cm^{-1}] versus $l_{min} = l_m - d_r$; (b) relative uncertainty $\delta g_s / g_s$ (in %) of track density measurement versus l_{min}; × – IGLD; • – AGL; ∘ – ABL; ▽ – MGD; △ – TL

of condensation. This relation is very sensitive to the measurement uncertainties of the track width. Therefore such a method is suitable only for comparison of particle tracks with clearly differing ionizing powers; for example, for separation of recoil protons and deuterons produced in the reactions n + p → p + n and n + p → d + π^0 in a hydrogen high-pressure diffusion cloud chamber (Schluter 1954). The widths of track images are measured on a microscope with a high magnification factor or by the photometric method (Fig. 5.29).

The optical density of a track image determined by the photometric method yields a better estimate of the relative ionizing power than measurement of the track width. This method, developed long ago, is still successfully applied (Balestra et al. 1987). In this case the photometric current excited in the photosensitive element of the microphotometer by a light beam is recorded. The light beam traverses a section of the track image limited by a special slit of width S_i. Measurements are performed by the integral or the differential method. In the former, a "broad" slit, $S_i > d$, is used, which is first projected onto the track (photocurrent i_t) and then onto the nearest blank region of the photograph (photocurrent i_b), where the background blackening is determined (Fig. 5.29). Thus, the measured optical density of the track $B_t = 1 - i_t / i_b$ is independent of

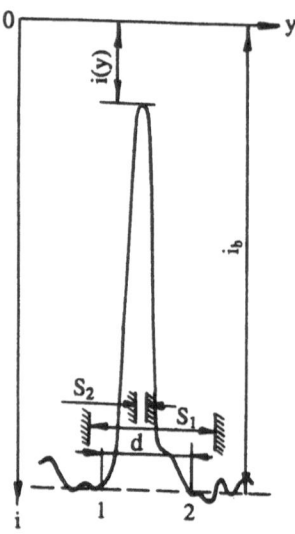

Fig. 5.29. Typical photogram of a dense track image: d – track width; S_1 – broad slit; S_2 – narrow slit; dashed line – average level of background; $i_t(y)$ and $i_b(y)$ are photo-electric currents corresponding, respectively, to the optical density of a track section and to the background optical density during scanning by the narrow slit

the light intensity and the sensitivity of the microphotometer. In the differential method a "narrow" slit, $S_i = (0.1–0.2) \cdot d$, is used which allows the profile of the track blackening to be determined by scanning the track in the transverse direction (along the y-axis). In this case the blackening is given by the area of the photogram

$$B_t = 1 - \int i_t(y)\, dy \Big/ \int i_b(y)\, dy \;,$$

where the integrations are carried out inside equal intervals with and without the track. The integral method of photometric evaluation of optical density is simpler, but owing to fluctuations of the background, it is less accurate than the differential method. The dependence between the relative blackening and the specific ionization, plotted on the basis of particle tracks of known ionizing powers, is linear, at least for an ionization density $g < 4 \cdot g_{min}$ where g_{min} corresponds to minimum ionizing particles (Bjornerud 1956).

Automation of Ionization Measurements in Track Detectors. The problem of automation of ionization measurements in track detectors can be solved in two ways: 1) by automation of the analysis of track images like in the case of bubble chamber pictures; 2) by using filmless recording methods involving subsequent recognition and analysis of the images of particle tracks.

Automation of the Analysis of Track Pictures. For the automation of ionization measurements in an analysis of streamer chamber pictures, the same principles and devices are applied as for bubble chamber photographs. A certain additional complication is a higher spread of the diameters of avalanches and streamers than is observed in bubble chambers.

280

The procedure of automatic track analysis includes the following steps:

a) searching for events of interest by conventional or automatic scanning of the film material;

b) determination of the coordinates of the track for its reconstruction in space and momentum measurement,

c) photometry of the track using automatic or semi-automatic measuring devices. It is performed by moving the track image across a narrow slit ($s_i \ll d$) transmitting light onto a sensitive photometer (Glassneck, Peter 1975), or by the scanning with a light beam (Eckardt et al. 1977; Basile et al. 1981b);

d) representation of analog information in digital form and its recording on a magnetic tape;

e) filtration of the film veil and background for reliable identification of blobs and gaps;

f) estimation of ionization parameters of the track and of the true ionization density g.

On the average, the complete automatic analysis of a single picture takes about 15–20 s. Here the efficiency η of finding tracks is independent of their number n_M in the picture ($\eta > 99\%$ for $n_M \leq 10$), of the track density [$\eta = 85\%$ even for old tracks with a density $0.25\,\text{cm}^{-1} < 0.1$ of $(g_s)_{\min} \simeq 3.5\,\text{cm}^{-1}$]; and of the position of the track in the streamer chamber (apart from the edges). A very critical stage in the analysis is the estimation of the discrimination level of the background and veil, which can be carried out in an objective manner (Glassneck, Peter 1975). The inability of automatic devices to distinguish weak streamers may lead to systematic underestimation of the measured ionization density. If, however, the results of ionization measurements are normalized, say, to the beam particles, the above effect does not affect the reliability of particle identification.

The sequence of automatic analysis of tracks when scanned with a light beam is shown in Fig. 5.30. Here the ionization density was determined in two ways: by direct avalanche counting and from the slope of the integral gap length distribution. In the first case a cluster was identified as an individual avalanche if it had a small size and round shape. The number of overlapping avalanches inside an elongated cluster was determined by analyzing its configuration under the assumption that all avalanches are round. The transverse spread of the avalanche centers with respect to the reconstructed particle trajectory was also determined. In the second case only the gap lengths between the clusters were measured, while their shapes were not analyzed in detail; g_s was determined from the slope of the integral gap length distribution. The results of the two measurements are compared in Fig. 5.31. It can be seen that they yield the same results for $g_s < 25\,\text{cm}^{-1}$. As g_s increases, the accuracy of the direct avalanche counting falls, and saturation sets in when $g_s \simeq 130\,\text{cm}^{-1}$.

The results of conventional and of automatic analysis procedures are in good agreement, apart from those cases when weak streamers were present on the track

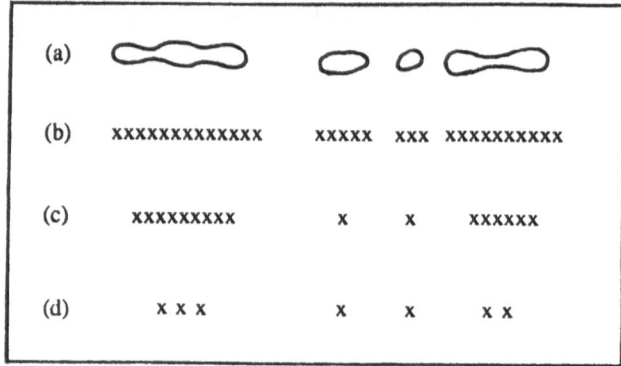

Fig. 5.30. Steps of automatic analysis procedure of a track section in a streamer chamber (Eckardt et al. 1977): (a) – track section; (b) – results of digitization; (c) – correction taking into account the spread of streamer images; (d) – streamers used for the track density estimation

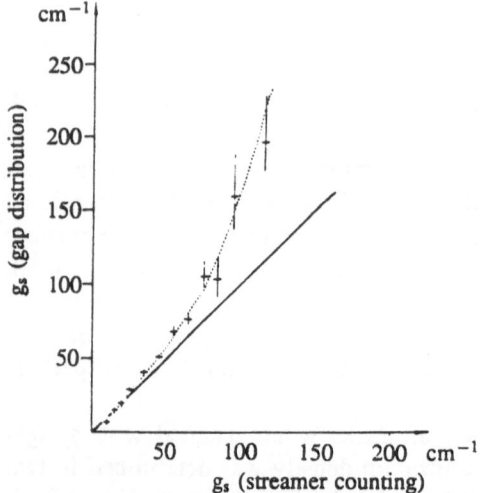

Fig. 5.31. Comparison of results of streamer density analysis applying direct streamer counting (X-axis) and integral gap length distribution (Y-axis) methods (Basile et al. 1981b)

(for example, in the periphery of the chamber). Investigation of the reproducibility of the results of atomized ionization measurements reveal that, independently of the value of g_s, systematic uncertainties do not exceed 5 % (Basile et al. 1981b).

The analysis of multiparticle events in a streamer chamber is time consuming because the recorded information amounts to hundreds of megabytes. Thus, for example the analysis of a single nuclear reaction $^{16}O + ^{197}Au$ at an energy of 200 GeV per nucleon in the NA–35 CERN experiment takes about 10 hours, even without ionization measurements. For a drastic enhancement of the processing facilities of CERN it has been proposed that computers of new architecture, the so-called associated string processors, should be employed, which would permit reduction of the film analysis time by approximately a factor of 10^3 (Rohrbach 1988).

(a)　　　　　　　　　　　　　　　　　　　　　　　　　　(b)

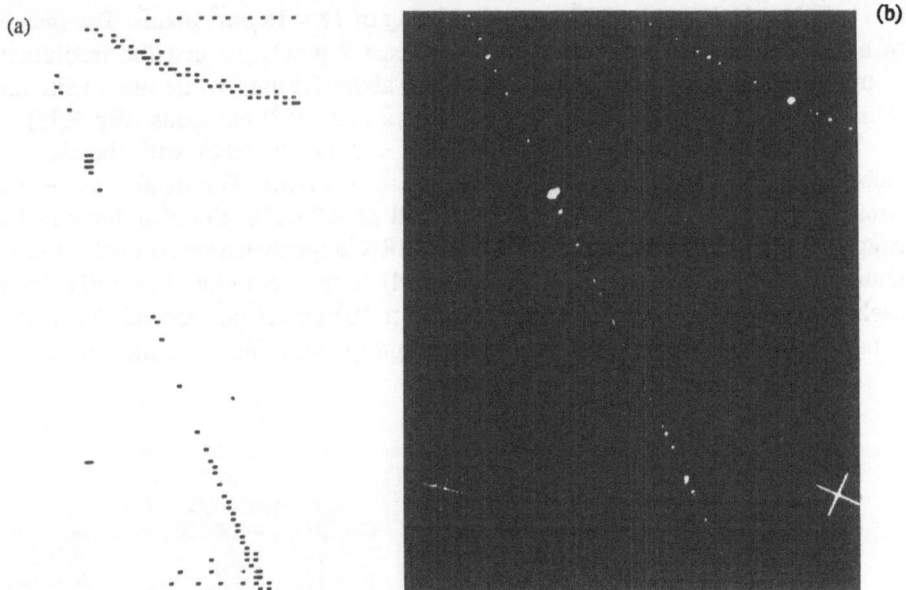

Fig. 5.32. Track recording in a SLAC streamer chamber (Villa, Wang 1977): (a) – using a CCD matrix; (b) – by direct photography

Filmless Track Recording. There exists in principle a new and extremely promising method for the automation of ionization measurements: filmless recording and analysis of tracks with the aid of optical-electronic systems operating on line with a computer. The method of filmless track recording has been tested on spark, discharge-condensation and streamer chambers, with the aim of automation of coordinate measurements. To this end transmitting television tubes (TTT) of high sensitivity were applied (Somov 1975) or solid-state light-sensitive matrices, i. e. charge coupled devices (CCD). Both methods exhibit good resolution, broad dynamic range and high speed of operation. In addition, the CCD matrices are insensitive to external magnetic fields. The reconstructed images of tracks in a streamer chamber repeat the track structure (Asatiani et al. 1972; Villa, Wang 1977) (Fig. 5.32), indicating the possibility, in principle, of automation of ionization measurements. Clearly, it would be reasonable to combine a solid-state image with light amplifiers. These methods are promising, also, for laser recording tracks in a streamer chamber.

An extremely important practical step in this direction was made recently by Tincknell et al. (1987), who suggested an automatized system for recording and analyzing multitrack events in a streamer chamber placed in the magnetic field. The CCD Supervision System developed by the authors makes use of industrial CCDs produced by Texas Instruments Co.. The system was tested in the NA–35 experiment at CERN using the large $2.0 \times 1.2 \times 0.72\,\mathrm{m}^3$ magnetic streamer chamber, viewed by two electron-optical cameras, each containing three lenses, a two-stage image intensifier with a 3×10^6 amplification, a controlled shutter, and a 1024×1024 pixel CCD array consisting of $18 \times 18\ \mu m^2$ pixels. The image of an individual streamer was fixed by at least 9 pixels, so that the resolution of track reconstruction in space amounted to about 2.5 mm, while the maximum charge released in a pixel corresponded to $2 \times (10^4\text{--}10^5)$ electrons (Fig. 5.33).

The NA–35 magnetic streamer chamber operates together with the shower calorimeter and is triggered by a chosen class of events. The dead time of the streamer chamber is about 1 s, so for a spill of 4.2–7.4 s, the chamber can be triggered several times during the 14.4 s CERN Supersynchrotron cycle. Information from the CCD array (9 bit per pixel) is read out simultaneously from each electron-optical camera at a rate of 5×10^5 pixels per second. Then the data are transmitted through conventional multiplexing interface into the com-

(a)

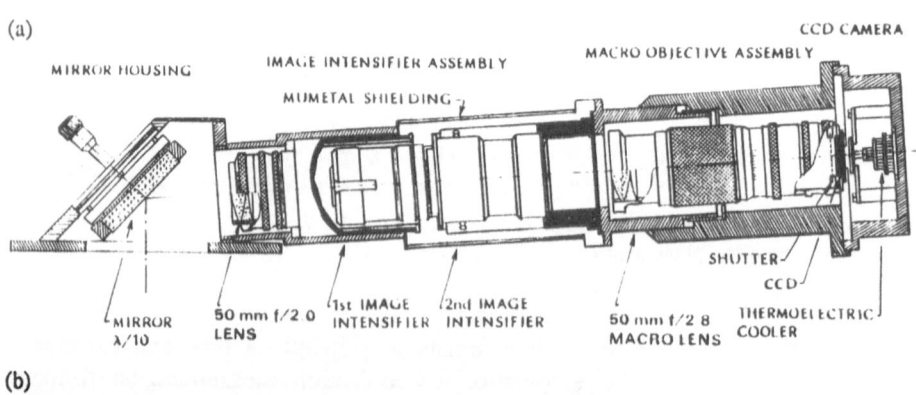

(b)

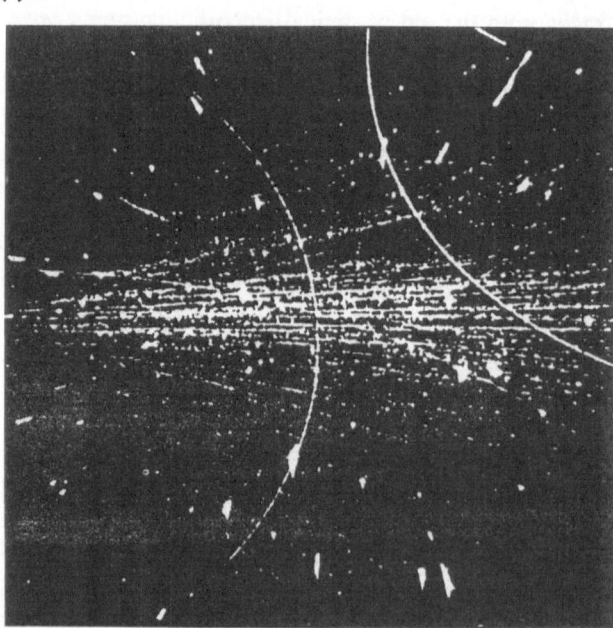

Fig. 5.33. Supervision CCD camera for CERN NA–35 streamer chamber (a), and CCD picture obtained with the aid of the Supervision system (b) (Tincknell et al. 1987)

puter system, which consists of a high speed integer array processor (Mercury ZIP 3216), a host processor (MicroVax II with 11 Mbyte RAM and 71 Mbyte disk memory), and a color display station. The developed software provides for recording, storage and pseudo-visualization of the streamer chamber tracks, and for their subsequent geometrical and physical off-line analysis.

The CCD super-video system was used in studies of nucleus-nucleus collisions, at an energy of 200 GeV/nucleon, at the first stage of the NA-35 experiment for on-line estimation of the event multiplicity on the basis of an analysis of the image profiles. The ultimate goal of this project is to implement a three-camera system that will record three views of a complex multiparticle event onto fast high volume digital media, and perform off-line stereoscopic reconstruction of the particle trajectories in space. Efficient handling of large volumes of data will be achieved by employing a gigabyte optical disk or videocassette tape storage systems. Algorithms have been developed for reliable image analysis in order to identify tracks in the pictures and to follow them through the streamer chamber, and to describe them by fitting curves in two dimensions along pixels with maximum charge. In principle, it is possible to measure the distribution of streamers along a particle track in order to determine the primary ionization, i. e. the charge and velocity of the particle, and then to identify it by its track curvature. Thus, automized filmless recording and analysis of streamer tracks opens up new opportunities for experimentation in high energy physics.

References

Aganianz A. O., Leksin G. A., Tiunchik V. M. 1969: Pribori Tekhn. Exp. No. 3, 55 (English transl.: Instr. Exp. Techn. R 12, 596). Sect. 5.2

Alexanian A. S., Asatiani T. L., Dayon M. I., Gasparian A. O., Jirova L. A., Ivanov V. A., Kamensky A. D., Kayumov F. F., Megrabian G. K., Mkrtchian G. G., Pikhtelev R. N. 1977: Nucl. Instr. Meth. 140, 473. Sect. 5.2

Allaby J., Bianchini G., Diddens A., Dobinson R., Gygi E., Hartung R., Klovning A., Miller D., Sakharidis E., Shlupmann K., Shneider F., Shtahlbrandt C. A., Vetherell A. 1969: "Streamer chamber in a quark experiment at CERN Proton Synchrotron" in Proc. of Symposium on Filmless Spark and Streamer Chambers. Dubna, JINR 13–452. (Dubna). Sect. 5.3

Allkofer O. C. 1969: *Spark Chambers* (Thiemig München). Sects. 5.2, 3

Ananyin P. S., Sidulenko O. H., Fotin E. V., 1979: Pribori Tekhn. Exp. No. 2 (English transl.: Instr. Exp. R 22, 1260). Sect. 5.2

Ananyin P. S., Stibunov V. N. 1983: Pribori Tekhn. Exp. No. 3, 45. Sect. 5.2

Anisimova N. Z., Davidenko V. A., Dolgoshein B. A., Somov S. V., Staroseltsev V. N. 1971: Pribori Tekhn. Exp. No. 2, 70 (English transl.: Instr. Exp. R 14, 408). Sect. 5.3

Anisimova N. Z., Davidenko V. A., Dolgoshein B. A., Somov S. V., Staroseltsev V. N. 1972: J. Exp. Teor. Fiz. 63, 21 (English transl.: Sov. Phys. JETP 36, 10, 1973). Sect. 5.3

Artamonov A. A., Guschin E. M., Zhukov V. V., Lebedev A. N., Somov S. V. 1984: Pribori Tekhn. Exp. No. 4, 70 (English transl.: Instr. Exp. R 27, 858). Sect. 5.3

Asatiani T. L., Gorokhov V. P., Ivanov V. A., Petrakov A. V. 1972: Pribori Tekhn. Exp. No. 4, 64. Sect. 5.4

Asoskov V.S., Grishin V.M., Ermilova V.C., Kotenko L.P., Merson G.I., Chechin V.A. 1982: Trudi FIAN **140**, 3 (in Russian). Sect. 5.1

Balestra F., Bussa M.P., Bussol L., Fava L., Ferrero L., Panzieri D., Piragino G., Tosello F., Bendiscioli G., Fumagalli G., Rotondi A., Salvini P., Zenoni A., Guaraldo C., Maggiora A., Batusov Yu.A., Falomkin I.V., Pontecorvo G.B., Sapozhnikov M.G., Lodi Rizzini E. 1987: Nucl. Instr. Meth. A **257**, 114. Sect. 5.3

Ballario C., De Marco A., Fortune R.D., Verkerk G. 1961: Nuovo Cimento **19**, 1142. Sect. 5.1

Bannikov A.V., Bom A., Vertogradov L.S., Grishkevich Ya.V., Javrishvili A.K., Krumshtein Z.V., Lomtadze T.A., Merekov Yu.P., Petrukhin V.I., Pishka K., Zkhadadze E.G., Shafarik K., Shelkov G.A. 1986: JINR Short Communications. No. 18, 48 (Dubna). Sect. 5.3

Barnaveli T.T., Grigalashvili N.S., Javrishvili A.K., Mandjavidze Z.Sh. 1975: Nucl. Instr. Meth. **125**, 89. Sect. 5.1

Basile M., Berbiers J., Contin A., Favale L., Rohrbach F., Zichichi A., Cara Romeo G., Cifarelli L., D'Ali G., Di Cesare P., Guisti P., Massam T., Palmonari F., Sartorelli G., Valenti G., Esposito B., Spinetti M., Susinno G., Votano L., Laakso I., Nania R., Rossi V. 1980: Lett. Nuovo Cimento **29**, 251. Sect. 5.3

Basile M., Berbiers J., Cara Romeo G., Cifarelli L., Contin A., D'Ali G., Di Cesare P., Esposito B., Favale L., Guisti P., Laakso I., Marrian C., Massam T., Nania R., Palmonary F., Rohrbach F., Rossi V., Sartorelli G., Spinetti M., Sasinno G., Valenti G., Votano L., Zichichi A. 1981a: Physica Scripta **23**, 743. Sect. 5.3

Basile M., Cara Romeo G., Castelwetri A., Cifarelli L., Contin A., Laakso I., Marrian C., Massam T., Nania R., Palmonari F., Rohrbach F., Rossi V., Sartorelli G., Spinetti M., Susinno G., Valenti G., Votano L., Zichichi A. 1981b: Physica Scripta **23**, 754. Sect. 5.3

Bjornerud E.K. 1956: Rev. Sci. Instr. **26**, 836. Sect. 5.1

Bolotov V.N., Bydanov G.A., Devisheva M.N., Devishev M.I., Karpov I.I., Mikhailov O.I., Semenov V.K. 1970: Pribori Tekhn. Exp. No. 3, 114 (English transl.: Instr. Exp. R 13, 761). Sect. 5.2

Bulos F., Odian A., Villa F., Yount O. 1967: *Streamer chamber development*, Stanford Linear Accelerator Technical Report, SLAC–74, UC–28. Sect. 5.3

Chikovani G.E., Mikhailov V.A., Roynishvili V.N. 1963: Phys. Lett. **6**, 254. Sect. 5.3

Cowan E.W. 1954: Phys. Rev. **94**, 161. Sects. 5.1, 4

Davidenko V.A., Dolgoshein B.A., Semenov V.K., Somov S.V. 1968a: J. Exp. Teor. Fiz. **55**, 426 (English transl.: Sov. Phys. JETP **28**, 223). Sect. 5.3

Davidenko V.A., Dolgoshein B.A., Somov S.V. 1968b: J. Exp. Teor. Fiz. **55**, 435 (English transl.: Sov. Phys. JETP **28**, 227). Sect. 5.3

Davidenko V.A., Dolgoshein B.A., Somov S.V. 1969a: J. Exp. Teor. Fiz. **56**, 3 (English transl.: Sov. Phys. JETP **29**, 1). Sect. 5.3

Davidenko V.A., Dolgoshein B.A., Somov S.V., 1969b: Nucl. Instr. Meth. **75**, 277. Sect. 5.3

Davidenko V.A., Dolgoshein B.A., Somov S.V., Staroseltsev V.N. 1969c: J. Exp. Teor. Fiz. **57**, 84 (English transl.: Sov. Phys. JETP **30**, 49, 1970). Sect. 5.3

Davidenko V.A., Dolgoshein B.A., Somov S.V., Staroseltsev V.N. 1970a: J. Exp. Teor. Fiz. **58**, 130 (English transl.: Sov. Phys. JETP **31**, 76). Sect. 5.3

Davidenko V.A., Dolgoshein B.A., Somov S.V., Staroseltsev V.N. 1970b: Proc. Int. Conf. on Instrumentation for High Energy Physics, Dubna, Sept. 1970, **1**, 339 (Dubna, 1971). Sect. 5.3

Dayon M.I., Dolgoshein B.A., Efremenko V.I., Leksin Yu.A., Lubimov V.A. 1967: *Spark Cambers* [*Iskroviye Kamery* (in Russian)] (Atomizdat, Moscow). Sect. 5.2

Dayon M.I., Yegorov O.K., Krylov S.A., Pozharova E.A., Smirnitsky V.A., Chechin V.A. 1970: Pribori Tekhn. Exp. No. 5, 64, (English transl.: Instr. Exp. R 13, 1217). Sect. 5.3

Dolgoshein B.A., Luchkov B.I. 1964: Nucl. Instr. Meth. **26**, 345. Sect. 5.3

Eckardt V., Ladage A. 1968: *A Streamer Chamber with Liquid Targets for the DESY Bubble Chamber Magnet*, DESY Report No. 68–31, Hamburg. Sect. 5.3

Eckardt V., Seyboth P., Derado I., Gebauer H.-J., Odian A., Pretzl K.P. 1977: Nucl. Instr. Meth. **143**, 235. Sect. 5.3

Eckardt V. 1970: *Storing Tracks in Streamer Chambers*, Report No 70/60; DESY, Hamburg. Sect. 5.3

Eckardt V., Lecoq P., Wenig S., Wiatrowsky E. 1983: Nucl. Instr. Meth. **225**, 651. Sect. 5.3

Eggert K., Gürich W., Smetan E. 1973: *"Ionization measurements with a streamer chamber in a quark experiment"* in Proc. Int. Conference on Instrumentation for High Energy Physics, Frascati, Italy, pp. 117–127. Sect. 5.3

Eggert K. 1974: *Messung der Ionisation von relativistischen Teilchen mit einer Streamerkammer*, Ph. D. Thesis 1974, Physikalisches Institut, Aachen. Sect. 5.3

Ermilova V. C., Kotenko L. P., Merson G. I., Chechin V. A. 1969: J. Exp. Teor. Fiz. **56**, 1608 (English transl.: Sov. Phys. JETP **29**, 861). Sect. 5.3

Falomkin I. V., Kulyukin M. M., Pontecorvo G. B., Shcherbakov Yu. A. 1967: Nucl. Instr. Meth. **53**, 266. Sect. 5.3

Falomkin I. V., Kulyukin M. M., Lozansky E. D., Lyashenko V. I., Nguen Minh Kao, Pontecorvo G. B., Troshev T. M., Shcherbakov Yu. A. 1975a: Nucl. Instr. Meth. **131**, 431. Sect. 5.3

Falomkin I. V., Kulyukin M. M., Lyashenko V. I., Nguen Minh Kao, Pontecorvo G. B., Troshev T. M., Shcherbakov Yu. A., Busso L., Piragino G. 1975b: Lett. Nuovo Cimento **13**, 427. Sect. 5.3

Fortune R. D. 1969: *Identification of Relativistic Particles in the Streamer Chamber*, Report No. CERN–ISR/DI/69–73 (CERN, Geneva). Sect. 5.3

Fretter W. B., Hansen L. 1960: Phys. Rev. **118**, 812. Sect. 5.1

Galaktionov Yu. V., Yech F. A., Lubimov V. A. 1965: Nucl. Instr. Meth. **33**, 353. Sect. 5.2

Gebauer H. J. 1977: Nucl. Instr. Meth. **144**, 477. Sect. 5.2

Ghosh C. K., Jones G. M., Wilson J. G. 1954: Proc. Phys. Soc. A **67**, 331. Sect. 5.1

Glassneck G.-P., Peter G. 1975: *Automatic Measurement of the Primary Ionization in a Streamer Chamber*, PHE (Physik Höheren Energien) 75–9 (Berlin-Zeuthen). Sect. 5.4

Gorbunov A. N., Spiridonov V. M., Cherenkov P. A. 1957: Pribori Tekhn. Exp. No. 2, 29. Sect. 5.1

Grigalashvili H. C., Javrishvili A. K., Varsimashvili T. V., Mandjavidze Z. Sh. 1975: Pribori Tekhn. Exp. No. 5, 46 (English transl.: Instr. Exp. Techn. **18**, 1373, 1975). Sect. 5.1

Gromova I. I., Nikanorov V. I., Peter G., Pisarev A. F. 1965: Pribori Tekhn. Exp. No. 1, 64. Sect. 5.3

Guschin E. M., Lebedev A. N., Somov S. V. 1984: Pribori Tekhn. Exp. No. 3, 7. Sect. 5.3

Gygi E., Schneider F. 1966: *Isotropic Spark Chamber Development in the AR Division. Some Practical Results* (Status Report) Report No. 66–14, CERN, Geneva. Sect. 5.3

Kalimov A. G., Kozlov V. S., Lobanov O. V., Stabnikov M. V., Tarakanov V. I., Tombak M. A. 1978: *Laser Track Recording in the Proton Beam of the LINP Synchrocyclotron*, Report No. LINP–407, Leningrad. Sect. 5.3

Kalimov A. G., Kozlov V. S., Stabnikov M. V., Tarakanov V. I., Tombak M. A., Anizoy E. I., Lobanov O. V., Lysenko V. V., Miroshkin V. V., Daschuk B. B., Tverskoy M. G. 1979: Pisma J. Exp. Teor. Fiz. **30**, 460 (English transl.: Sov. Phys. JETP Lett. **30**, 429). Sect. 5.3

Kalimov A. G., Kozlov V. S., Stabnikov M. V., Tarakanov V. I., Tombak M. A., Budziak A., Ivanov I. C., Shcherbakov Yu. A. 1981: Nucl. Instr. Meth. **185**, 81. Sect. 5.3

Kanofsky A., Shoen N. 1969: Rev. Sci. Instr. **40**, 921. Sect. 5.3

Kepler R. E., d'Andlau C. A., Fretter W. B., Hansen L. F. 1958: Nuovo Cimento **7**, 71. Sect. 5.1

Krider E. P., Bowen T., Kalbach R. M. 1970: Phys. Rev. D **1**, 835. Sect. 5.2

Lecoq P., Rohrbach F., Cifarelli L., Guisti P., Heusse P. 1979: *Avalanche Chambers at LEP*, Report of ECFA/LEP, Working group SSG/11/5/27–3, CERN, Geneva. Sect. 5.3

Lomanov M. F., Chirikov B. V. 1957: Pribori Tekhn. Exp. No. 5, 22. Sect. 5.4

Lozansky E. D., Firsov B. M. 1976: *Spark Theory* [Teoriya Iskry (in Russian)] (Atomizdat, Moscow). Sect. 5.2

Maimudar D., Mazumder K., Mundra J. P., Sen Choudry P. K. 1976: Nucl. Instr. Meth. **138**, 151. Sect. 5.2

Mandjavidze Z. Sh., Roynishvili V. N. 1967: Phys. Lett. **24**, 392. Sect. 5.1

Merson G. I. 1972: Yad. Fiz. **15**, 278 (English transl.: Sov. J. Nucl. Phys. **15**, 157). Sect. 5.1

Merson G. I., Sitar B., Budagov Yu. A. 1983: Fiz. Elem. Chastits At. Yadra **14**, 648 (English transl.: Sov. J. Part. Nucl. **14**, 270). Sect. 5.3

Merson G. I. 1990: *Streamer chamber as an identifier of relativistic hadrons* (in Russian). New Methods and New Trends in Physics of Elementary Particles, p. 34 (Metsniereba, Tbilisi) Sect. 5.3

Mikhailov B. A., Ilyin N. S. 1969: Pribori Tekhn. Exp. No. 3, 204. Sect. 5.3

Mikhailov V. A., Lomtadze T. A. 1975: Nucl. Instr. Meth. **130**, 61. Sect. 5.3

Morrison R. W. 1972: *"Ionization Measurements with Streamer Chambers"* in Streamer Chamber Technology, Proc. of the 1st Int. Conf. on Streamer Chamber Technology, Sept. 14–15, 1972. Report No. ANL–8055, pp. 150–166, Argonne 1972. Sect. 5.3

Nikitin V. A., Nomofilov A. A., Sviridov V. A., Slepets L. A;, Sitnik I. M., Strunov L. M. 1965: Yad. Fiz. **1**, 183 (English transl.: Sov. J. Nucl. Phys. **1**, 127). Sect. 5.1

Pavlovsky F. A. 1966: Pribori Tekhn. Exp. No. 1, 184. Sect. 5.2

Puchkov V. S. 1973: *Precision Ionization Measurements in the Relativistic Rise Region and the Determination of π/p Ratio in Cosmic Ray Flux at Mountain Altitudes*, Ph. D. Thesis P. N. Lebedev Physical Institute. Sect. 5.1

Rice-Evans P. 1974: *Spark, Streamer, Proportional and Drift Chambers*, (Richelieu, London). Sect. 5.3

Rochester G. D., Wilson J. G. 1952: *Cloud Chamber Photographs of the Cosmic Radiation.* (Pergamon, London). Sect. 5.1

Rohrbach F., Bonnet J. J., Cathenoz M. 1973: Nucl. Instr. Meth. **111**, 485. Sect. 5.3

Rohrbach F. 1988: *"Streamer Chambers at CERN During the Last Decade"*; Report No. CERN/EP 88–17, CERN, Geneva, in Sect. 5.3

Sandweiss J. 1982: Phys. Repts **83**, 39. Sect. 5.3

Schluter R. A. 1954: Phys. Rev. **96**, 734. Sect. 5.1

Schneider F. 1969: *Precision of Spark Chamber Track Parameters*, Report No. ISR–GS/69–2, CERN, Geneva. Sect. 5.4

Schroeder L. S. 1979: Nucl. Instr. Meth. **162**, 395. Sect. 5.3

Somov S. V. 1975: *Streamer Chambers in Experiments in Particle Accelerators* (in Russian), (Moscow Institute of Physics and Engineering, Moscow). Sect. 5.3

Stabnikov M. V., Tombak M. A. 1971: in Proc. Int. Conf. on Instrumentation for High Energy Physics, Dubna, Sept. 1970, p. 382. Sect. 5.3

Tan B. C., Shmied H., Rousset A., Rohrbach F., Piuz F., Grey-Morgan C., Cathenoz M. 1970: *Etude et Development des Chambres a Dards a Hydrogene*, Report No. TC–L/Int. 70–7, CERN, Geneva. Sect. 5.3

Tincknell M. L., Chase S. I., Dinh T., Harris J. W., Teitelbaum L. 1987: Optical Engineering **26**, 1067. Sect. 5.4

Tombak M. A. 1979: *Formation of Laser Shadowgrams on Streamers*, Report No. LINP 499, Leningrad Institute of Nuclear Physics. Sect. 5.3

Valley G. E., Vitale J. H. 1949: Rev. Sci. Instr. **20**, 411. Sect. 5.4

Vardenga G. L., Zhuravleva M. S., Makarov L. G., Lukstinsh Yu. P., Okonov E. O., Khorozov S. A. 1970: in Proc. Int. Conf. on Instrumentation for High Energy Physics, Dubna, Sept. 1979, **1**, 362. Sect. 5.3

Villa F., Wang L. C. 1977: Nucl. Instr. Meth. **144**, 529. Sect. 5.4

6. Alternative Methods of Ionization Measurement

There exist several devices for ionization measurements which, although not universal, are indispensable in a number of applications: for instance, pulsed ionization chambers, "low efficiency" gas discharge detectors (counters, spark chambers, discharge tubes), have been in use for a long time in studies or estimations of fast particle ionization powers. Other detectors, for example, the so-called light, or gas-scintillation proportional chambers are used, when it is expedient for whatever reason, to have information in the form of light, instead of electric signals.

6.1 Application of Ionization Chambers for Measuring High Ionization Densities

Characteristics of Pulsed Ionization Chambers. Pulsed ionization chambers, the output signal of which results from collection of the charge produced in the gas by ionizing radiation, are applied in high-energy physics for the measurement of high ionization densities, caused by flows of fast charged particles or by individual strongly ionizing slow particles. Examples are provided by experiments in which ionization calorimeters are used for measuring the energy of ultra-relativistic shower cosmic ray particles (Danilova et al. 1982), or by experiments involving the spectrometry of recoil fragments produced in elastic nuclear reactions in the gas (Vorobyov et al. 1972, 1974; Burq et al. 1981). Ionization chambers are also utilized for monitoring high-intensity external accelerator beams (Blair et al. 1984) and for measuring the beam profiles.

Ionization chambers have a number of important advantages over other gas detectors. These are reliability and cheapness, which is a decisive factor in their mass application, for example, in ionization calorimeters and burst detectors for studies of cosmic radiation, and for their use as beam monitors. Ionization chambers exhibit an enormous dynamic range (up to 10^7) and are free from fluctuations related to gas amplification. They are extremely simple in construction, and many gases free from electron affinity (H_2, He, N_2, Ne, Ar, Kr, Xe and their mixtures at pressures $P \gtrsim 10^5$ Pa) can serve as fillings. The typical time of charge collection in an ionization chamber amounts to 10–$20 \, \mu s$ (for electron collection), which is quite acceptable for recording events with a frequency up to $10^4 \, s^{-1}$, although it does impose a limit on the concentration of electro-negative admixtures (e. g.,

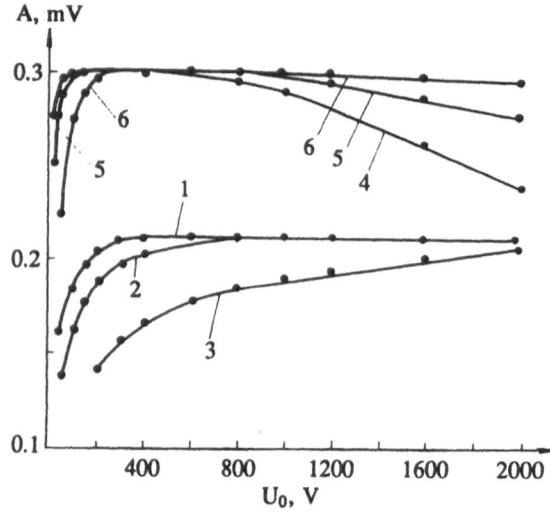

Fig. 6.1. Pulse height characteristics of a cylindrical ionization chamber 9.5 cm in diameter and 4 m in length (C_i = 54 pF) filled with N_2 (1–3) and Ar (4–6) and exposed to 5.15 MeV α-particles (Danilova et al. 1982): $P = 10^5$ Pa (1 and 4); $P = 1.5 \cdot 10^5$ Pa (2); $P = 2 \cdot 10^5$ Pa (3 and 5); $P = 3.5 \cdot 10^5$ Pa (6). The gas purity is 99.99 %. Reduction of the pulse height with voltage U_0 in the argon chamber is due to the cross-section of electron capture by the O_2 admixture increasing with the electron temperature. The drop of the electron temperature as the argon pressure increases leads to an expansion of the voltage plateau (6). The rise of the saturation voltage with pressure is caused by the enhancement of recombination

less than 10^{-2} % for O_2). When the voltage U_0 applied to the electrodes of an ionization chamber is higher than the saturation voltage U_s, (Fig. 6.1), total charge collection is provided. Therefore, the amplitude A of the output signal is almost independent of where the particle actually traverses the chamber, and is proportional to the energy released in the gas. The independence of A upon U_0 in the region of the voltage plateau permits using a common power supply for a large number of ionization chambers and reduces the requirements on its stability.

Owing to the absence of gas amplification, a single relativistic particle produces a very weak charge signal in an ionization chamber, $\Delta Q = e\,(dN_t/dx)lP = 10^{-5}–10^{-3}$ pC. Thus, for a typical input capacitance $C_i \gtrsim 10^2$ pF of an ionization chamber preamplifier, the voltage pulse height arriving at its input is merely $A = \Delta Q/C_i \simeq 10^{-7}–10^{-5}$ V. This imposes severe restrictions on the sensitivity of the preamplifier and on the minimum number of relativistic particles to be recorded. The threshold sensitivity A_t (the input noise level) of a high-quality preamplifier is close to $10\,\mu$V. Accordingly, the energy noise equivalent is $\Delta_t = A_t C_i w/e \simeq 0.2$ MeV ($w \sim 30$ eV, Table 1.3). The quantity Δ_t characterizes the lower limit of the sensitivity range of the detector. For an argon ionization chamber with $lP = 20 \cdot 10^5$ Pa, the value $\Delta_t = 0.2$ MeV corresponds to about 30 relativistic particles traversing the chamber simultaneously. Fluctuations of ionization, which are inevitable, actually raise this limit to an even higher level.

Ionization Chambers in Experiments at Accelerators. The first attempts to use ionization chambers for measuring the ionizing powers of fast charged particles were related to studies of the energy losses of relativistic electrons in gases (Barber 1955, 1956). In theses experiments the relative ionization produced by beams from an electron accelerator was measured for beam energies $E_e \leq 35$ MeV.

To reduce the influence of electron-ion recombination the chambers were filled with light gases (H_2 and He at $P = 10^5$–10^6 Pa), and the results of measurements in beams of intensity $(1$–$6) \cdot 10^6$ electrons per pulse were extrapolated to zero intensity. The decrease of the relativistic rise of ionization, due to the medium density effect, was clearly demonstrated; this coincided with the calculated restricted energy losses within a 5 % uncertainty. In a similar experiment, involving ionization chambers filled with H_2, He, Xe and a He + Xe mixture, and carried out at electron energies $\varepsilon_e = 150$–550 MeV, measurements of dN_t/dx were performed and the existence of a Fermi plateau in H_2 was established at $P = 10^5$ Pa and $\gamma > 10^2$ (Aggson, Fretter 1962). Studies of the energy loss distribution for cosmic muons in argon ($lP = 1.75 \cdot 10^5$ cm · Pa) revealed that it is close to the distribution predicted by the Landau theory (Hall 1959).

In 1972 a group from the Leningrad Institute of Nuclear Physics (LINP) proposed an ionization chamber filled with H_2, D_2 or He at $P = 10^6$ Pa, for the investigation of elastic scattering of high energy hadrons at very small angles, corresponding to the region of Coulomb-nuclear interference (Vorobyov et al. 1972, 1974; Burq et al. 1978). In these experiments the ionization chamber served simultaneously as the gas target and as the spectrometer of recoil nuclei. Their energy in the 1–10 MeV interval was measured according to the total ionization released before they stopped in the gas. The outgoing recoil nucleus angle was determined according to the rise time of the anode pulse, while the coordinate of the interaction vertex was derived from the delay of this pulse relative to the passage of the incident particle. This approach is adequate for the physical problem under investigation, since the energy range of recoil nuclei corresponding to the Coulomb-nuclear interference depends very weakly on the momenta and masses of the incident particles. To solve this problem the ionization spectrometer IKAR was constructed, and studies were performed of the elastic scattering of protons in H_2, D_2 and He at the Gatchina synchrocyclotron (Vorobyov et al. 1972, 1974) and of π^-p-scattering ($\varepsilon_\pi = 30$–140 GeV) at the IHEP and CERN accelerators (Burq et al. 1978, Vorobyov 1979).

The IKAR spectrometer (Fig. 6.2) contains six identical sections of ionization chambers. Each section consists of anode and cathode planes perpendicular to the longitudinal axis of the chamber and situated 12 cm from each other; in between the two planes there is a grounded wire grid. The incident beam is directed along the axis of the spectrometer. The recoil nucleus produced as a result of an elastic nuclear reaction in the gas travels at an angle close to $\pi/2$, nearly parallel to the electrode plane. These events are accompanied by anode pulses with very short

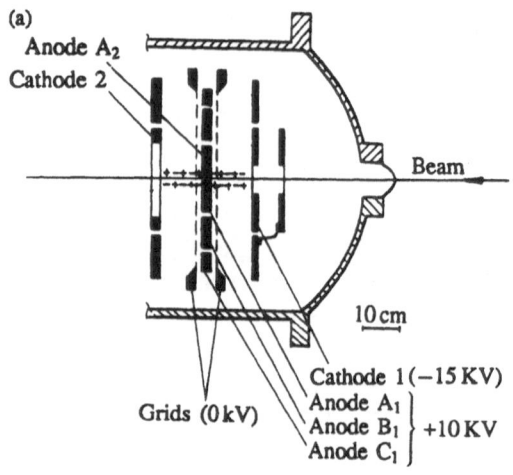

(a)

Anode A_2

Cathode 2

Beam

10 cm

Grids (0 kV)

Cathode 1 (−15 KV)

Anode A_1
Anode B_1 } +10 KV
Anode C_1

Fig. 6.2. Schematic layout of two sections of the IKAR ionization spectrometer for recoil nuclei (a); front edge of the anode pulse, and time positions of the control and the main strobing pulses S_c and S_m, respectively, of length 0.1 μs (b) (Vorobyov et al. 1974; Burq et al. 1978)

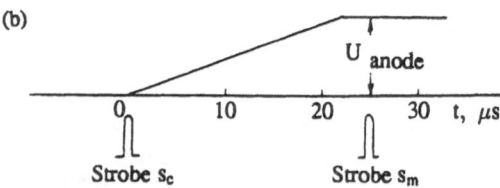

(b)

U_{anode}

0   10   20   30   t, μs

Strobe s_c Strobe s_m

rise times. The drift time of electrons during charge collection at the anodes does not exceed 20 μs.

If the range of a recoil nucleus is smaller than the radius of the anode B (Fig. 6.2), which is recognized by the absence of any signal arriving from the external circular anode C, then the sum of the signal amplitudes at the anodes A and B determines the energy of the recoil nucleus. Long-range particles leave only part of their energy in the gas. The stability of the gas composition is controlled by the position of spectrometric signals from α-sources used for calibration of the energy scale. In working with helium, a stabilizing admixture of 10 % H_2 has to be introduced.

With the aid of the IKAR ionization spectrometer the constancy of the Fermi plateau was demonstrated for ionization energy losses up to $\gamma = 1.8 \cdot 10^3$ (Burq et al. 1981) (Fig. 6.3). This is convincing evidence against theoretical predictions made by Zytovich (1962), according to whom the taking into account of radiative corrections should lead to a fall of the Fermi plateau level for $\gamma > 10^2$. The average total ionization measured in this experiment corresponds to an energy release of 0.24 MeV (Fig. 6.4).

Ionization Chambers in Calorimeters for Investigation of Cosmic Radiation.
The most important application of ionization chambers in modern high energy physics is their use in ionization calorimeters for studies of ultra-relativistic cosmic particles. Such calorimeters contain hundreds of ionization chambers of

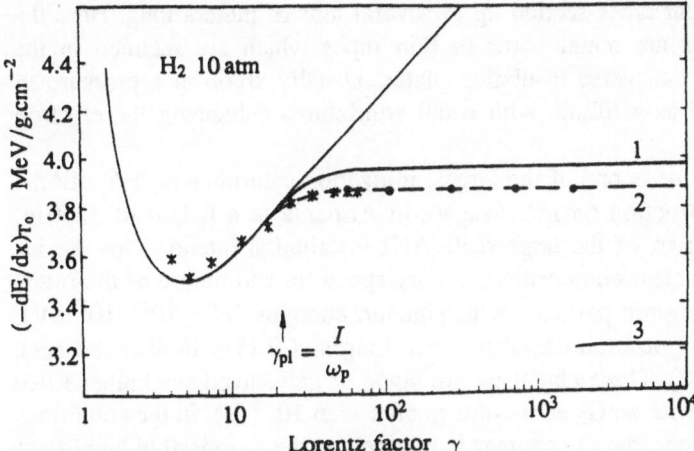

Fig. 6.3. Relativistic rise of restricted energy losses and the Fermi plateau measured in ionization chambers: $P = 10^6$ Pa of H_2; (∗) Barber (1956); (•) Burq et al. (1981); *smooth curves* are (*1*) calculation using (1.50) at $T_0 = 0.12$ MeV; (2) approximation of experimental data; (3) predictions of Fermi plateau level according to Zytovich (1962). The difference between *curves 1* and 2 is due to energy losses related to Vavilov-Čerenkov radiation, not recorded in the experiment

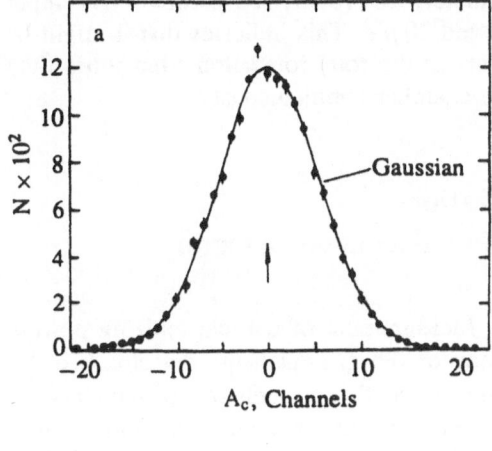

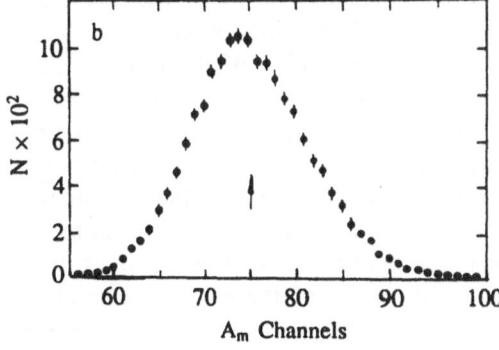

Fig. 6.4. Distribution of amplitudes A of the total signal (16707 events) from six ionization chambers of the IKAR spectrometer, placed in a 100 GeV/c proton beam (Burq et al. 1981): (a) for the control strobe S_c ($\overline{A}_c = 0.1 \pm 0.17$, $\sigma_a = 22.10 \pm 0.13$ arbitrary units); (b) for the main strobe S_m ($\overline{A}_m = 76.5 \pm 0,2$ arbitrary units). The *solid line* is a Gaussian

rectangular or circular cross-section up to several tens of meters long. Here the collecting electrodes are metal wires or thin pipes which are retained in the central position by transverse insulating plates. Usually argon at a pressure of $3–5 \cdot 10^5$ Pa is used as a filling, with small admixtures enhancing the electron drift velocity.

An example of this is one of the largest ionization calorimeters, $1.6 \cdot 10^3$ m^2 in area, under construction on Mt. Aragatz in Armenia at a height of 3250 m. This calorimeter is part of the large-scale ANI installation intended for the investigation of the nuclear composition, energy spectrum and nature of the interaction products of cosmic particles with gigantic energies ($E = 10^{15}–10^{17}$ eV). It consists of many ionization chambers 40 m long and 0.25 m in diameter each (Danilova et al. 1982). These chambers are made of galvanized steel tubes filled with pure nitrogen (with an O_2 admixture smaller than 10^{-3} %). In the conditions of the ANI experiment, the O_2 content is retained at the permissible level (less than 10^{-3} %) by utilizing industrial air purifiers involving a N_2 flow of 4.2 liters per hour per chamber. Owing to a reduced specific ionization ($dN_t/dx \simeq 60$ cm^{-1} in N_2 compared to 97 cm^{-1} in Ar) the signal in an ionization chamber filled with nitrogen is about 2 times weaker than in an argon chamber (Fig. 6.1). Calibration 5.15 MeV α-particles from a radioactive ^{239}Pu source produce in the gas of such a chamber ($C_i = 600$ pF) a saturation signal with an amplitude of about 38 μV, which corresponds to approximately 10^2 relativistic particles. The input noise of the preamplifier does not exceed 20 μV. This indicates that it might be feasible to make reliable measurements of the total ionization from more than 10^2 relativistic particles traversing the chamber simultaneously.

6.2 Determination of the Ionization from the Efficiency of a Gas-discharge Detector

Physical Principles of the Method. Measurement of particle ionizing powers on the basis of the recording efficiency of the gas-discharge detector (i. e. by the mean probability of a particle to be detected) represents one of the classical methods for determining ionization; for instance, the specific primary ionization. With the aid of low-efficiency Geiger-Müller counters in the 1940s and 1950s, measurements were performed of the primary specific ionization in H_2, He, Ne and Ar produced by relativistic particles, in the region of its minimum, and of the beginning of the logarithmic rise (McClure 1953; for earlier references see Price 1955). The choice of gases was later greatly extended by Rieke and Prepejchal (1972). Using narrow-gap spark chambers, including wire chambers at low pressures, instead of low-efficiency counters, made possible the investigation of the relativistic rise of the primary specific ionization in inert gases up to $\gamma \simeq 10^3$, corresponding to the Fermi plateau (Blum et al. 1974; Asoskov et al. 1975, 1977, 1979, 1982a; Sochting 1979). Some results of these measurements are presented in Table 1.3. The relative particle ionizing power was also determined by the operation efficiency of a track spark chamber (Grieder 1966; Ashburn et al. 1980) and of hodoscopic gas-discharge tubes (Coxell and Wolfendale 1960; Ashton et al. 1971). Attempts have also been made to use the multigap projection spark chamber for these purposes (Golovin et al. 1982).

The method considered here is based on measuring the efficiency η of a detector, which depends on the number N_1 of ionizing collisions in the gas; this follows a Poisson distribution with a mean value $\overline{N}_1 = lP\,(dN_1/dx)$. If at least one free electron in a gas volume is sufficient to trigger the gas discharge detector, then its efficiency numerically equals the probability for one or more $(N_1 \geq 1)$ ionizing collisions to occur in the gas:

$$\eta = 1 - \exp(-\overline{N}_1) = 1 - \exp[-lP\,(dN_1/dx)]\,. \tag{6.1}$$

The standard deviation of the efficiency, $< \delta\eta > = [\eta(1 - \eta)/n]^{1/2}$ (where n is the number of measurements or the number of independent layers of a multilayer detector) corresponds to the binomial distribution, taking into account only two alternative possibilities – a count or a miscount of the detector. The relative measurement uncertainty of the primary ionization,

$$\frac{< \delta\,(dN_1/dx) >}{dN_1/dx} = - \left[\frac{\eta}{n(1 - \eta)}\right]^{1/2} \Big/ \ln(1 - \eta)$$

achieves a minimum at $N_1 = 1.6$, i.e., for $\eta = 0.8$. At $n = 10^3$ it amounts to 3.9 % (within the interval $0.5 \leq \eta \leq 0.95$ this uncertainty does not exceed 4.6 %). If it is necessary to separate particles of differing ionizing powers, then the optimum magnitude of η will differ somewhat from 0.8 (Tiapkin 1968).

The required detector efficiency $\eta < 1$ can be achieved by the following methods [see (6.1)]:

1) by appropriate choice of the gas composition or pressure, and of the detector thickness. For $l = 1\,\mathrm{cm}$, a gas pressure lower than the atmospheric pressure (from $P = 0.5 \cdot 10^5\,\mathrm{Pa}$ for He, down to $P = 0.04 \cdot 10^5\,\mathrm{Pa}$ for Xe) corresponds to the optimum conditions for primary ionization measurements. Reduction of the detector size to $l < 0.5$–$1.0\,\mathrm{cm}$ is undesirable, because the detector capacitance will rise, and also because of mechanical difficulties in keeping the distance between the electrodes constant to a high precision. Thus, for measurements of the primary ionization, low-efficiency Geiger-Müller counters (sometimes operating in the proportional mode) and low-pressure spark chambers with $l \simeq 1\,\mathrm{cm}$ and $P \lesssim 10^5\,\mathrm{Pa}$ are applied;

2) by reduction of the mean number of ion pairs in the gas of a triggerable detector, delaying the high-voltage pulse. In this case some of the ionization electrons are precipitated into the detector walls, owing to diffusion in the gas, or are captured by electro-negative admixtures and lose their ability to initiate a discharge. Here, miscounts of the detector are possible, even when several ion pairs are produced initially in the detector volume; so the measured ionization is no longer identical to the primary one. This operation mode is utilized in spark chambers containing electro-negative admixtures, and in hodoscopic

gas-discharge tubes. In these detectors, particles are identified by their relative ionizing powers;

3) by discrimination of weak signals with amplitudes lower than the preset threshold in multilayer proportional chambers. This method is very promising because information on the coordinates and on the ionizing powers of the particles under investigation are obtained simultaneously.

Measurement of the Primary Ionization. Determination of the primary ionization from the operation efficiency of a gas-discharge detector is possible only when for the development of a discharge under the influence of the electric field in the gas the presence of a single free electron is sufficient. This condition was verified for a spark chamber experimentally, and it is valid for all gas-discharge detectors exhibiting a plateau of the efficiency (Fig. 6.5), where it is constant, in spite of the exponential rise of the Townsend ionization coefficient α_T with increasing the electric field strength (Chap. 2). Another important condition for the method to be correct is the absence of δ-electrons from the walls of the detector; this is verified by measurements of the primary ionization at various gas pressures with subsequent extrapolation of the results to $P = 0$ (Price 1955; Asoskov et al. 1977).

Low-efficiency counters exhibit several major drawbacks, e.g. long pulse duration and bad spatial resolution. However, spark chamber ionization measurements are also subject to certain disadvantages. Some of the ionization electrons land on the electrodes before a discharge is initiated, because of electron diffusion in the gas during the high-voltage pulse delay time, and because of drift in the rising pulsed electric field before the discharge is developed. Thus, these electrons may lose the capability to initiate a gas breakdown. Such effects are especially noticeable in gases with high diffusion coefficients and electron mobilities, and are enhanced when the pressure is decreased. They can be significantly weakened by the use of a spark chamber with wire electrodes. In this case, electron diffusion from the inter-electrode gap is almost compensated by the backward diffusion of electrons produced outside the gap. As a result, it is possible to measure the primary ionization, even in pure inert gases with high diffusion coefficients: this is difficult to achieve by other methods. In a wire spark chamber the influence of δ-electrons from the walls, and the escape of ionization electrons due to their drift toward the anode before initiation of the discharge, are significantly lower than in a chamber with continuous electrodes. Nevertheless, to determine the absolute values of dN_1/dx it is necessary to perform measurements for various time delays τ_d of the high-voltage pulse, and then to extrapolate the results to $\tau_d = 0$ (Asoskov et al. 1977). In the case of relative ionization measurements with the aim of particle identification, this procedure is no longer required, while normalization is performed using the results obtained for the ionization minimum or for the Fermi plateau.

Electron and ion collisions occurring during the transport processes in the gas of a spark chamber may distort the measured primary ionization (Chap. 2). These processes, however, like the presence of admixtures in the gas, have almost

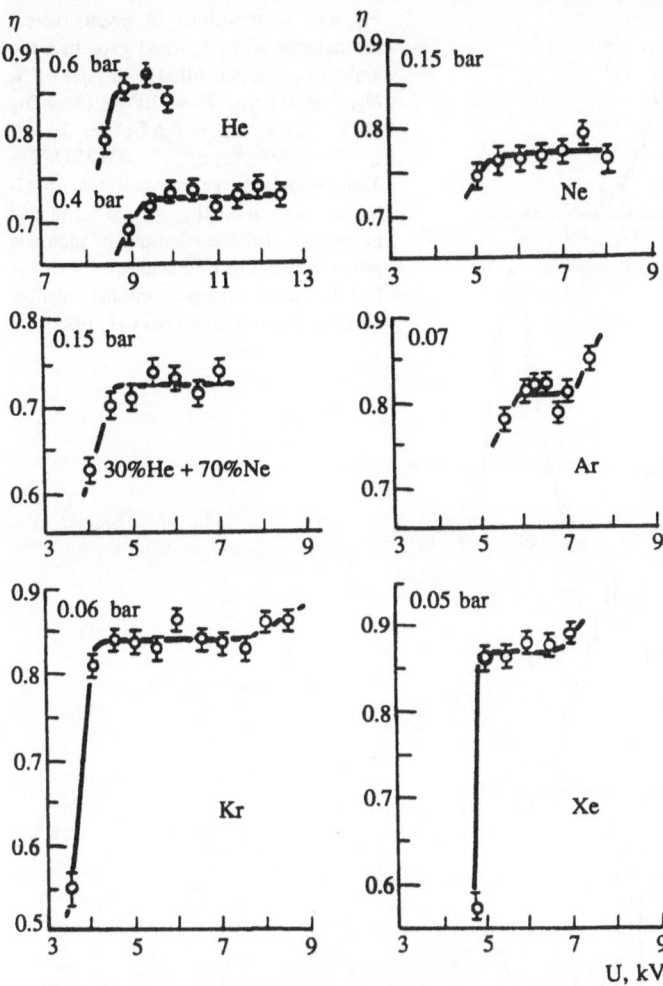

Fig. 6.5. Dependence of the efficiency η of a spark chamber, filled with a noble gas, on the high-voltage pulse amplitude U (Asoskov et al. 1977): $l = 1\,cm$; $\tau_d = 0.2\,\mu s$; $RC = 0.2\,\mu s$. The figures near the curves indicate the gas pressure P in bar. The existence of an efficiency plateau confirms that a single electron is sufficient for discharge formation

no influence on both the results of relative ionization measurements and on the relativistic rise of primary ionization. If such measurements are used for particle identification, the presence of molecular admixtures together with a weak electric field may turn out to be useful (Sochting 1979). Because of the Penning effect, they remove the excitation of metastable atoms produced in the spark discharge, thus reducing the dead time of the spark chamber.

Particle identification from the operation efficiency of a multilayer detector involving low-efficiency narrow-gap spark chambers have been performed in experiments at accelerators (Fig. 6.6a–c) and with cosmic rays (Fig. 6.7). To provide the ionization resolution $\sigma_n = 3\,\%$ required for reliable separation of hadrons in

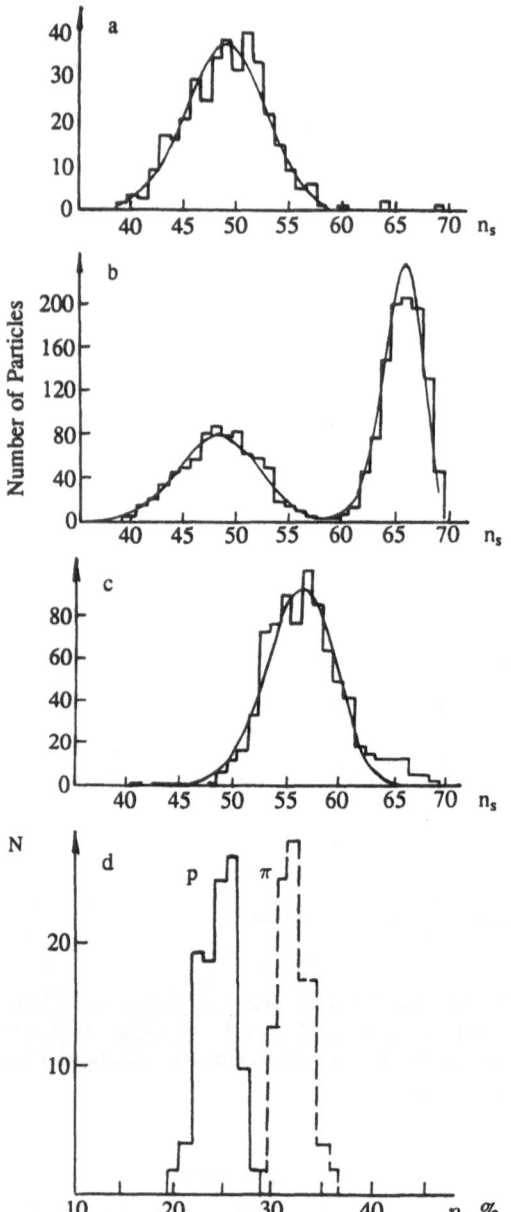

Fig. 6.6. Distributions of events versus the number n_s of sparked gaps in a 69-gap spark detector filled with He + 0.7 % N_2, l = 0.8 cm, P = 10^5 Pa (Sochting 1979): (a) π^-, p = 0.5 GeV/c; (b) π^+, p = 0.5 GeV/c; (c) e$^-$, p = 0.25 GeV/c. The *smooth curves* are binomial distributions approximating the experimental histograms. (d) Distributions of pion and proton recording efficiencies at 4 GeV/c in a 100-layer wire proportional chamber with l = 2 mm (Gurzhiev et al. 1989)

the region of the logarithmic rise of primary ionization (γ = 10–10^2), a detector with a number of layers $n \geq 10^3$ is necessary, which represents quite a difficult technical problem. But such a multilayer detector would require quite inexpensive electronics, should exhibit a high operation speed and should have a simple algorithm for the analysis of information, which would amount to only 1 bit per channel.

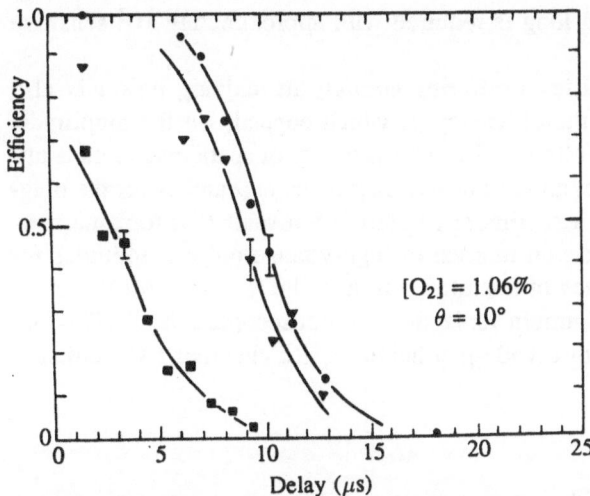

Fig. 6.7. The efficiency of a spark chamber filled with He + 1.06 % O_2 versus the delay time τ_d of the high voltage pulse (Ashburn et al. 1980): (■) cosmic ray particles, $(-dE/dx) = 0.56$ keV/cm; (▼) protons, $(-dE/dx) = 7.08$ keV/cm; (•) protons , $(-dE/dx) = 10.01$ keV/cm; $l = 4.9$ cm; $P = 10^5$ Pa; $\mathcal{E}_0 = 5.76$ kV/cm; $RC = 0.82\,\mu$s; $\theta = 10°$

Detectors with Controllable Efficiency. Discrimination or identification of high-energy particles using the efficiency of a gas-discharge detector does not necessarily require measurement of primary ionization. This aim can also be achieved by determination of the *relative ionizing powers* of the recorded particles with the aid of low-efficiency multilayer proportional and spark chambers, or of hodoscopic discharge tubes (neon flash tubes). In doing so the optimum detector efficiency can be achieved by varying the voltage supply or signal discrimination level, and in pulsed detectors, by variation of the amplitude or the time delay of the high-voltage pulse.

Identification of relativistic hadrons using the mean efficiency of a multilayer proportional chamber was tested by a group from IHEP (Protvino) (Gurzhiev et al. 1989). In the experiment, a single-layer 8×8 cm^2 cell was used, with two high-voltage cathode planes made of 15 μm aluminum-coated mylar, situated at a distance of 2 mm from each other. In between them forty 25 μm wires were stretched parallel to the cathode planes with a 2 mm spacing and were combined into a single channel. The proportional chamber was filled with an Ar + 20 % CH$_4$ or Ar + 20 % CO$_2$ mixture. Its efficiency was regulated by variation of the gas pressure or of the high-voltage applied to the cathodes. The ratio of the registration efficiencies for pions and protons, η_π/η_p increases as the voltage supply decreases and as η_π falls. η_π/η_p is almost independent of the gas pressure, which favors operation at atmospheric pressure. Fig. 6.6d demonstrates the separation of pions and protons of momentum 4 GeV/c with $\eta_\pi = 20$ % ($\eta_\pi/\eta_p \simeq 1.2$; the separation coefficient is $S_n = 6$). Each event in Fig. 6.6d corresponds to the passage of 10^3 pions or $\sim 5 \cdot 10^2$ protons. To provide separation of individual particles a detector about 2 m long is required with approximately 10^3 sensitive planes.

The discrimination of particles differing strongly in ionizing power is also possible in spark chambers, the efficiency of which depends on the amplitude, length and delay of the high-voltage pulse. This property of spark chambers is utilized in studies of recoil nuclei and of nuclear fragments, in searches for the magnetic monopole, and in other experiments. Figure 6.8 reveals that for a narrow-gap argon chamber there exists an interval of high-voltage pulse amplitudes for which the registration efficiency of α-particles from ^{210}Po ($\varepsilon_\alpha = 5.3$ MeV) equals 100 %, while electrons of minimum ionization are not recorded at all. This operating mode was also used in a wide-gap helium spark chamber (Alexanian et al. 1977).

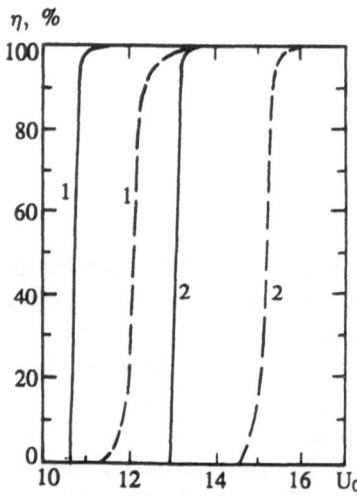

Fig. 6.8. Dependence of the efficiency η of an argon spark chamber on the high voltage pulse amplitude U_0 (Grieder 1966): (———) α-particles from Po^{210}, (- - - -) β-particles of minimum ionization: (1) $RC = 0.7\,\mu s$; (2) $RC = 0.28\,\mu s$; the clearing field is 10 V/cm, $l = 1$ cm

The efficiency of a spark chamber with electro-negative admixtures present in the gas filling, is conveniently regulated by variation of the time delay τ_d of the high-voltage pulse, instead of the amplitude. In such a spark chamber the recording efficiencies of various particles depend on their specific energy losses (Fig. 6.7). This permits particle discrimination in a multigap spark chamber by the number of discharge gaps in which breakdowns occur.

In the presence of electro-negative admixtures the average number $\overline{N}_e$ of electrons available in the gas up to the moment when the high-voltage pulse is applied is

$$\overline{N}_e = \kappa_e w^{-1}(-dE/dx)_{T_0} \exp[-(\tau_d - \tau_i)/\tau_a] .$$

Here $\kappa_e < 1$ is the "survival" coefficient of an electron, i. e., the probability that it will avoid capture by admixtures, and will not precipitate on the anode or the detector walls in the course of thermalization and diffusion during the time τ_d as well as of drifting during the rise time of the high-voltage pulse; while τ_i and τ_{at}

are the average thermalization time and capture time of an electron, respectively. Using (6.1) we find that

$$\ln(1 - \eta) = [-(\tau_d - \tau_t)/\tau_{at}] \ln[\kappa_e l(-dE/dx)_{T_0}/w] .$$

Consequently, in the case of two particles of differing ionizing powers, the same registration efficiency is achieved for values of τ_{d_1} and τ_{d_2} related to each other as follows:

$$\ln \{[(-dE/dx)_{T_0}]_1 / [(-dE/dx)_{T_0}]_2\} = (\tau_{d_2} - \tau_{d_1})/\tau_a .$$

This makes it possible to determine the expected dependence $\eta(\tau_{d_2})$ for particles of specific energy loss $[(-dE/dx)_{T_0}]_2$ if $\eta(\tau_{d_1})$ for particles of specific energy loss $[(-dE/dx)_{T_0}]_1$ is known.

Regulation of the detector efficiency by variation of the time delay of the high-voltage pulse is also applied when working with hodoscopic flash tubes. The latter are hermetically sealed glass or plastic (sometimes flexible) tubes of a diameter of $d_t = 5$–20 mm, retorts or channel plates usually filled with neon at a pressure of $(0.6$–$3) \cdot 10^5$ Pa, and placed between the high-voltage parallel plane electrodes. [See the review article by Conversi and Brosco (1973)]. If the gas inside a tube is ionized, then a high-voltage pulse ($\mathcal{E}_0 = 6$–9 kV/cm, $\tau_p = 3$–$10 \,\mu$s) is applied to the electrodes, and a discharge is initiated, which propagates throughout the entire volume of the tube. This discharge can be photographed through the transparent end, or recorded with the aid of a wire electrode introduced into the tube and terminated by a simple electronics channel. Neon discharge tubes have acquired wide spread circulation because of their simplicity, low cost and reliability. They serve as a basis for the construction of multilayer hodoscopic "discharge chambers" involving up to tens or hundreds of thousands of channels, and used for visualization of trajectories, coordinate measurement and particle identification, in cosmic rays studies, experiments, searching for the proton decay, and so on. However, a large sensitivity time ($\tau_s \simeq 100 \,\mu$s) and long dead time ($\tau_m \simeq 10^{-1}$ s) limit the application of flash tubes in experiments performed at accelerators.

The characteristics of discharge formation in a flash tube depends on its diameter d_t and on the rise time τ_R of the high-voltage pulse. When $d_t \gtrsim 1$ cm and $\tau_R \ll 1 \,\mu$s, the discharge is initiated, as it is in a spark chamber, directly by the ionization electrons (which are produced by the charged particle in the gas inside the tube). When $d_t < 1$ cm and $\tau_R \gtrsim 1 \,\mu$s, the ionization electrons land on the walls, because they drift during the rise time of the high-voltage pulse and are not able to initiate the discharge. In this case the discharge is caused by photo-ionization of the walls in the course of diffusion of the resonance radiation (which is emitted by atoms excited in collisions with ionization electrons accelerated by the pulsed electric field).

An important property of flash tubes is the decrease of their efficiency as τ_d is increased; this is explained by diffusion of thermalized free electrons during the delay time of the high-voltage pulse and their landing on the walls of the tube. This mechanism for the fall of efficiency follows from the universality of

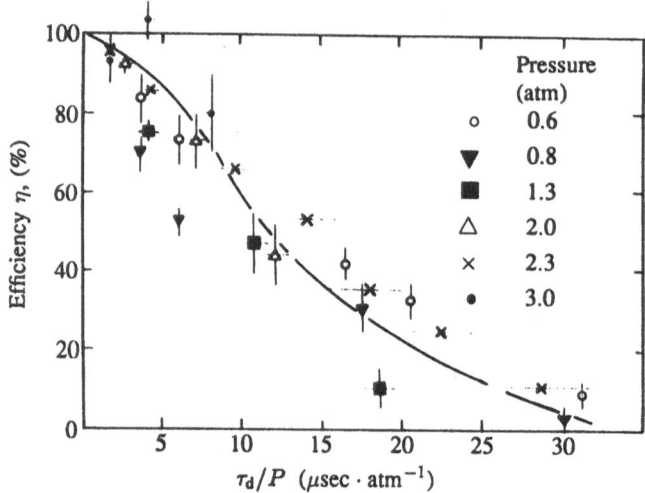

Fig. 6.9. Dependence of the efficiency of a neon flash tube upon $\tau_{\rm d}/P$ measured by Coxell and Wolfendale (1960); the experimental data are obtained for $P = (0.6\text{--}3.0) \cdot 10^5$ Pa, flash tube diameter $d_{\rm t} = 5.5$ mm, high voltage pulse duration $\tau_{\rm p} = 3.5\ \mu$s

the observed efficiency dependence versus the $\tau_{\rm d}/P$ ratio (Fig. 6.9). Indeed, in accordance with (2.12) the diffusion displacement of thermal electrons during the time $\tau_{\rm d}$ is

$$l_{\rm d} = (2D_0\tau_{\rm d}/P)^{1/2}\ , \tag{6.2}$$

where D_0 is the electron thermal diffusion coefficient under normal conditions. Therefore in the absence of electro-negative admixtures, the number of electrons arriving at the detector walls during the time $\tau_{\rm d}$ and, consequently, the detector efficiency, depend on the ratio $\tau_{\rm d}/P$.

By varying $\tau_{\rm d}$ one can choose the best conditions for the identification or discrimination of particles of different ionizing powers. A similar method was applied, for example, in the search for particles with fractional charge $z = 1/3$ in cosmic radiation [see, for instance, Ashton et al. (1971)]. To optimize $\tau_{\rm d}$, the shape of the $\eta(\tau_{\rm d})$ dependence can be computed; it is determined by the "effective" number of electrons $N_{\rm e}$ produced by the charged particle along a path $r_{\rm t} = d_{\rm t}/2$ in the gas (Lloyd 1960):

$$N_{\rm e} = \kappa_{\rm e}r_{\rm t}(-dE/dx)_{\rm T_0}/w\ .$$

The results of such a calculation for cosmic muons with an average momentum $\overline{p} = 2.1$ GeV/c, are in good agreement with experimental data (Fig. 6.10), which permits the calculation of the expected $\eta(\tau_{\rm d})$ dependence for particles of arbitrary z, for example, $z = 1/3$.

The high-voltage pulse amplitude utilized for the operation of neon flash tubes usually corresponds to the efficiency plateau: $\eta(U_0) = $ const. Regulation of the efficiency of flash tubes by variation of the high-voltage pulse amplitude is not applied, since in this case very fine tuning is required.

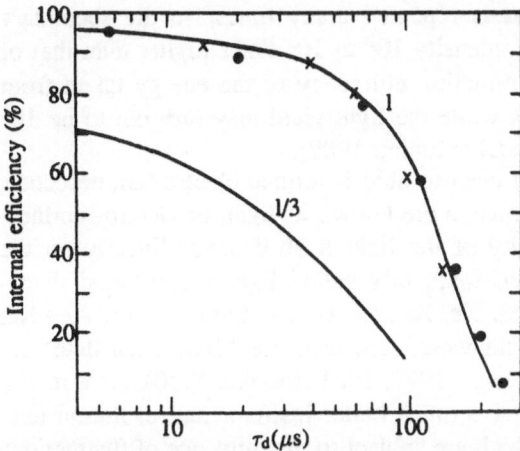

Fig. 6.10. The efficiency η of recording singly charged and fractionally charged relativistic particles by neon discharge tubes versus the time delay τ_d of the high-voltage pulse (Ashton et al. 1971): *smooth curves* – results of calculation by Lloyd (1960) for $N_t = 12$ ($z = 1$) and $N_t = 1.33$ ($z = 1/3$); experimental points correspond to (•) exponential ($RC = 10\,\mu s$) and (×) to rectangular ($\tau_p = 14\,\mu s$) high-voltage pulses; $\mathcal{E}_0 = 8\,kV/cm$.

6.3 Possibilities of Ionization Measurement in Gas Scintillation Detectors

The phenomenon of radio-luminescence, or scintillation under the influence of ionizing radiation, has been known since the early days of radioactivity science. The noble gases and their mixtures occupy a special place among the substances exhibiting this property. These gases are characterized by a high radiation yield [$\sim 10^3$ photons per MeV of released energy (Birks 1964)]. Upon passage of a fast charged particle through a noble gas, a short ($\lesssim 10\,ns$) flash of light arises with an intensity proportional to the energy liberated in the gas. The spectrum of this primary scintillation lies mainly in the vacuum UV range, so that for its registration, one must use light amplifiers with quartz windows and light wave length shifters.

For the same energy release, the recorded light yield of scintillations in noble gases is 10–40 times lower than in a NaI(Tl) crystal. Therefore gas scintillation detectors exhibit a low energy resolution (Policarpo 1977, 1981),

$$\sigma_1 \simeq 1.3/\Delta^{1/2} \,, \tag{6.3}$$

where Δ (keV) is the energy release in the gas. Thus their application is restricted to the spectrometry of slow heavy particles stopping in the gas, – e.g. recoil nuclei, fission fragments and neutrons.

If a reduced electric field $\mathcal{E}/P = 0.1$–$1.0\,kV/cm \cdot 10^5\,Pa$, somewhat lower than the gas amplification threshold, is applied to the gas, then the free electrons being accelerated become capable of exciting the atoms of the gas, without producing additional ionization. This process is repeated many times, so the secondary scintillation flash produced has an intensity 10^2 to 10^3 times higher than that of the primary flash. Here the transformation efficiency of the energy taken from the electric field is close to 100 %, while the light yield may turn out to be 10^2 times higher than in a NaI(Tl) crystal (Monich 1980).

The scintillation of a gas in an electric field is termed electro-luminescence, and detectors based on this phenomenon are known as light, or electro-luminescence, detectors. Since the intensity of the light flash depends linearly on the energy release in the gas, they are frequently called light proportional detectors. Usually the heavy noble gases, Xe, Kr and Ar, and the mixtures, Ar + Xe, or Xe + N_2, which exhibit high light yields, are used for filling such detectors (Dolgoshein, Rodionov 1969; Policarpo 1977, 1981; Monich 1980). As a result, gas light proportional detectors have a much better instrumental resolution than ordinary proportional detectors, which are subject to the influence of fluctuations of the gas amplification. Thus, for example, in recording X-ray quanta emitted by the isotope ^{55}Fe (E_γ = 5.9 keV), the energy resolutions of these two types of detectors are, respectively, 8–10 % and 20–22 %. Owing to the absence of any space charge in the gas, light proportional detectors are capable of handling intensities up to $10^6\,s^{-1}$, and are limited only by the rise time of the light pulse ($\lesssim 1\,\mu s$). These detectors have become wide spread in the spectroscopy of X-rays and of soft gamma radiation, as well as in α- and β-spectroscopy (particularly in experimental studies of outer space), where because of their large apertures and high resolution they compete successfully with semiconductor detectors. In high-energy physics gas light proportional chambers have been used quite recently and, as a rule, only in two cases: either when optical readout of information is expedient [in this case the energy resolution is not so important, so gas amplification is used for enhancement of the light yield (Golovkin 1979; Poliakov, Rykalin 1988; Volnov et al. 1988)]; or when the energy resolution and the counting rate capability have to be improved [here the electric field strength is somewhat lower than the gas discharge threshold (Charpak et al. 1975; Herold et al. 1983; Grishin, Merson 1988)].

A light proportional chamber usually involves two functionally different gaps: a drift gap and a light gap, separated by a transparent wire grid (Fig. 6.11). Electrons produced by an ionizing particle in the gas of the drift gap move in the constant drift field ($\mathcal{E}_0/P \lesssim 1\,kV/cm \cdot 10^5\,Pa$) to the light gap, where a significantly higher field ($1 \leq \mathcal{E}_0/P \leq 7\,kV/cm \cdot 10^5\,Pa$) is applied. Here, undergoing acceleration up to energies a bit lower than the first ionization potential of the gas, they excite its atoms repeatedly in inelastic collisions, thus allowing a secondary scintillation to be produced. Further, upon re-emission by the wavelength shifter deposited on the walls of the chamber, or on its transparent window, the light arrives at the photocathode of the photomultiplier and is transformed into an electric signal. The rise time of the electro-luminescence pulse τ_R depends on how long the electron traverses the light gap, i.e. on the width of the gap, as well as on the gas composition, gas pressure and the electric field strength. A typical value is $\tau_R = 0.1–1.0\,\mu s$.

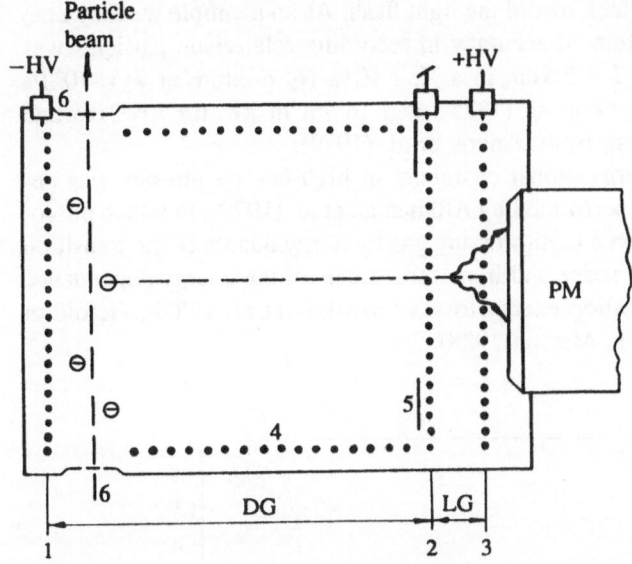

Fig. 6.11. Schematic layout of a light proportional chamber: DG, LG – drift and light gaps; PM – photomultiplier; (1–3) wire planes; (4) potential wires; (5) mask; (6) input and output windows

If a particle travels along the grid of the light gap, then an analysis of the rise time of the pulse permits to distinguish events with compact or elongated ionization clusters, and thus to discard particles accompanied by delta-electrons. This, for instance, results in contraction of the ionization distribution produced by charged particles in the gas (Fig. 6.12), i. e. in the improvement of the ionization resolution (Herold et al. 1989). Particle coordinates in the light proportional chamber can be measured by the drift time or by the relative pulse heights in

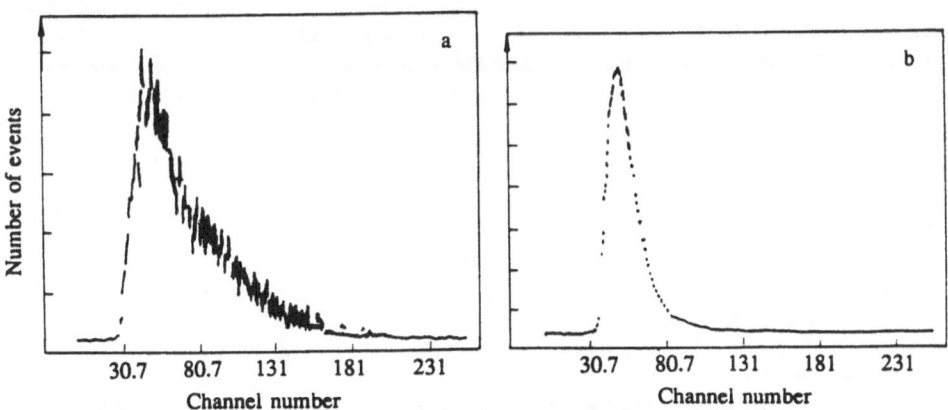

Fig. 6.12. Fluctuations of ionization produced by relativistic pions ($\beta = 0.92$) in a xenon light proportional chamber (Herold et al. 1983): (a) events with long pulse rise time accompanied by δ-electrons; (b) events with short pulse rise time (without δ-electrons)

several photomultipliers which record the light flash. As an example we may note that the coordinate measurement accuracy in recording relativistic particles was 0.3 mm, for a drift length $l = 2.5$ cm in a Xe + 10 % N_2 mixture at $P = 10^5$ Pa in an experiment by Charpak et al. (1975), and 16 μm in Xe, for $l = 2$ cm and $P = 10^6$ Pa in an experiment by Baskakov et al. (1979).

Application of light proportional chambers in high-energy physics was demonstrated in experiments performed by Alikhanian et al. (1979), in which photoelectron clusters produced in a high-pressure gas by X-ray quanta of the transition radiation were separated in space and in measurements of the energy dependence of the most probable ionization energy losses (Asoskov et al. 1982b; Herold et al. 1983 (Fig. 6.13); Grishin, Merson 1988).

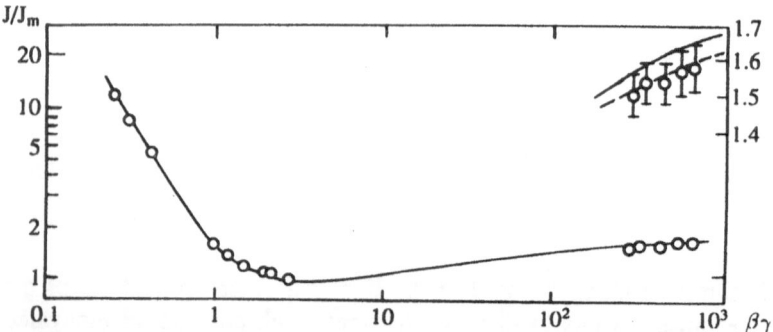

Fig. 6.13. The relative most probable ionization, measured in a xenon light proportional chamber, versus $\beta\gamma$: $l = 9.1$ cm, $P = 10^5$ Pa (Herold et al. 1983); (———) calculation according to (1.83); (- - - -) Monte Carlo calculation (Allison and Cobb 1980)

References

Aggson T., Fretter W. B. 1962: Suppl. Nuovo Cimento **23**, 75. Sect. 6.1

Alexanian A. S., Asatiani T. L., Dayon M. I., Gasparian A. O., Jirova L. A., Ivanov V. A., Kamensky A. D., Kayumov F. F., Megrabian G. K., Mkrtchian G. G., Pikhtelev R. N. 1977: Nucl. Instr. Meth. **140**, 473. Sect. 6.2

Alikhanian A. I., Baskakov V. I., Cherniatin V. K., Dolgoshein B. A., Fedorov V. M., Gavrilenko I. L., Konovalov S. P., Kozodajeva O. M., Lebedenko V. N., Majburov S. N., Muravjev S. V., Pustovetov V. P., Romanjuk A. S., Shmeleva A. P., Vasiliev P. S. 1979: Nucl. Instr. Meth. **158**, 137. Sect. 6.3

Allison W. W. M., Cobb J. M. 1980: Ann. Rev. Nucl. Sci. **30**, 253. Sect. 6.3

Ananyin P. S., Sidulenko O. A., Fotin E. V., 1979: Pribori Tekhn. Exp. No. 5, 68 (English transl.: Inst. Exp. R 22, 1260). Sect. 6.2

Ashburn M. J., Caldwell T. G., Grey R. N. C., Stuteley P. C., Yook P. C. M. 1980: Nucl. Instr. Meth. **169**, 121. Sect. 6.2

Ashton F., Coats R. B., King J., Tsuiji K., Wolfendale A. W. 1971: J. Phys. A **4**, 895. Sect. 6.2

Asoskov V. S., Blazhenkov V. V., Grishin V. M., Kotenko L. P., Merson G. I. 1975: *Application of a Low Pressure Spark Chamber for Primary Specific Ionization Measurements in Gases*, Report No. 45 FIAN Moscow 1975. Sect. 6.2

Asoskov V.S., Blazhenkov V.V., Grishin V.M., Kotenko L.P., Merson G.I., Pervov L.S. 1977: J. Exp. Teor. Fiz. **73**, 146 (English transl.: Sov. Phys. JETP **46**, 75). Sect. 6.2

Asoskov V.S., Blazhenkov V.V., Grishin V.M., Ermilova V.K., Kotenko L.P., Merson G.I., Pervov L.S. 1979: J. Exp. Teor. Fiz. **76**, 1274 (English transl.: Sov. Phys. JETP **49**, 646). Sect. 6.2

Asoskov V.S., Grishin V.M., Ermilova V.C., Kotenko L.P., Merson G.I., Chechin V.A. 1982a: Trudy Fian **140**, 3. Sect. 6.2

Asoskov V.S., Grishin V.M., Ermilova V.K., Kotenko L.P., Merson G.I., Pismenny O.G. 1982b: *Electroluminescence Proportional Drift Chamber for Measurement of Relativistic Charged Particle Energy Loss in a Gas*. Report No. 214, FIAN Moscow 1982. Sect. 6.3

Barber W.C. 1955: Phys. Rev. **97**, 1071. Sect. 6.1

Barber W.C. 1956: Phys. Rev. **103**, 1281. Sect. 6.1

Baskakov V.I., Cherniatin V.K., Dolgoshein B.A., Levedenko V.N., Romanjuk A.S., Fedorov V.M., Gavrilenko I.L., Konovalov S.P., Majburov S.N., Muravjev S.V., Pustovetov V.P., Shmeleva A.P., Vasiliev P.S. 1979: Nucl. Instr. Meth. **158**, 129. Sect. 6.3

Birks J.B. 1964: *The Theory and Practice of Scintillation Counting* (Pergamon, Oxford). Sect. 6.3

Blair B., Jin B., Macfarlane D., Messner R.L., Novikoff D.B., Purohit M.V., Auchincloss P.S., Schuilli F., Shaevitz M.H., Edwards D., Edwards H., Fiak H.E., Fukushima Y., Kerns O.A., Kondo T., Rapidis P.A., Segler S.L., Stefanski R.J., Theriot D., Yovanovich D., Bodek A., Colleman R., Marsh W., Fackler O., Jenkins K.A. 1984: Nucl. Instr. Meth. Phys. Res. A **226**, 281. Sect. 6.1

Blum W., Sochting K., Stierlin U. 1974: Phys. Rev. A **10**, 491. Sect. 6.2

Burq J.P., Chemarin M., Chevallier M., Denisov A.S., Dore C., Ekelof T., Grafstrom P., Hagberg E., Ille B., Kashchuk A.P., Korolev G.A., Kullander S., Lambert M., Martin J.P., Mauri S., Paumier J.L., Querrou M., Shchegelsky V.A., Spiridenkov E.A., Tkach I.I., Vorobyov A.A. 1978: Phys. Lett. B **77**, 438. Sect. 6.1

Burq J.P., Chemarin M., Chevallier M., Denisov A.S., Grafstrom P., Hagberg E., Ille B., Kashchuk A.P., Kulikov A.V., Lambert M., Martin J.P., Mauri S., Querrou M., Shchegelsky V.A., Tkach I.I., Vorobyov A.A. 1981: Nucl. Instr. Meth. **187**, 407. Sect. 6.1

Charpak G., Majewsky S., Sauli F. 1975: Nucl. Instr. Meth. **126**, 381. Sect. 6.3

Conversi M., Brosco G. 1973: Ann. Rev. Nucl. Sci. **23**, 75. Sect. 6.2

Coxell M., Wolfendale A.W. 1960: Proc. Phys. Soc. (L) **75**, 378. Sect. 6.2

Danilova T.V., Dunaevsky A.M., Yerlikin A.D., Kozina N.P., Kulichenko A.K., Machavariani S.K., Nikolskaya N.M., Nikolsky S.I., Romakhin V.A., Slavatinsky S.A., Subbotin B.V., Tukish E.I., Feinberg E.L., Shaulov S.B., Avakian V.V., Avakian K.M., Azarian M.O., Akopian C.K., Amatuni A.Z., Arzumanian S.A., Asatiani T.L., Babajan H.P., Bagdasarian G.A., Bojadzian N.G., Bujukian S.P., Gariaka A.P., Yeganov B.S., Zazian M.Z., Ivanov V.A., Mamidzhanian E.A., Martirosov R.M., Matinian S.G., Mnatsakanian E.A., Ter-Antonian S.V., Hodzhamirian A.Yu., Chilingarian A.A. 1982: Izv. Ak. Sci. Armenian SSR, Ser. Fiz. **17**, 129. Sect. 6.1

Dolgoshein B.A., Rodionov B.U. 1969: *"The Mechanism of Noble Gas Scintillation"* in Elementary Particles and Cosmic Rays No. 2 (Atomizdat, Moscow). Sect. 6.3

Golovin B.M., Gornushkin Yu.A., Nadezhdin V.S., Petrov N.I. 1982: Pribori Tekhn. Exp. No. 5, 47 (English transl.: Instr. Exp. R **25**, 1088). Sect. 6.2

Golovkin S.V. 1979: *The Charged Particle Counter on the Basis of Electroluminescence*, Report No. 79–83 IHEP Serpukhov. Sect. 6.3

Grieder P.K.T. 1966: Rev. Sci. Instr. **37**, 80. Sect. 6.2

Grishin V.M., Merson G.I. 1988: *The Measurement of the Relativistic Rise of Electron Ionization Energy Loss in Thin Noble Gas Layers by the Electroluminescence Proportional Drift Chamber*, Report No. 53 FIAN Moscow. Sect. 6.3

Gurzhiev A.N., Kryshkin V.I., Kurchaninov L.L. 1989: *Particle Identification by Low Efficiency "Thin" Proportional Chambers*, Report No. 89–84, IHEP Serpukhov. Sect. 6.2

Hall G. 1959: Can. J. Phys. **37**, 189. Sect. 6.1

Herold W.-D., Egger J., Kaspar H., Pocar F. 1983: Nucl. Instr. Meth. **217**, 277. Sect. 6.3

Lloyd J.L. 1960. Proc. Phys. Soc. (L) **75**, 387. Sect. 6.2

McClure G. W. 1953: Phys. Rev. **90**, 796. Sect. 6.2

Monich V. A. 1980: Pribori Tekhn. Exp. Nos. 5, 7 (English transl.: Instr. Exp. R **23**, 1061). Sect. 6.3

Polyakov V. A., Rykalin V. I. 1988: *"Primary Ionization Recording in the Gas of an Electroluminescence Chamber with Longitudinal Electron Drift by the Cluster Counting Method"* in Coordinate Detectors, Proc. on Int. Symp. on Coordinate Detectors in High-Energy Physics, Dubna, p. 223. Sect. 6.3

Policarpo A. J. P. L. 1977: Space Sci. Instrum. **3**, 77. Sect. 6.3

Policarpo A. J. P. L. 1981: Physica Scripta **23**, 539. Sect. 6.3

Price B. T. 1955: Rep. Prog. Phys. **18**, 52. Sect. 6.2

Rieke F., Prepejchal W. 1972: Phys. Rev. A **6**, 1507. Sect. 6.2

Sochting K. 1979: Phys. Rev. A **20**, 1359. Sect. 6.2

Tiapkin A. A. 1968: *Optimal Conditions of the Low Efficiency Counter Method: Application for Particle Separation by Primary Ionizing Ability*, Report No. JINR-1-3686. Sect. 6.2

Volnov A. D., Zalikhanov B. Zh., Komissarov E. V., Serdjuk V. Z., Sidorkin V. V., Filimonov I. S. 1988: *"The Possibility of Electroluminescence Signal Recording in Wire Chambers with a Plastic Wave Shifter"* in Coordinate Detectors, Proc. of Int. Symp. on Coordinate Detectors in High-Energy Physics, Dubna. Sect. 6.3

Vorobyov A. A. 1979: Elastic Scattering in Coulomb-Nuclear Interference Region. Proc. 19th Int. Conf. High Energy Phys. Tokyo 1978, p. 24. Sect. 6.1

Vorobyov A. A., Denisov A. S., Zalite Yu. K., Korolev G. A., Korolev V. A., Kovshevny G. G., Maev Ye. M., Medvedev V. I., Sokolov G. L., Solyakin G. Ye., Spiridenkov E. M., Tkach I. I., Shchegelsky V. A. 1972: Phys. Lett. B **41**, 639. Sect. 6.1

Zytovich V. M. 1962: J. Teor. Exp. Fiz. **43**, 1782 (English transl.: Sov. Phys. JETP **16**, 1260, 1963). Sect. 6.1

List of Names and Abbreviations

ABL	– average blob length
ADC	– analog to digital converter
AFS	– detector used on ISR collider at CERN
AGL	– average gap length
ALEPH	– detector installed on e^+e^- collider LEP at CERN
AMY	– detector installed on e^+e^- collider TRISTAN at KEK
ANI	– large scale installation for cosmic ray studies at Mount Aragatz in Armenia
ARES	– detector installed in muon beam of JINR, Dubna
ARGUS	– detector installed on e^+e^- collider DORIS II at DESY
ASTRON	– detector built at IHEP, Serpukhov
BEBC	– Big European Bubble Chamber at CERN
BES	– detector built on e^+e^- collider at Beijing
BNL	– Brookhaven National Laboratory
CCD	– charge coupled device (line or matrix of integrated semiconductor capacitors)
CDC	– cylindrical drift chamber
CDF	– Collider Detector of Fermilab (detector installed on $p\bar{p}$ collider at FNAL)
CERN	– European Organization for Nuclear Research
CESR	– e^+e^- collider at Cornell University
CLEO	– detector installed on e^+e^- collider CESR
CMD-2	– detector installed on e^+e^- collider VEPP at Novosibirsk
CMS	– center-of-mass system
CPC	– cylindrical proportional chamber
CRISIS	– Considerably Reduced ISIS (detector installed at Fermilab)
DC	– drift chamber
DEA	– diethylaniline
DELPHI	– detector installed on e^+e^- collider LEP at CERN
DESY	– Deutsches Electronen Synchrotron at Hamburg
DIOGENE	– detector installed on SATURN accelerator

DMA	– dimethylaniline
DME	– dimethylether
DORIS	– e^+e^- collider at DESY
DPI	– drift precision imager
DT	– drift tube
DTR	– drift time recorder
DØ	– detector installed on pp collider at Fermilab
EHS	– European Hybrid Spectrometer at CERN
ELC	– electroluminescence (light proportional) chamber
ELDC	– electroluminescence (light proportional) drift chamber
EMC	– European Muon Collaboration at CERN
EPI	– External Particle Identifier (detector used behind bubble chamber BEBC at CERN)
E-128	– experiment at Serpukhov 70 GeV proton synchrotron
FADC	– flash analog-to-digital convertor
FAST BUS	– fast electronic system
FHS	– Fermilab Hybrid Spectrometer
FNAL	– Fermi National Accelerator Laboratory at Batavia
fwhm	– full width at half maximum
HERA	– electron–proton collider at DESY
HRS	– High Resolution Spectrometer at SLAC
HVPG	– high voltage pulse generator
HYPERON	– spectrometer installed in the secondary beam of U-70 accelerator at Serpukhov
H1	– detector installed on ep collider HERA at DESY
ICS	– Ionization Coordinate Spectrometer at JINR, Dubna
IDC	– induction drift chamber
IGLD	– integral gap length distribution
IHEP	– Institute of High Energy Physics at Serpukhov
IKAR	– recoil nucleus ionization spectrometer
ISIS	– Identification of Secondaries by Ionization Sampling (detector installed on EHS at CERN)
ISR	– Intersecting Storage Ring (pp collider at CERN)
ITEP	– Institute of Theoretical and Experimental Physics at Moscow
JADE	– detector used on e^+e^- collider PETRA at DESY
JINR	– Joint Institute for Nuclear Research at Dubna
KEDR	– detector to be built on e^+e^- collider VEPP at Novosibirsk

KEK	– Japanese institute of high energy physics
LEAR	– Low Energy Accumulator Ring at CERN
LEP	– Large Electron Positron collider at CERN
LHC	– Large Hadron Collider (pp collider project at CERN)
LINP	– Leningrad Institute of Nuclear Physics
LS mode	– limited streamer mode
LTD	– LEP time digitizer
L3	– detector installed on LEP at CERN
MAC	– detector used on e^+e^- collider at SLAC
MARK II	– detector used on PEP collider at SLAC
MARK II/SLC	– updated detector MARK II installed on SLC at SLAC
MARK III	– detector installed on e^+e^- collider SPEAR at SLAC
MC	– multistep chamber
MDC	– multilayer drift chamber
MGD	– mean gap density
MHTDC	– multihit time-to-digital converter
MPC	– multilayer proportional chamber
MSC	– modified streamer counting
MSD	– multistrip silicon detector
MWPC	– multiwire proportional chamber
NA-5	– experiment carried out at CERN
NA-9	– experiment carried out at CERN
NA-28	– experiment carried out at CERN
NA-35	– experiment carried out at CERN
NA-36	– experiment carried out at CERN
OBELIX	– detector installed on LEAR at CERN
OMEGA	– fixed target detector installed at CERN
OPAL	– detector installed on LEP at CERN
PDC	– planar drift chamber
PEP	– e^+e^- collider at SLAC
PEP-4	– detector (called also TPC) used on PEP collider at SLAC
PETRA	– e^+e^- collider at DESY
PION	– detector developed for identification of cosmic particles at Aragatz Laboratory, Armenia
PPAC	– parallel plate avalanche chamber
ppm	– parts per million
PS-179	– experiment being carried out at CERN
PVC	– polyvinilchloride

RC	– output time constant of an electronic device
RDC	– radial drift chamber
RISK	– experiment carried out at 70 GeV Serpukhov proton synchrotron
r.m.s.	– root mean square
RTL	– roughened track lacunarity

SATURN	– proton synchrotron at Saclay
SCDC	– drift chamber with symmetric cells
SCIFI	– scintillating fiber detector
SIDC	– silicon drift chamber
SLAC	– Stanford Linear Accelerator Center
SLC	– Stanford Linear Collider
SLD	– Stanford Linear Detector
SPEAR	– e^+e^- collider at SLAC
SPS	– Super Proton Synchrotron at CERN
SQS mode	– self-quenching streamer mode
SRINP	– Scientific Research Institute of Nuclear Physics, Moscow State University
SSC	– Superconducting Super Collider (pp collider built at Texas); Stereo Superlayer Chamber
STP	– standard temperature and pressure (0°C, 1 atm)

TDC	– time-to-digital converter
TEA	– triethylamine
TEC	– time expansion chamber
TL	– track lacunarity
TMAE	– tetradimethylaminoethylene
TOF	– time of flight
TOPAZ	– detector installed on e^+e^- collider TRISTAN at KEK
TPC	– time projection chamber; detector used on PEP collider at SLAC
TRED	– fixed target detector at Fermilab
TRISTAN	– e^+e^- collider at KEK
TRIUMF	– accelerator center at Vancouver, Canada
TTT	– transmitting television tube

UA-1	– detector installed on pp̄ collider at CERN
UA-5	– experiment carried out at CERN
UA-5/2	– experiment carried out at CERN
UCD	– Universal Calorimetric Detector being planned for pp collider UNK at Serpukhov
UNK	– accelerator and collider complex being built at IHEP, Serpukhov

VEPP-2M	– e^+e^- collider at Novosibirsk
WA-44	– quark search experiment carried out at CERN
ZEUS	– detector installed on ep collider HERA at DESY

List of Symbols

A	– atomic (molecular) weight; amplitude of a signal
$\boldsymbol{A}$	– vector potential
A_b	– constant in (4.12)
A_c, A_m	– ionization spectrometer signal amplitudes for the control and the main strobes, respectively
A_t	– threshold signal amplitude corresponding to the detector noise level; threshold detector sensitivity
A_0	– constant in (1.50)
A_1, A_2	– constants in (1.66a) and (1.71)
a	– concentration
a_a	– concentration of an admixture with great electron affinity
a_c	– correlation parameter associated with a crosstalk
a_d	– correlation parameter associated with the electron drift in a gas
a_j	– concentration (relative partial pressure) of the j-th gas component
a_k	– concentration of a catalyzer
a_m	– constant in (3.29)
a_0	– Bohr atomic radius
A_t	– threshold detector sensitivity
B	– magnetic induction
B_t	– optical density of the track image (track image blackening)
b	– parameter in Table 1.6
b_c	– capacitance coupling parameter
b_m	– constant in (3.29)
C	– capacity, capacitance; parameters in (1.57), and in Tables 1.2 and 1.6
C_e	– constant in (2.1)
C_F	– constant in (2.20)
C_i	– input capacity
C_w	– linear wire capacity

$C_0, C_1, C_2, \ldots$	– coefficients in (1.84), (3.10), (3.16) and Table 2.10
c	– velocity of light in vacuum
c_m	– constant in (3.29)
D, D	– electron diffusion coefficient; D-meson
$D(\pi/K)$	$= \Delta_0(\pi) - \Delta_0(K)$
D_a	– diffusion coefficient for atoms
D_i	– diffusion coefficient for ions
D_j	– diffusion coefficient for the j-th component
D_L	– longitudinal diffusion coefficient
D_M	– electron diffusion coefficient in a magnetic field
D_T	– transverse diffusion coefficient
D_0	– thermal diffusion coefficient for electrons at STP
D_+	– diffusion coefficient for positive ions
D_-	– diffusion coefficient for negative ions
d	– track width, diameter of a track structure element; constant
$-dE/dx$	– average specific total energy loss
$(-dE/dx)_{\check{C}}$	– energy loss due to Vavilov–Čerenkov radiation
$(-dE/dx)_{T_0}$	– average specific restricted energy loss
dJ/dx	– observed specific ionization effect
dN/dx	– average number of particle collisions per unit length
$d^2N/dxd\omega$	$= n_a(d\sigma/d\omega)$
dN_t/dx	– average total specific ionization
dN_1/dx	– average primary specific ionization
$(dN_1/dx)_{\min}$	– minimum value of dN_1/dx
dS/dt	– flux of electromagnetic energy
$d\sigma_k/dq$	– differential cross section for atomic excitation to a quantum level with the energy E_k
$d\sigma/d\omega$	– differential collision cross section
$d^2\sigma/dqd\omega$	– doubly differential collision cross section
d_a	– diameter of the electron avalanche in a gas
d_c	– mean distance between ionization clusters
d_r	– reconstructed diameter of a track structure element
d_s	– diameter of a streamer
d_t	– flash tube diameter
d_0	– diameter of a cylindrical detector
E	– total particle energy
E_e	– total electron energy
E_k	– energy of k-th excitation atomic level
E_γ	– photon energy

E_0	– energy of atomic ground state
$\mathcal{E}$	– electric field strength
$\mathcal{E}_0$	– strength of the applied electric field
$\mathcal{E}_\tau$	– additional electric field strength produced by a streamer
e	– electron charge
e, e^-	– electron
F	– Fano factor
$F(\varepsilon_e)$	– electron kinetic energy distribution function
$F_e(\varepsilon_e, \theta_e)$	– electron energy-angular distribution function
F_L	– Lorentz force
F_p	– lens aperture number
$F_0(\Delta_j / \Delta_0)$	– shape of the ionization fluctuation curve
$F_0(\varepsilon_e), F_1(\varepsilon_e) \ldots$	– electron energy distribution functions (2.17–2.21)
$\mathcal{F}(\omega) = \int_0^\omega f(\omega') d\omega'$	– integral oscillator strength
$\mathcal{F}_s$	$= \mathcal{F}(\omega = \omega_s)$
f	– frequency
$f(x, \Delta)$	– energy loss distribution
$f(\omega)$	– optical distributed atomic oscillator strengths
$f(\omega, q)$	– generalized distributed atomic oscillator strengths
$f_e(v_e)$	– electron velocity distribution function
f_k	– optical discrete atomic oscillator strengths
$f_k(q)$	– generalized discrete atomic oscillator strengths
f_L	– Larmor frequency
f_M	– relative variance of the gas gain
f_{m_1}, f_{m_2}	– energy loss distribution functions for particles of masses m_1, m_2
f_s	– recording frequency of events
f_1, f_2	– recording frequency of a recording device
G_H	– correction of the transport coefficients for an influence of the magnetic field (2.35, 2.37, 2.39)
g	– average ionization density along the track in a visual detector
g_b	– mean blob density along the track
g_g	– mean gap density along the track
g_{min}	– average ionization density along the track for a minimum ionizing particle
g_s	– streamer or track structure element density along the track
g_π, g_p	– average ionization density along the pion (proton) track

H	– magnetic field strength
H_s	– spark track brightness (luminosity)
$\hbar$	– reduced Plank constant
h_s	– streamer brightness (luminosity)
I	– mean logarithmic ionization pontential of matter; the electric current strength
$I_{K,L,M\ldots,j}$	– ionization potentials of $K, L, M \ldots, j$ atomic shells
I_m	– potential of a metastable atomic state
I_r	– resonance potential of an atom
I_1	– first ionization potential of an atom
i_b	– background photocurrent in a track image photometry
i_t	– track image photocurrent in track image photometry
J	– ionization effect
J_e	– diffusion current of electrons
J_0	– the most probable value of the ionization effect
J_+	– positive ion diffusion current
j_e	– electron diffusion current density
j_+	– positive ion diffusion current density
K	– amplification coefficient
K	– kaon
$K_s, (K_l)$	– short (long) lived kaon
$K_0(x), K_1(x)$	– modified Bessel functions
K_1	– constant in (4.11)
k	– Boltzmann constant; energy level of the atom
k_a	– electron capture probability (attachment probability)
k_i	– constant in (2.74)
k_p	– optical magnification factor
k_s	– constant in (5.1)
k_1	– relative dispersion of the single electron spectrum
L	– multilayer detector length; total track length
L'	– effective track length
L_a	– electron capture length (attachment length)
L_d	– maximum drift distance (drift gap length)
L_m	– molecular mean free path in a gas
L_w	– sense wire length
L_+	– positive ion mean free path in a gas

$\mathcal{L}$	– stopping number; likelihood function; track lacunarity
$\mathcal{L}_0$	– likelihood function containing Δ_0 as a parameter; roughened track lacunarity
$\mathcal{L}_{12}$	– likelihood ratio
l	– layer thickness; single layer detector length; anode–cathode gap length; root of the eq. $\varepsilon(il) = \beta^{-2}$
l_a	– avalanche size along the electric field direction
l_b	– blob length
l_c	– average distance between ionizing collisions, or centers of adjacent ionization clusters on a track
l_D	– electron displacement due to diffusion
l_d	– total electron displacement due to thermalization and diffusion
l_{dec}	– decay length
l_e	– mean free path between elastic electron–atom collisions
l_g	– distance between adjacent track structure elements; gap length
l_k	– mean free path between inelastic collisions accompanied with excitation of an atom into the level k
l_i	– electron mean free path for inelastic collisions accompanied by excitation and/or ionization
l_j	$= \left(\omega_j^2 + f_j \omega_p^2 / Z\right)$
l_m	– resolution distance of adjacent track structure elements
l_{min}	– minimum distance between track structure elements excluding their interaction
l_{mt}	– mean momentum transfer length (transport length)
l_s	– streamer length along the electric field direction
l_t	– thermalization displacement of an electron
l_0	– cell size in roughened track lacunarity analysis method

M	– gas amplification coefficient (gas gain)
M_a	– mass of an atom (molecule)
M_+	– positive ion mass
m	– particle mass; constant in Table 2.10
m_e	– electron mass
$m_0 = m_e m / (m_e + m)$	– reduced mass

N	– total number of collisions; number of events
N_c	– number of counted clusters on a track
N_{ch}	– charged particle multiplicity

N_{cr}	– critical number of electrons in an avalanche
N_d	– number of droplets in a cluster on a cloud chamber track
$(N_d)_{max}$	– maximum magnitude of N_d
N_e	– number of electrons in an ionization cluster (in an avalanche, streamer, spark gap)
N_p	– number of photons emitted by an avalanche or by a streamer
N_s	– increase of the electron number due to multiplication; total number of streamers on a track in the streamer chamber
N'_s	– effective number of streamers on a track (i.e. number of streamers situated at a distance greater than l_m from the preceding counted streamer)
N_t	– total number of ion pairs produced by a charged particle (total ionization)
$N_0(x)$	– the most probable value of the ionization distribution in a sample of thickness x
N_1, N_2	– primary and secondary ionizations
N_Δ	– number of ionization electrons for energy release $\leq \Delta$
n	– number of layers in a multilayer detector; number of samples
$n(0)$	– number of electrons (ions) at initial point at time $t = 0$
n_a	– number of atoms per cm^3
n_b	– number of cells in a CCD matrix
n_e	– number of electrons per cm^3
n_{exp}	– number of points on a track used for ionization measurements in experimental conditions
n_L	– Loschmidt constant
n_M	– collision multiplicity; number of tracks in the visual detector
n_r	– refraction index
n_s	– number of spark gaps of the multigap spark chamber
n_t	– number of points on a track used for ionization measurements in test conditions
n_v	– number of valence electrons of an atom
n_0	– number of "empty" cells or spark gaps
n_α	– number of samples selected after truncation
P	– pressure
P_k	– Legendre polynomial
P_l	– left pick-up electrode
P_r	– right pick-up electrode
P_α	– signal difference on pick-up electrodes

$\mathcal{P}$	– probability
$\mathcal{P}(\omega)$	– weight function (local response function)
p	– particle momentum; parameter of Laplace transformation
p	– proton
p_{max}	– maximum particle momentum of the reliable particle identification
p_T	– particle transverse momentum
Q	– electric charge
Q_e	– charge density on the drift electrode
Q_j	– electric charge after amplification
Q'_j	– electric charge transformed by electronics (including crosstalk between channels)
Q_l	– total charge on the left pick-up electrode
Q_r	– total charge on the right pick-up electrode
Q^+, Q^-	– total charge of positive and negative ions, respectively
q	– momentum transfer
q_j	– electric charge before amplification
q_w	– linear charge density on a sense wire
R	– relativistic rise of ionization; electric resistance
R_c	– relativistic rise of recorded number of clusters
R_E	– relativistic rise of energy loss
R_J	– relativistic rise of the observed ionization effect
R_N	– relativistic rise of the collision number
R_{N1}	– relativistic rise of the primary ionization
R_p	– particle range
R_s	– electric resistance of the spark channel
R_t	– radius of a particle trajectory in a magnetic field
R_{th}	– ultimate resolution
R_0	– amplitude resolution
R_{10}	– amplitude resolution after noise subtraction
r	– radius of a detector; distance between two points; parameter in (1.66)
r_a	– characteristic dimension of an atom; anode wire radius; electron avalanche radius
r_b	– cylindrical cathode radius
r_c	– critical radius
r_m	– molecular rotational constant
r_s	– streamer radius
$r_t = d_t/2$	– radius of a cylindrical flash tube
r_0	– distance from the anode of an electron–ion pair at time $t = 0$

S	$= \mathcal{E}/P$ (reduced electric field)
$S(\omega, q)$	– dynamical factor of the medium
S_a	– magnitude of S on the anode surface
S_c	– control strobe pulse
S_{col}	– collision integral (2.15)
S_{cr}	– magnitude of S at the point corresponding to the critical radius
S_i	– slit width in the track photometry
S_m	– main strobe pulse
S_n	– separation coefficient of the n-layer detector (in σ_n units)
S'_n	– separation coefficient of the n-layer detector (in W_n units)
S_{n0}	– maximum separation coefficient of the n-layer detector
S_{nt}	– separation coefficient of the n-layer detector in test measurements
S_p	– sensitivity of the optical recording system
S_0	– constant in Table 2.10
s	– anode wire pitch; parameter in (1.66)
T	– absolute temperature
T_e	– electron temperature
T_m	– maximum energy transferred in collisions
T_0	– upper instrumental restriction for energy transfers in individual collisions
t	– time
$t°C$	– temperature in degrees centigrade
t_c	– characteristic time (2.93)
t_d	– electron drift time
t_s	– signal saturation time (2.91)
t_{s1}	– signal saturation time in the absence of gas amplification (2.89)
t_y	– signal propagation time along the delay line
t_0	– time of particle passage through a detector
t_1	– time of electron swarm arrival at the anode
U	– electric potential; voltage
U_a	– anode potential
U_c	– critical potential
U_d	– drift gap potential
U_g	– grid voltage
U_l	– light gap potential
U_s	– saturation potential

U_0	– voltage applied to the electrodes; high voltage pulse amplitude
U_1	– anode potential in the absence of gas amplification
V	– volume
v	– particle velocity
v_a	– atom velocity in a gas
v_e	– velocity of atomic electrons; electron velocity in a gas
v_g	– velocity of an avalanche (streamer) growth in a strong electric field
v_+	– velocity of a positive ion in a gas
$W = \text{fwhm}$	– ionization resolution (full width at half maximum) of a single layer detector
W_a	– instrumental resolution (fwhm)
W_c	– constant in (3.32)
W_g	– energy stored in HVPG
W_J	– resolution (fwhm) of a total ionization distribution
W_n	– ionization resolution (fwhm) of the n-layer detector
$W_{n,\exp}$	$= W_n$ obtained in a physical experiment
W_{nt}	$= W_n$ obtained in a test experiment
W_t	– total ionization resolution (fwhm)
W_R	– energy released in the resistance R
$W_1(\omega)$	$= (1/\sigma)d\sigma/d\omega$
$\mathcal{W}_i$	– probability that the particle has the mass m_i
$\mathcal{W}_{f1}$	$= f_1(p)\mathcal{W}_1$
$\mathcal{W}_0$	– probability
w	– mean energy spent to create an ion pair in the gas
w_d	– electron drift velocity
w_L	– longitudinal drift velocity
w_M	– electron drift velocity in a magnetic field
$w_R(\omega)$	– Rutherford collision spectrum
w_T	– transverse drift velocity
w_y	– signal propagation velocity in the delay line
w_+	– positive ion drift velocity
$w(\omega)$	$= x\,d^2N/dx\,d\omega$
X	– x-coordinate; symbol for an atom or a reaction product
X^*	– symbol for an excited atom
$X_{\text{pl}} = \lg(\beta\gamma)_{\text{pl}}$	– parameter corresponding to the relativistic rise saturation of the ionization effect
X_0	– radiation length

X_0, X_1	– parameters in Tables 1.2 and 1.6
x	– gas layer thickness (in g/cm^2); electron drift length (in cm)
x_1	– path length of the first electron arriving at the anode of a drift detector
x_2	– double track resolution along the x-direction
Y	– y-coordinate; symbol of an atom or of a reaction product
Y^*	– symbol for an excited atom
y_1	– distance between the first electron and the coordinate axis in a drift detector
y_2	– double track resolution along the y-direction
Z	– z-coordinate; number of electrons; atomic number; symbol of an atom or of a reaction product
z	– particle charge expressed in electron charge units
α	– cut-off coefficient (truncation parameter)
α_c	– angle of cluster incidence on the sense wire
α_M	– angle between directions of electron motion in magnetic field and of electric field lines
α_s	– tilt stereo angle of layers in multilayer drift chamber
α_T	– first Townsend ionization coefficient
β	$= v/c$
Γ	– gamma function
γ	– Lorentz factor (gamma factor)
γ_f	– contrast coefficient of photographic film
γ_{pl}	– Lorentz factor corresponding to ionization curve saturation and its transition to the Fermi plateau
γ_T	– second Townsend ionization coefficient
Δ	– particle energy loss; energy release as the result of a number of collisions; deviation of particle trajectory in electric or magnetic field
Δ_{eff}	– effective energy loss
Δ_{ij}	– energy loss of the i-th particle in the j-th layer of a multilayer detector
Δ_j	– energy loss in the j-th layer of a multilayer detector
Δ_t	– noise energy equivalent corresponding to the threshold sensitivity of an amplifier
Δ_{th}	– threshold magnitude of energy loss

Δ_α	– truncated mean value
$\Delta_{\alpha m}$	• $= \Delta_\alpha$ for minimum ionizing particles
Δ_0	– the most probable value of the ionization energy loss
Δ_{0m}	$= \Delta_0$ for minimum ionizing particles
Δ_{0p}	$= \Delta_0$ corresponding to the Fermi plateau
δ	– Dirac delta function; full width at half maximum (fwhm) of a distribution
δ_E	– density effect correction to the energy loss
δ_L	– fwhm of Landau distribution
δ_{ls}	– streamer length spread
δ_{N1}	– density effect correction to the primary ionization
δ_n	– fwhm of Δ_0 distribution for an n-layer identifier
$\delta_{\tau f}$	– spark formation time difference for a particle under investigation and the minimum ionizing particle
δ_z	– depth of the photographic field
ε	– dielectric constant
$\varepsilon(\omega, q)$; $\varepsilon(\omega) \equiv \varepsilon(\omega, 0)$	– dielectric permeability of a medium
ε_c	– characteristic energy
ε_e	– kinetic electron energy
ε_{ef}	– "effective" dielectric constant
ε_g	– dielectric constant of a gas
ε_δ	– delta-electron energy
ζ	– fraction of collisions accompanied by excitation
η	– detector efficiency; relative part of "full"cells, containing track structure elements; η-meson
η_a	– electron attachment coefficient
η_b	– efficiency of the beam particle recording
η_c	– cluster counting efficiency
η_g	– ratio of the gas density to its density in normal conditions
η_s	– streamer detection (counting) efficiency
$\eta_\pi/\eta_K/\eta_p$	– efficiency of $\pi/K/p$ separation
η_+	– efficiency of positive ion transportation through a grid
Θ	– parameter of Poya distribution (2.82)
θ	– track inclination angle (angle between the track and the normal to the sense wire plane)
θ_e	– angle between the direction of electron motion and a reference direction

θ_H	– angle of the particle trajectory to the magnetic field direction
θ_s	– track inclination angle to the streamer chamber electrode plane
$\kappa = \xi / T_{max}$	– parameter determining the shape of the energy loss distribution
κ_e	– electron survival coefficient
Λ	– mean electron energy in a gas in the presence of an electric field
λ	– Landau variable (1.82); wavelength
μ	– muon; mass photoabsorption coefficient
μ_e	– electron mobility
μ_j	– ion mobility for the j-th component of a gas mixture
μ_+	– positive ion mobility
μ_-	– negative ion mobility
ν	– frequency
ν_a	– frequency of electron capture in a gas
ν_e	– electron collision frequency in a gas
ν_f	– relative r.m.s. deviation of spark formation time in a single gap spark chamber
ν_s	– natural frequency of a harmonic oscillator
ξ	$= 2\pi x n_a Z e^4 / m_e v^2$
ξ_e	$= \varepsilon_e / kT$ (2.26)
$\prod$	– multiplication symbol
π	– pion
ϱ	– mass density of a medium; impact parameter
ϱ_w	– specific resistance of a wire
ϱ_x	– spatial resolution of a cluster counting method
ϱ_y	– transversal density of ions on a particle track
ϱ_0	– mass density of a gas at STP
ϱ_+	– positive charge volume density
σ	– total collision cross section; r.m.s. deviation (standard deviation)
σ_A	– r.m.s. deviation of amplitude measurements
σ_a	– r.m.s. instrumental error
σ_{at}	– attachment (capture) cross section for thermal electrons

σ_c	– r.m.s. deviation of cluster number distribution
σ_D	– r.m.s. spread of the electron swarm diameter due to diffusion (2.14)
$\sigma_{D'}$	$= \sigma_D$ value for a realistic detector
σ_e	– electron elastic collision cross section
$\sigma_{ex} = \sum_k \sigma_k$	– total excitation cross section
σ_G	– r.m.s. error of a charge centroid position measurement (4.7)
σ_i	– ionization cross section
σ_{in}	– electron inelastic scattering cross section
σ_J	– ionization measurement resolution
σ_j	– spatial resolution of the j-th element of an optical recording system; r.m.s. spatial resolution corresponding to arrival of the j-th electron to the anode
σ_k	– cross section of inelastic collisions accompanied by atomic excitations to a level k
σ_L	– r.m.s. spread of the electron swarm due to longitudinal diffusion (2.28)
$\sigma_L(1)$	– r.m.s. spread due to single electron longitudinal diffusion (4.6)
σ_M	– r.m.s. spread of the gas gain coefficient
σ_m	– momentum transfer cross section
σ_{ma}	– mass measurement resolution
σ_{N_t}	– r.m.s. spread of the total ionization measurement
σ_{N_Δ}	– r.m.s. deviation of N_Δ distribution
σ_n	– ionization measurement resolution of an n-layer identifier
$\sigma_{ni,exp}$	– ionization resolution obtained in a physical experiment for a particle with mass m_i
σ_{nt}	– ionization resolution in a test experiment
$\sigma_{n\alpha}$	– r.m.s. deviation for Δ_α-distribution
σ_{opt}	– optical resolution
σ_P	– cross section of Penning reaction
σ_p	– r.m.s. resolution of momentum measurements
$\sigma_{pi} \equiv \sigma_\gamma$	– photoionization cross section
σ_R	– Rutherford collision cross section
$\sigma_{r\varphi}$	– r.m.s. accuracy of the azimuthal coordinate ($r\varphi$) measurement
σ_s	– r.m.s. spread of electrons along the track (4.5); spatial resolution of streamers in a streamer chamber (5.3)
σ_t	– time measurement resolution
σ_{tr}	– spatial accuracy of track reconstruction
σ_x	– spatial resolution of X-coordinate measurements

σ_y	– spatial resolution of Y-coordinate measurements
σ_0	– reference cross section (2.26); constant in (4.12)
σ_1	– ionization measurement resolution of a single layer detector (a single gap spark chamber)
σ_{1e}	– spatial resolution obtained by the measurement of the arrival time of the first electron to the anode
σ_{12}	– spatial resolution of two adjacent tracks
$\sigma_\gamma(\omega)$	– photoabsorption cross section
τ_A	– characteristic time of associative ionization reaction
τ_a	– lifetime of free electrons in a gas (attachment time)
τ_c	– time constant
τ_d	– delay time of the high-voltage pulse formation
τ_e	– mean time between successive elastic collisions of the electron in a gas
τ_{ex}	– lifetime of an excited state
τ_f	– gas discharge formation time
τ_{fm}	– gas discharge formation time for minimum ionizing particles
τ_M	– electron mean free path time in crossed electric and magnetic fields
τ_m	– dead time (recovery time) of a detector
τ_p	– pulse duration
τ_R	– pulse rise time
τ_r	– repetition time (duty cicle) of a detector
τ_s	– sensitivity (memory) time of a detector
τ_t	– electron thermalization time
τ_0	– spark discharge formation time constant (5.6)
τ_+	– drift time of a positive ion
Φ	– particle flux; phi-meson
Φ_γ	– photon flux
φ	– asimuthal angle
$\varphi(\lambda)$	– Landau distribution
χ_i	– ionization efficiency
$\chi^2_{m_i}$	– chi-squared value defined by (3.30) for particle mass determination
ψ	– stopping power for electrons in a gas
ω	– energy transfer in an individual collision
ω_a	– binding energy of the electrons in an atom

328

$\omega_{\check{C}}$	– energy transfer for Vavilov–Čerenkov radiation
ω_{in}	– total excitation energy
ω_j	– energy of the j-th oscillator in the harmonic oscillator model
$\omega_{K,L,\ldots}$	– energy of the $K, L, \ldots$ absorption edge
$\omega_k = E_k - E_0$	– excitation energy of an atom
ω_m	– minimum energy transfer in an individual collision
ω_p	– plasma energy of a medium
ω_r	– excitation energy of the molecular rotational level
ω_s	– energy transfer for the natural frequency of a harmonic oscillator
ω_{th}	– energy transfer with threshold energy
ω_v	– excitation energy of the molecular vibrational level

Subject Index

Springer Tracts in Modern Physics

---

* denotes a volume which contains a Classified Index starting from Volume 36